NATURAL RESOURCE MANAGEMENT AND POLICY

EDITORIAL STATEMENT

There is a growing awareness to the role that natural resources such as water, land, forests and environmental amenities play in our lives. There are many competing uses for natural resources, and society is challenged to manage them for improving social well being. Furthermore, there may be dire consequences to natural resources mismanagement. Renewable resources such as water, land and the environment are linked, and decisions made with regard to one may affect the others. Policy and management of natural resources now require interdisciplinary approach including natural and social sciences to correctly address our society preferences.

This series provides a collection of works containing most recent findings on economics, management and policy of renewable biological resources such as water, land, crop protection, sustainable agriculture, technology, and environmental health. It incorporates modern thinking and techniques of economics and management. Books in this series will incorporate knowledge and models of natural phenomena with economics and managerial decision frameworks to assess alternative options for managing natural resources and environment.

In an era where water quantity and quality problems lead to an increased number of domestic and international conflicts, empirical experiences are quite valuable. Readers of *Braving the Currents* would benefit from reading this book because it provides, in addition to having a sound theoretical framework, a set of cases that altogether indicate how conflicts emerged, how tensions rose, how mediators were brought in and how the mediators designed and implemented processes that brought the conflict to some sort of resolution. *Braving the Currents* is likely to be very influential among those lawyers, mediators, academics and policy makers concerned with the resolution of public policy disputes generally and environmental conflicts in particular.

The Series Editors

Recently Published Books in the Series

Moss, Charles B., Rausser, Gordon C., Schmitz, Andrew, Taylor, Timothy G., and Zilberman, David
Agricultural Globalization, Trade, and the Environment
Haddadin, Munther J.
Diplomacy on the Jordan: International Conflict and Negotiated Resolution
Renzetti, Steven
The Economics of Water Demands
Just, Richard E. and Pope, Rulon D.
A Comprehensive Assessment of the Role of Risk in U.S. Agriculture
Dinar, Ariel and Zilberman, David
Economics of Water Resources: The Contributions of Dan Yaron
Ünver.I.H. Olcay, Gupta, Rajiv K. IAS, and Kibaroğlu, Ayşegül
Water Development and Poverty Reduction

BRAVING THE CURRENTS

BRAVING THE CURRENTS
Evaluating Environmental Conflict Resolution in the River Basins of the American West

TAMRA PEARSON d'ESTRÉE
University of Denver
BONNIE G. COLBY
University of Arizona

Kluwer Academic Publishers
Boston/Dordrecht/London

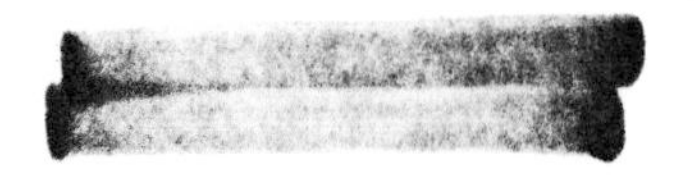

Distributors for North, Central and South America:
Kluwer Academic Publishers
101 Philip Drive
Assinippi Park
Norwell, Massachusetts 02061 USA
Telephone (781) 871-6600
Fax (781) 871-6528
E-Mail <kluwer@wkap.com>

Distributors for all other countries:
Kluwer Academic Publishers Group
Post Office Box 322
3300 AH Dordrecht, THE NETHERLANDS
Telephone 31 78 6576 000
Fax 31 78 6576 474
E-Mail <orderdept@wkap.nl>

Electronic Services <http://www.wkap.nl>

Library of Congress Cataloging-in-Publication

BRAVING THE CURRENTS
Evaluating Environmental Conflict Resolution in the River Basins of the American West
by Tamra Pearson d'Estrée and Bonnie G. Colby
ISBN: 1-4020-7503-0
1-4020-8132-4
1-4020-8129-4 (e-Book)

Printed on acid-free paper.

Printed in the United States of America

Dedication

To dedicated professionals in the public, private, and non-governmental sectors who have committed their skills and experience to creatively addressing environmental conflicts – T.P.E. & B.G.C.

To my parents, Donnette and Marlyn Pearson, for wisdom, guidance, and childhood camping trips that engendered a respect and love for the natural world; and to my husband Claude and sons Jean-Marc, Alec, and Daniel, for their support and sacrifices in the name of "the book" – T.P.E.

To my parents, Mason and Patricia Colby, with gratitude for daily examples of a graceful and collaborative approach to life – B.G.C.

Contents

Foreword

During the past twenty-five years there has been an exponential growth in the number of environmental and natural resource conflicts. They occur at local, regional, national and international levels and encompass an ever increasing range of issues. Many have proven to be very difficult to successfully resolve. A look at just a few arenas where conflicts have emerged illustrates this trend.

Many significant conflicts have emerged over water, an essential element for the life of human beings and other species. Water disputes have developed over surface and groundwater development, use, allocation and quality. The conflicts have involved diverse competing users – agriculturalists, environmentalists, municipalities, companies, tribes and non-human species. The disputes have often been highly contentious, protracted and costly in terms of dollars, personnel and use of other limited resources.

Resource and land-use issues - conversion of agricultural land, management and protection of forests and open spaces, urban growth and suburban sprawl - have increasingly resulted in intense conflicts where competing parties strive to their up-most to achieve their desired ends. As populations have shifted, development patterns changed, and employment opportunities declined or grown, there has been an increase in competition over how land and associated resources are used, what development will or will not occur, and how quality of life can be protected and preserved.

In response to the growing number of disputes, parties - governmental agencies, the private sector, agriculturalists, environmentalists, other public interest groups, tribes, and citizens - have explored and utilized a range of approaches to address and try to resolve them. Procedures have included lobbying and legislative action, lawsuits and judicial rulings, administrative hearings and decisions, mediated or facilitated negotiations, unassisted direct talks, and nonviolent and violent direct action. Some approaches seem to have been more successful than others, and proponents of specific processes have often been quick to advocate the superiority of their preferred methods, and have frequently been highly critical of others, often with little concrete data to back up their claims.

Over the years, as environmental disputes have either been settled by various approaches or continue to fester and periodically re-surface as parties try yet again to resolve issues in dispute, participants, conflict management professionals (in the broadest sense), and the general public have become more concerned about the kinds of outcomes and impacts that result from specific dispute resolution procedures. Specifically, they want to know 1) which procedures are the most successful, and 2) whether it is possible to predict success in the resolution of specific kinds of cases if one or another dispute resolution procedure is utilized.

These two sets of questions related to "success" and "prediction", are critical for conducting careful analysis and evaluation of specific dispute resolution processes. However, the first, that related to success, is the foundation for all future assessment. Without a clear understanding of what constitutes success, it is very difficult to compare dispute resolution processes, and impossible to try to predict which ones will are more likely to result in the outcomes desired by the involved parties.

Since the middle 1980's users of various environmental conflict management procedures, dispute resolution practitioners and researchers have debated the merits and costs of various approaches. They have called for the development of, rigorous means to evaluate the viability, success rate and costs of various dispute resolution procedures. However, until the completion of *Braving the Currents*, no comprehensive assessment process to define the components of success, and operationalize them, has been developed.

After more than a decade of research on evaluation methodologies, and extensive testing of their "success criteria" on a number of actual environmental cases, d'Estrée and Colby have developed the most comprehensive set of criteria to date. Their framework for evaluating the outcomes and success of various dispute resolution procedures can be utilized by users, conflict management practitioners and researchers to evaluate the outcomes of various disputes, and the procedures used to resolve them. In their Guidebook, an appendix to this book, they outline step-by-step procedures for assessing various types of "success" at various stages in the evolution of a dispute and its termination.

What is so significant about d'Estrée and Colby's work and development of a framework for categories of success, measurement approaches and outcomes? Why is the development of a standardized and operational set of success criteria important?

First, their work enables us to develop a highly sophisticated understanding of dispute outcomes and the meaning of success. Their approach moves beyond the simplistic debate regarding whether settlement or non-settlement should be the measure of success, and provides a rich set of variables and dimensions that can be used to determine the various types of successes that are possible.

Second, the framework provides an assessment methodology that can be used across a wide variety of cases and procedures to effectively compare outcomes. While the authors developed and tested the approach on a fairly small number of water disputes in which a range of resolution approaches were applied, the framework is equally applicable to other public disputes.

Third, if the evaluation framework and approach for assessing outcomes and success is adopted widely by researchers, dispute resolvers and parties, and data is collected on a significant number of conflicts, it will enable us to begin to address questions of predictability - what kinds of outcomes and successes can be expected when specific procedures are applied to various types of disputes. This information will be of critical importance to government agencies as they strive to build effective public policies and dispute resolution systems, and to parties as they deliberate about approaches that will best enable them to satisfy their interests.

Braving the Currents is an invaluable contribution to the field of dispute resolution evaluation. It should be on the bookshelf of every researcher, dispute resolution practitioner and program administrator. It is also a valuable resource for parties engaged in conflicts to help them decide what kind of outcomes they really want and which strategies to pursue to address and resolve serious differences.

Christopher W. Moore
Partner, CDR Associates,
and environmental and water mediator

Acknowledgments

This research was supported by a grant from the Udall Center for Public Policy at The University of Arizona to both authors. Directors Steven Cornell and Robert Varady, and Program Director Kirk Emerson (now at the U.S. Institute of Environmental Conflict Resolution) have been extremely supportive of our efforts.

Eric Abitbol, Connie Beck, Annette Hanada, Kristina Jones, Shawn MacDonald, Kathryn Mazaika, Erin McCandless, and Therese Suomi all contributed to the research and writing of various sections of our larger project which led to the development of this evaluation framework. Framework development benefited from interviews with multiple unnamed practitioners. Framework revisions emerged from discussions among audiences of practitioners and scholars stimulated by its public presentation. We acknowledge our debt to these creative and helpful audiences. We also appreciate the constructive comments provided by Lawrence Susskind, Christopher Moore, Gregory Sobel, Gayle Landt, Sarah Connick, Ariel Dinar, and anonymous reviewers. All errors of omission and commission are the sole responsibility of the authors. Monica Jakobsen, Tamara Zizzo, Nancy Bannister, and Shay Bright provided competent and tireless editing of our many manuscript revisions. Anne Enderby ushered the final manuscript into its publishable form. Finally, we thank Marilea Fried, our editor at Kluwer, for her support, encouragement, and unswerving patience.

Authors

Bonnie G. Colby is Professor of Agricultural and Resource Economics at The University of Arizona, where she has been a faculty member since 1983. Colby's PhD is from the University of Wisconsin. Her expertise is in the economics of natural resource policy and disputes over water, the public lands and environmental regulation. She has authored over one hundred journal articles and five books, including the books *Water Markets in Theory and Practice* and *Indian Water Rights: Negotiating the Future*. She has provided invited testimony on these matters to tribal councils, state legislatures and to Congress. She served on the National Research Council's Committee on Western Water Management and on the National Academy of Science committee investigating use of economic methodology by the Army Corps of Engineers, involving billion-dollar proposed projects on US waterways. Dr. Colby advises public and private sector organizations on managing natural resources and resolving environmental disputes.

Tamra Pearson d'Estrée, PhD in Social Psychology, Harvard University, is Henry R. Luce Professor of Conflict Resolution at the University of Denver. She has also held faculty appointments at the Institute for Conflict Analysis and Resolution (ICAR) at George Mason University, and the Psychology Department at the University of Arizona. Her research bridges conflict resolution and social psychology, encompassing work on social identity, intergroup relations, and conflict resolution processes, as well as on evaluation research and reflective practice. She engages in conflict resolution through leading trainings and facilitating interactive problem-solving workshops in various intercommunal contexts, and she has directed and/or evaluated projects aimed at conflict resolution capacity- and institution-building in several U.S. and international communities.

* * *

Contributors

Connie J. A. Beck, Ph.D., is an assistant professor of psychology in The Psychology, Policy and Law Program at the University of Arizona. Her work focuses on how the legal system creates or exacerbates psychological distress and how it can be adjusted or restructured to minimize that distress. She is currently focused on issues surrounding domestic violence in couples mandated to attend divorce mediation.

Kristine Crandall has a M.S. in natural resource economics, with an emphasis in water resources, from the University of Arizona. She has worked as a consultant in the area of water policy and valuation for over 10 years, assessing the benefits of water for instream uses including outdoor recreation and habitat protection. She has also evaluated the costs and benefits of proposed water projects. Since 1998, she has worked as a Research & Writing Specialist for the Roaring Fork Conservancy, a non-profit watershed conservation organization. She actively follows water quality and quantity issues for the Conservancy, and is helping steer a collaborative project to strategically plan for water management within the Roaring Fork Watershed.

Annette Pfeifer Hanada completed her BA in Cultural Anthropology at DePaul University, and then studied Environmental Science and Policy at George Mason University where she earned an MS. With a National Science Foundation dissertation grant she conducted field research in Germany and Japan, which lead to a doctoral degree in Environmental Science and Policy. She was born in Germany and lived for many years in Japan. She lives and works in the Washingtron DC area.

Landon Hancock holds a Ph.D. in Conflict Analysis and Resolution from George Mason University, and a BA and MA in international relations from San Francisco State University. His research interests center around comparative ethnic conflict causes, processes, intervention strategies and long-term resolution, focusing on the elements of identity that drive these conflicts. His articles have appeared in journals such as *Civil Wars* and

International Studies Perspectives. He has taught at George Mason University and American University, as well as one year teaching high school in Japan.

Kathryn Mazaika, who works to protect natural resources both as a professional, and volunteer in national parks around the San Francisco Bay Area, brings her knowledge of conflict analysis to help resolve problems related to natural resources. She holds advanced degrees in Conflict Analysis and Resolution from George Mason University and Environmental Law from Vermont Law. She also works as an independent researcher while completing doctoral studies in conflict analysis and resolution. She is focusing on natural resource issues, having gained insights from her earlier work on transportation and air quality planning, forestry, and Superfund investigations with the U.S. Environmental Protection Agency. In her recent work, she has studied conflicts surrounding endangered species, water in the American West, and how culture influences intractable conflicts.

Erin McCandless is a scholar, practitioner and activist living in Zimbabwe. She is founder and co-Executive Editor of the refereed, international *Journal of Peacebuilding and Development*, a lecturer at Africa University in Mutare and a co-director of/trainer with the South North Centre for Peacebuilding and Development in Harare. She is a doctoral candidate with American University in Washington D.C.

Part I

DEVELOPING THE FRAMEWORK

Chapter One

UNDERSTANDING 'SUCCESS'

WHY ANALYZE SUCCESS?

Resource conflicts in the American West, and water conflicts in particular, are complex, difficult and costly. The Yakima River adjudication currently ongoing in the state of Washington involves 4,000 claims and 40,000 land owners, and by the time it concludes will have taken over 30 years.[1] The Arizona general stream adjudication involves 27,000 persons asserting 77,000 water rights claims. The Big Horn Case involving Wind River in Wyoming has been to the U.S. Supreme Court once and the Wyoming Supreme Court five times.[2] These conflicts represent huge social costs and drains on communities, industries, and societies. Formal judicial dispute resolution processes have led to less than efficient and enduring solutions; consequently, alternative dispute resolution (ADR) processes have also been tried, with varying levels of success. Though examples of successful resolution exist, we more often hear of protracted litigation, frozen legislative bodies, and defensive agencies and utilities.

Even with the enormous costs of conflicts, relatively little reflection exists on what would be considered to be successful resolution, and even less work exists on how one might document and demonstrate it. Collectively, we are not yet clear on what we are trying to achieve when we resolve these costly and protracted conflicts and what might be the most effective way for achieving these goals. What we lack is cumulative wisdom on the nature of successful environmental conflict resolution. In order to identify "best practices" for addressing disputes, success must be defined, and a framework for organizing, comparing, and learning from past experiences is required.

Thus the impetus for this book. Systematic evaluation can help to develop information about what works. Evaluation methods have as their primary goal the provision of information that is not only useful, but which will actually be used for decision-making and planning.[3] Evaluation methodologies provide useful tools for capturing what actually happened during a given process, for systematically assessing where goals were met and where shortfalls occur or modifications are needed, and for actually documenting changes that have taken place. We strive in this book to define success, to outline a framework and method for documenting it, to use this method to compare cases, and then to realistically evaluate the process of evaluation itself. We conclude with reflections for policy and process in environmental conflict.

The Call for Comparative Analysis

In the last decade, the call for comparative analysis and for lessons on "best practices" in resolving environmental conflicts has become more urgent. Those interested in ECR clearly have begun to critically consider "success" and noteworthy evaluations of particular cases exist.[4] However, little effort has been made to systematically evaluate the relative effectiveness of various ECR processes using comparative analysis. Most writers on dispute resolution start from an assumption that either litigation or ADR is preferable, and then cite cases to support this argument. Fortunately, recent reviewers of the field have noted this lack and have called for comparative analysis. For example, Innes[5] calls for true comparative analysis *within* conflict resolution methods (e.g., different consensus-building designs), *across* conflict resolution methods (e.g.: consensus-building and litigation), and even compared to other methods of influencing environmental conflicts (e.g., lobbying and media campaigns).

Similarly emphasizing the need for such analyses, Emerson[6] notes that research in ECR suffers from two principal shortcomings. First, the research is narrow and lacks effectively integrated relevant research from other fields. Secondly, it is unfocused in that it draws on research related to any non-litigation process rather than addressing well-defined forms of ECR. Emerson also observes that the existing body of work includes very little comparison across cases or dispute resolution methods, and argues that research on how to evaluate ECR processes and outcomes is one of the essential tasks facing ECR research. Emerson finds that most research has been performed at the "micro" level, examining particular cases, and that "theory building" has focused on improved technique and better understanding of why specific dispute resolution processes either achieve negotiated agreements or not. Macro concerns such as the role of government agencies, the efficacy of outcomes and the representativeness of the process are only beginning to be evaluated. She notes, further, that environmental dispute resolution research has not addressed the relationships between mediation and more institutionalized forums, such as the courts and administrative processes. Emerson argues that after two decades of experience with ADR, the time is ripe for more sophisticated analyses and that the rich diversity of cases provides fertile ground for analyses that can provide a better understanding of ECR and policy processes.

Finally, funding agencies that support varied work in conflict resolution echo this concern. In their final report on the activities of the Fund for Research on Dispute Resolution, the National Institute for Dispute Resolution[7] notes that workshops convened to discuss dispute resolution research highlighted a pre-occupation with how the outcome of dispute resolution processes should be evaluated and a strong concern for how evaluation measures would be used. Despite widespread interest among practitioners and researchers, the report notes that research to date has not systematically examined how best to measure success. The report to the Hewlett Foundation on their projects funded from 1984 through 1994 also noted the lack of research and consensus regarding what constitutes "success" in resolving disputes. It reports that only one study among the 18 Hewlett-funded centers focused on definitions of quality in dispute resolution over that period.[8]

Recognizing that the success or failure of a particular conflict resolution effort depends on what one considers to be its goals and priorities, scholar-practitioners have called for more systematic criteria for analyzing success and more comparative case study analyses. In the field of dispute resolution, multiple views on what constitutes "success" exist, and prior research has not systematically examined how to measure it. In short, voices from numerous quarters have concluded that additional substantive efforts are needed to further our understanding of conflict resolution success, how it should be measured, and how the results of evaluation can best be used to improve dispute resolution.

THE NEED FOR A FRAMEWORK

When we began this project in the mid-1990's, little work had been done on defining success in conflict resolution generally, much less environmental conflict resolution in particular, and even less work had been done on how to assess it. Scattered in the literature were several diverse, but often overlapping, frameworks for considering success.[9] Using several strategies including reviewing multiple literatures, interviewing practitioners and researchers, and drawing from evaluation work in related fields, we worked empirically to assemble a comprehensive list of the varying ways that 'success' in conflict resolution has been conceptualized. This work is documented in Chapter 2.

These criteria lent themselves very naturally to conceptual categories: (1) Outcome[10] Achieved, (2) Process Quality, (3) Outcome Quality, (4) Relationship of Parties to Outcome, (5) Relationship Quality Among Parties, and (6) Social Capital [see Figure 2.1]. Our added conceptual organization stimulated further reflection, discussion with practitioners and scholars, and the addition of more criteria.

Even with these multiple definitions of success present in the field, very few case reports address even a fraction of these many criteria. Typically, case studies are written with a specific focus for a specific audience, and thus may contain much information on certain criteria of successful resolution but almost none on others. For one to make any comparisons among cases in order to draw larger lessons becomes almost impossible, as one is forced to compare case documentations that are, figuratively, apples and oranges. For true comparison, cases must document similar criteria.

A clear need exists for a comprehensive framework and method for documenting and reflecting on cases. Both stakeholders and practitioners who toil daily with environmental conflicts, and researchers who observe and study their patterns, need a framework for organizing and analyzing case information. Potential users of such a framework include policymakers, practitioners, agencies, researchers, educators, and students.

Policy makers must stretch limited budgets among budding ADR programs, agency litigation efforts and other activities related to environmental conflicts. They would benefit from comparative work that evaluates the effectiveness of different dispute resolution processes.

Alternative Dispute Resolution (ADR) practitioners and case managers face demanding caseloads which can preclude them from writing the type of reflective case analyses that could provide a cumulative evaluation of effective conflict resolution. A repository of consistently documented cases would encourage more reflection and provide critical information for training new generations of conflict resolution practitioners. As the field of alternative dispute resolution is increasingly called upon to justify itself and "prove" its results, a framework for case reporting and analysis is needed to improve rigor and consistency throughout the dispute resolution profession.

Individual practitioners also would find a comprehensive framework provides a useful vehicle for reflective practice and process design. Once a practitioner became familiar with such an evaluation framework and the many possible criteria for consideration, it would likely encourage reflection before and during future interventions. What would initially be used *ex post* to evaluate, could be used *ex ante* to design.

Many students who take courses on environmental conflict resolution and environmental policy will not ultimately become practitioners. However, many of them will work in public agencies, local or state government, or as part of other groups involved in environmental conflict resolution processes:

legislative staff, technical staff for environmental groups, engineers in corporations, etc. Understanding the multiple issues and dimensions of successful environmental dispute resolution will make them better users and participants in dispute resolution processes. A framework for thorough, yet succinct comparative case analysis would serve both university-level learning and students' future analysis efforts.

In sum, the need exists for an analytic frame that will allow for further reflection and accumulation of wisdom regarding successful ECR. This book introduces a framework that synthesizes much of the wisdom in the field about what constitutes success in ECR.

TESTING THE FRAMEWORK ON WATER CONFLICT CASES

Our task in this book is not only to address the need for accumulating ECR wisdom, by providing a framework for defining and discussing success, as well as a method for documenting it, but also to demonstrate the fruits of engaging in case analysis – single and comparative – using the common framework. We focus in particular on a domain of work that we authors share, the resolution of complex water disputes, but this framework and method could just as easily be applied, and has been applied, to other environmental and public policy conflicts.

Case studies are necessary, both for providing a context for substantive learning, but also for testing and refining the evaluation framework and method. It is through application to actual cases that abstract success concepts are transformed into concrete, measurable indicators. Through examining actual cases, we can determine where and how success actually "lives."

Water conflicts provide an ideal domain for considering both the complexities of environmental conflict resolution and the multiple possible aspects of success. Water conflicts, like many environmental and public policy conflicts, involve multiple parties and multiple issues. They are broad in scope, and often endure for decades. The may seem neverending, as parties return for multiple rounds of litigation. Parties vary in their stakes in the outcome, their resources, their power, and in their internal organization. Water conflicts inevitably involve large amounts of information that is often technically complex. These conflicts also often include a clash of cultures: between urban and rural interests, between ethnic cultures, between traditional resource users (mining, ranching) and new resource users (mountain bikers, hikers, and anglers), and between different values and worldviews on the relationship of humans to the environment.

After reviewing a larger set of water conflicts in the American West, eight cases were selected for analysis. These cases were chosen based on their geographic diversity across the southwest, the range of issues contributing to the conflict, the types of resolution mechanisms attempted, and preliminary evidence that adequate data sources would be available for researchers to analyze the cases. These cases included Big Horn River (Wyoming), Edwards Aquifer (Texas), Lower Colorado River (Arizona, California, Nevada), Mono Lake (California), Pecos River (Texas, New Mexico), Pyramid Lake (Nevada), Salt River (Arizona), and Snowmass Creek (Colorado). Figure 1 shows the location of these cases. The eight cases are described in more detail in chapter 3, and the complete analyses of four cases are reported in chapters 4-7.

These eight cases share the common thread of conflict over water resources in the arid West, but also represent the common categories or character of such conflicts. Cases include disputes over allocation of interstate rivers, conflicts between Native American tribes and non-Indian water users, and conflicts between instream needs and consumptive water uses. They involve endangered species and overlapping jurisdictions. They also provide examples of the multiple processes that may be used to resolve such disputes, including negotiations, court proceedings, administrative actions, and legislation. While it was necessary in the application of our framework for case study researchers to select a specific form of resolution (i.e., court ruling, legislation, etc.) and specific time period in the course of the dispute in order to focus their efforts, the histories of each case reveal a rich mixture of conflict management processes. Litigation was a vital motivating factor at some point in every case.

By examining and considering the contributions of various forms of resolution used to address these water conflicts, this study goes beyond prior ECR research comparing across negotiated agreements or comparing litigation and mediation. We felt it important to evaluate "success" across the variety of resolution mechanisms actually used by parties to attempt to bring closure to environmental conflicts, in order that we might best draw conclusions and venture new hypotheses about the relative merits and possible contributions of these various processes.

BENEFITS OF ANALYSIS AND REFLECTION

In the process of engaging in comparative case analysis using the common framework, recognition of patterns and the accumulation of lessons and learning can begin. We began to appreciate both the richness and difficulties in actually assessing success. We were surprised to learn how processes vary on what type of successful resolution they attempt to achieve, in other words, that not all processes are striving for the same goals. We began to understand more clearly the tradeoffs among criteria and tradeoffs among goals in conflict resolution. We were intrigued to note the unique and potentially complementary contributions of different conflict resolution methods. We were

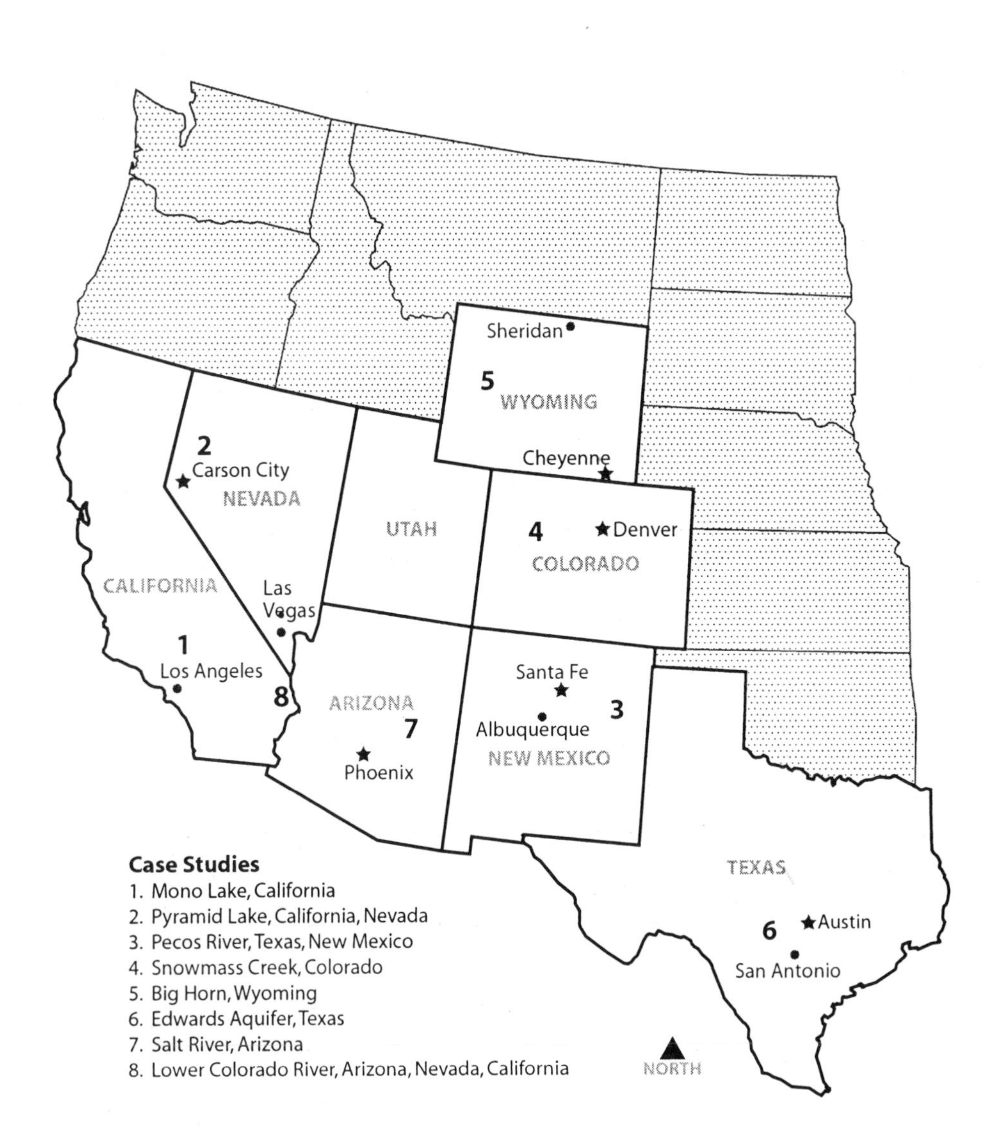

Figure 1.1 Case Study Locations

impressed by the various and creative ways that cases, and therefore communities, attempted to achieve successful resolutions and meet criteria. Finally, we were excited by the goldmine of comparative learning that is possible by noting both successful practices across cases, but also the important contribution of public policies to structuring the likelihood of communities' successfully resolving their conflicts over water and other shared trusts, and the larger lessons these suggest for public managers, policy makers, and concerned citizens.

Users of this book will find numerous benefits. Practitioners will benefit from the thoughtful reflection it encourages. Reflection on practice encourages the advancement of expertise and the development of mastery and artistry.[11] Even if one does not have time to thoroughly document a case, just reading through the Guidebook will make one aware of the many dimensions to consider. Though all criteria are seldom addressed in any one process, reflecting on the categories of what was achieved and what was not achieved will be useful both for subsequent summarizing and reflection, but also for influencing future practice. Practitioners also can be clearer with parties on exactly what may be traded off, given the various process choices possible.

Users will find the Guidebook useful not only for information on criteria and their documentation, but also as a roadmap for the many potential pitfalls and hidden snags on the road to resolution. For example, have all signatories ratified any official agreement? Have culturally effective methods for advertising public meetings and soliciting public input been explored? The accompanying water conflict cases provide illustration of both best practices and worst case scenarios, depending on the criterion of interest.

Students and scholars will learn more on the nature of evalution/measurement, case documentation, and comparative analysis. Are all criteria equally accessible and measurable? Are they equally reliable as measurement tools? What indicators in actual case documentation can be used to measure this criterion? When should it be measured over the life of the case? Does it apply to all types of conflict resolution processes (litigation, negotiation, etc.) or only to some?

Diverse groups of professionals such as Congressional aides, technical staff for environmental groups and corporations, and attorneys representing stakeholders will benefit from a better understanding of the multiple dimensions of successful environmental dispute resolution. This increased understanding can make them more effective participants in dispute resolution processes. Students of environmental policy, law, political science, and natural resource management will benefit from exposure to systematic evaluation of cases and conflict resolution processes. A framework for thorough, yet succinct, comparative case analysis will serve university-level learning, as well as agency and corporate efforts to tackle environmental conflicts.

STRUCTURE OF THIS BOOK

In this chapter, we have introduced the goals and general outline of this book. In Chapter 2, we consider past attempts to define and evaluate success, including explicit attempts to develop evaluation frameworks. We draw conclusions from this review of others' efforts to define and evaluate success. We then propose a framework (with six categories of "success") to organize the twenty-eight criteria used in the definition of success, and examine each criterion one by one. In closing, we discuss the implications of the inherent tradeoffs among criteria, their ties to values and policy, and the use of a common framework in comparative analysis. This chapter is co-authored with our colleague, Connie Allen Beck.

In Chapter 3, we review our methodology for applying these criteria to case studies and for evaluating the value of the criteria themselves as measurement devices. We discuss case study analysis as a methodology, and summarize our cases in further detail. We describe the process of operationalizing and testing each criterion on actual case data and developing the case analysis Guidebook. We review the application of the Guidebook to our cases and summarize our insights on systematic case study evaluation.

Chapters 4-7 present in full detail four of the eight case studies conducted for this project. These four cases were chosen to demonstrate the benefits of applying a common analytic framework (the same set of criteria) to various cases that involve different issues and conflict resolution processes. We include cases which use an administrative rule promulgated by a state agency (Mono Lake, authored by Kathryn Mazaika), a supreme court ruling (Pecos River, written by Annette Hanada and Landon Hancock), a mediated negotiated agreement followed by legislation (Pyramid Lake, written by Erin McCandless), and state legislation followed by a negotiated agreement (Snowmass Creek, authored by Kristine Crandall).

By applying a common analytic framework to the cases, we can thoroughly assess what was achieved in each case and begin to draw tentative comparisons between the varying conflict resolution methods used in those cases. In Chapter 8, we summarize the performance of each of our eight case studies based on each of the criteria contained in our evaluation framework. We note patterns across the cases, and the ways in which the parties and/or authorities have addressed the challenges posed by varying criteria for successful conflict resolution. We conclude Chapter 8 by comparing the cases and resolution processes, and discuss the limits of these comparisons.

In Chapter 9, we explore the limitations of the criteria themselves. Clearly, a wide array of success criteria exists from which to choose. As described in Chapters 2 and 3, the choice of criteria is ultimately a question of values and of the goals for a specific conflict resolution effort. However, certain criteria may be more accessible and measurable, while other criteria, though accessible, may be more difficult to measure *reliably*. This critical examination of the criteria is important in order to inform our goal of defining conflict resolution 'success.' In Chapter 9, we present our analysis of the accessibility, reliability and validity of the varying criteria, as they were operationalized for this analysis.

In Chapter 10, we examine how public policies interact with ECR and suggest directions for public agencies and elected officials to take toward improving ECR. We present examples of problem-solving and innovative strategies drawn from the cases. We highlight findings about the criteria for success, and we note that some criteria seem to have conflicting purposes, presenting tradeoffs for policy-makers, agencies, and communities (For instance, a thorough and inclusive conflict resolution process may rank high on Process Quality but low on Process Costliness). We discuss special uses of the criteria and applications that rely on subsets of the twenty-eight criteria. Finally, we suggest avenues for future research to further our understanding of ECR.

Appendices are included to facilitate further case analysis. Appendix A provides our case analysis Guidebook (also available in CDROM format from the first author). Appendices B and C provide tutorials on the U.S. federal policy framework and financial and economic criteria, respectively.

SUMMARY

This book showcases a new, comprehensive framework for evaluating multiple dimensions of success in water conflict resolution and in ECR more generally. The framework furthers the search for a more systematic understanding of how effectively different conflict resolution mechanisms perform in specific conflict cases. The reflection that the framework and analysis encourages should also further more informed practice and a more conscious and constructive integration of traditional and innovative resolution processes. In addition, the evaluation framework can potentially be extended and adapted to fit other research and assessment needs, such as conflict resolution within communities and organizations and between nations.

It is our hope that the evaluation framework and method presented here facilitates the standardized documentation of the processes used to manage and resolve environmental conflicts. This, in turn, can further the understanding of how different processes perform with respect to various components of success in resolving conflicts. Our ultimate hope is that our children and grandchildren's generations will have learned from the bitter and extended environmental conflicts of the late 20th century to devise more effective, durable and just methods for resolving conflicts over natural resources in the new century.

NOTES

[1] Jim Pharris, Mary Sue Wilson, and Alan Reichman, *Federal and Indian Reserved Water Rights: A Report to the Washington State Legislature by the Office of the Attorney General* (Olympia, WA: Attorney General of Washington, 2002).

[2] Mike Conner, *Congressional Briefing: Background and Status of Indian Land and Water Claims* (Department of Interior Ad Hoc Group on Indian Water Rights, 2001).

[3] AEA Task Force on Guiding Principles for Evaluators, "Guiding Principles for Evaluators," *New Directions for Program Evaluation* (Summer, 1995): 19-34. See also Patton, M.Q. *Utilization-Focused Evaluation* (Thousand Oaks, CA: Sage Publications, Inc., 1997)

[4] J.C. Neuman, "Run River Run: Mediation of a Water Rights Dispute Keeps Fish and Farmers Happy for a Time," *University of Colorado Law Review* 67 (Spring 1996): 259–339.; Innes (1999) cites case studies by Corburn (1996), Southeast Negotiation Network (1993), Kerwin and Langbein (1995), and the Consortium on Negotiation and Conflict Resolution (1994).

[5] Judith E. Innes, "Evaluating Consensus Building," in *Consensus Building Handbook*, 1999. For a recent example of comparative case study analysis in environmental conflict resolution, see Lawrence Susskind, Sarah McKearnon, and Jennifer Thomas-Larmer, eds.

[6] Kirk Emerson, *A Critique of Environmental Dispute Resolution Research,* Presentation to the Conflict Analysis and Resolution Working Group Seminar, University of Arizona, April, 1996.

[7] National Institute for Dispute Resolution, *Final Report: Fund for Research on Dispute Resolution* (Washington DC: Author, 1996).

[8] Robert A. Baruch Bush, *Report on the Assessment of the Hewlett Foundation's Centers for "Theory Building" on Conflict Resolution* (Hewlett Foundation, 1995).

[9] Gail Bingham, *Resolving Environmental Disputes: A Decade of Experience* (Washington, DC: The Conservation Foundation, 1986); E. Brunet, "Questioning the Quality of Alternative Dispute Resolution," *Tulane Law Review* 1, no. 62 (1987): 1-56; Leonard G. Buckle, & Suzann R. Thomas-Buckle, "Placing environmental mediation in context: Lessons from failed mediations," *Environmental Impact Assessment Review* 6 (March 1986): 55–70; Robert A. Baruch Bush, "Defining Quality in Dispute Resolution: Taxonomies and Anti-Taxonomies of Quality Agreements," *Denver University Law Review* 66 (1989): 335–380; Robert A. Baruch Bush, *Report on the Assessment of the Hewlett Foundation's Centers for "Theory Building" on Conflict Resolution*, 1995; A. Bruce Dotson, "Defining Success in Environmental and Public Policy Negotiations," Unpublished manuscript; H.T. Edwards, "Commentary: Alternative Dispute Resolution: Panacea or Anathema?" *Harvard Law Review* 99, (1986): 668-684; Judith E. Innes, "Evaluating Consensus Building," in *Consensus Building Handbook*, eds. Lawrence Susskind, Sarah McKearnon and Jennifer Thomas-Larmer, 631-675 (Thousand Oaks, CA: Sage Publications, 1999); James S. Kakalik, Terence Dunworth, Laural A. Hill, Daniel McCaffrey, Marian Oshiro, Nicholas M. Pace, and Mary E. Vaiana, *An Evaluation of Mediation and Early Neutral Evaluation Under the Civil Justice Reform Act* (Report produced by the Institute for Civil Justice, Rand, Santa Monica, CA, 1996); National Institute for Dispute Resolution, *Final Report: Fund for Research on Dispute Resolution* (Washington DC: Author, 1996); Dean G. Pruitt, Robert S. Pierce, Neil B. McGillicuddy, Gary L. Welton, and Lynn M. Castrianno, "Long-Term Success in Mediation," *Law and Human Behavior* 17 (1993): 313–330; Lawrence Susskind and Connie Ozawa, "Mediated Negotiations in the Public Sector," *American Behavioral Scientist* 27, no. 2 (1983): 255–79; William Ury, J. M. Brett, and S.B. Goldberg, *Getting Disputes*

Resolved: Designing Systems to Cut the Costs of Conflict (San Francisco: Jossey-Bass Publishers, 1988).

[10] An "Outcome" may be a negotiated agreement, a court ruling or a legislative or administrative resolution of a conflict. See Chapter 3 for a further discussion of terms.

[11] Donald A. Schön, *The Reflective Practitioner* (New York: Basic Books, 1983); M.D. Lang, and A. Taylor, *The Making of a Mediator: Developing Artistry in Practice* (San Francisco: Jossey Bass, 2000).

Chapter Two

DEFINING AND EVALUATING SUCCESS IN ENVIRONMENTAL CONFLICT RESOLUTION

With Connie J.A. Beck

In the world of complex and lengthy resource conflicts, parties have been compelled to try various paths for resolution. While numerous attempts have been made to claim benefits for these various conflict resolution processes, few systematic attempts have been made to outline criteria for defining success.

In this chapter we attempt to define success, and consider how it might also be measured. Building on a review of relevant work in several fields as well as practitioner insights, we identify and develop criteria for evaluating the merits of environmental conflict resolution (ECR) procedures and propose an organization for such criteria into six conceptual categories. This organization highlights the various categories possible for criteria that assess environmental conflict resolution and should assist practitioners seeking to identify "best practices" and researchers seeking a framework for comparative analysis.

Although the last decade has seen increasing interest and attempts to judge effectiveness in conflict resolution domains,[1] attempts to define and measure effectiveness in resource conflicts have been relatively rare. We consider first those works focusing on "success" within the domain of resource disputes, and then we review useful contributions from evaluations of conflict resolution processes in other domains.

RESOURCE DISPUTES

Susskind was one of the first to propose criteria for evaluating dispute resolution processes. In 1981, McCrory[2] summarized one of Susskind's earliest presentations on criteria "for judging the fairness of a mediation process and the quality of the mediation agreement." Susskind and his colleague Ozawa offered the following refined list of criteria in a subsequent article:[3]

1) the negotiated agreement is acceptable to the parties;
2) the results are perceived as fair by the community;
3) the results maximize joint gains;
4) the results consider past precedents;
5) the agreement is reached with minimal expenditure of time and money; and
6) the process improves relationships among the disputants.

In reviewing criteria such as these and others, Buckle and Thomas-Buckle[4] point out that such criteria should be evaluated not against some absolute standard, but in comparison to other options available – such as litigation or continuation of the dispute without resolution. They observe that the majority of environmental disputes presented for mediation do not result in a fully implemented agreement and thus they explore the value mediation might provide aside from achievement of an implemented agreement.

Buckle and Thomas-Buckle examined mediators' records and interviewed mediators and disputants in 81 cases. In ninety percent of the cases examined, no agreement was reached. They found that while mediators judged success on the basis of whether an agreement was reached, the disputants had a much broader view of the benefits of mediation. Many of these benefits were process related and included: (1) training in negotiation and communication skills; (2) more frank and open dialogue among the parties; and (3) generation of new options for problem solving. Regulatory agency representatives involved in the cases found the mediators helpful in clarifying the issues and interests and in assembling technical information, all of which was valuable when the cases proceeded to an administrative hearing. Many parties noted that mediation was valued as an indication of good faith that might serve them well when they appear before a judge. Buckle and Thomas-Buckle conclude that success should be defined far more broadly than just in terms of agreements achieved and implemented, and 'success' should incorporate the benefits that mediation sessions provide to participants in subsequent processes, such as judicial and administrative hearings.

Bingham's 1986 book[5] remains the most thorough examination of alternative approaches to resolving environmental disputes and the most explicit attempt to document and evaluate resolution attempts. Her data were based on interviews with mediators and other parties involved in over 160 environmental disputes. Bingham observed that parties care about the outcome of a dispute resolution process, specifically they care about the stability, longevity, and implementability of any agreement that was reached, and the extent to which it satisfies their own interests and what they perceive to be the public interest. Parties also care about the process itself—whether it is perceived as fair and legitimate, whether it is efficient, and whether they were able to influence the outcome. If the parties have an ongoing relationship, then they also care about whether the process enhances the quality of that relationship and their ability to communicate with one another.

Bingham compared 132 environmental dispute resolution cases with an index based on two criteria: whether or not an agreement was reached, and whether implementation was full, partial, or not at all. Of the 78% of cases that did reach an agreement, the majority were either fully or partly implemented. Bingham acknowledged that this analysis did not reflect other potential criteria such as satisfaction, fairness, and improved communications.

Bingham also attempted to examine the 'efficiency' of mediation as compared to litigation, focusing on time and costs expended to reach a decision as the measurable components of efficiency. She observed that systematic comparisons between litigation and mediation proved difficult, being confounded by three issues: lack of closely parallel cases that took the different tracks, the fact that many disputes involve a combination of litigation and negotiation, and the lack of comprehensive data on costs and time delays incurred by the various parties to a dispute.

More recent studies comparing mediation with litigation argue for both cost savings and higher satisfaction. In a recent national survey of environmental and natural resource attorneys' attitudes regarding ADR, O'Leary and Husar[6] found that attorneys considered ADR a source of significant savings for parties, with the average savings to clients who chose ADR over litigation estimated at $168,000. This finding echoes those of Kloppenberg,[7] who found that in a demonstration project on the use of ADR in federal district courts, attorneys reported parties' substantial cost savings of several thousands of dollars as well as satisfaction with the results. Similarly, in their study of 100 land disputes assisted in resolution by a professional neutral, Susskind, van der Wansem, and Ciccarelli[8] found that 81 percent of participants felt their mediation had consumed less time and money than would have been consumed by their same dispute had it been resolved through more traditional adjudicatory processes. Most attorneys in O'Leary and Raines[9] Environmental Protection Agency study felt that "ADR saved time and money." A study by the Oregon Department of Justice[10] for the Oregon legislature compared the costs of several forms of dispute resolution across diverse types of disputes, and found mediation averaged the cheapest at approximately $9,537, while taking a case through to trial was the most expensive, at $60,557. A summary of evidence for the cost-effectiveness of ECR can be found in Orr's briefing for the U.S. Institute for Environmental Conflict Resolution.[11]

O'Leary and Husar's[12] survey highlighted other benefits of ADR as well, which in turn suggest additional success criteria. These include fairer cost allocation, solutions more likely to benefit all parties, agreements on remedial measures, better understanding of other parties' interests, resolution of difficult technical issues, and long-term benefits such as environmentally beneficial projects and positive corporate-government relations. Even when the primary controversy is not settled, benefits mentioned include allowing hostile parties to communicate and share information that otherwise might not have been shared, better pretrial preparation and issue clarification, a 'reality check' for parties on options and possibilities, and commitment among parties to creating their own solution.

Yet another attempt to examine and frame effective environmental conflict resolution produced what came to be known as the "Park City Principles."[13] In 1991, the Western Governor's Association and the Western States Water Council held three workshops on Western water management in Park City, Utah. The workshops included federal, state, and local governments, Indian tribes, private users, and academics. The attendees wanted to improve Western water management by considering competing demands and attending to the public interest. The group authored a set of principles which are meant to serve as a guide to institutions developing and implementing effective water policies.

These principles included the following core components: recognition (both legal and administrative) of diverse interests and values; a holistic or systemic approach that encompasses all affected interests and needs within the "problemshed" (i.e., basin-wide, rather than within traditional political boundaries); a policy framework that is responsive and balances flexibility, adaptability, and predictability; a decentralized approach that recognizes the role of states and tribes; a focus on negotiation, incentives, and performance standards, rather than command and control; and encouragement of broad-based participation in both policy development and administration.[14]

Neuman's[15] case study on mediation in the Umatilla Basin (located in Oregon) focuses on satisfaction and on the quality of the substantive outcome, concluding that while mediation has much to recommend, it does not assure a superior substantive outcome when compared to litigation or other dispute resolution approaches. She measured the disputants' satisfaction with the process and the outcome and, in addition, evaluated both the mediation process and the substantive outcome.

Neuman defines several products that dispute resolution processes can provide: private goods, public goods, and hybrid goods. Private goods are those benefits provided to the disputants for which they are willing to expend time, money, and other resources. Private goods include getting the problem solved, the satisfaction of being heard and being treated fairly, and other psychological satisfaction.

Public goods are those benefits that extend beyond the disputants themselves. These include an outcome that is consistent with laws, policies, and public values and also "just" in the broad sense of the right thing being done. Therefore, public goods encompass the production of a fair process and outcome in a manner that builds confidence in dispute resolution mechanisms. This helps create constructive precedents to guide future disputes. Public goods also include the enhancement of the overall credibility and legitimacy of the social system.

A hybrid category of products includes those that combine public and private good aspects: resolution of the dispute in an expedient manner without excessive costs or delays; provision of a forum for creative and flexible solutions that address the array of the parties' needs and interests; and building of commitment and legitimacy for the outcome by giving the parties control over the process (as contrasted with giving the courts control and having legitimacy stem from court authority). The most important of the hybrid goods is the ability of dispute resolution to create, preserve, and enhance working relationships among the parties and, therefore, build the parties' capabilities for solving problems together.

In order to evaluate the process and outcome independently of the participants' evaluations, the author examined four components of the process and outcome: the mediation process itself, the resulting natural resource allocation, the implications for resolving public policy disputes, and the implications for regional water management.

With regard to providing the public goods mentioned earlier, Neuman suggests that a conflict resolution process be evaluated as to whether a reasoned outcome is created, consistent with public policies and values. In addition, the outcome should be perceived as just, enhance confidence in the government and law, and build legitimacy and consensus about working within the system. Neuman's final evaluative measure is that the outcome sets a constructive precedent for future similar disputes.

In June 1997 two organizations, Resolveand the National Institute for Dispute Resolution, [16] convened a roundtable in Washington, D.C., for researchers and practitioners. Criteria discussed as characterizing a successful process were: including all relevant parties, promoting problem-solving, de-escalating the conflict, building empathy among participants, empowering and giving voice to less powerful interests, generating a decision (which may or may not constitute a consensus agreement) that is well-informed, building institutions for future policy decisions, setting a good public policy precedent, contributing to a good track record for ADR, and positively influencing future interactions. Group dialogue emphasized a diversity of approaches to defining success and the need for systematic research on evaluating and measuring various components of success.

Dotson (undated)[17] both embraces and rejects a signed agreement as the critical criterion for success. He considers the 60 cases he and his colleagues have so far facilitated for the Institute for Environmental Negotiation at the University of Virginia, and found half of the cases explicitly strove for this goal, while the other half had primary objectives other than a signed agreement. He considers cases with signed agreements to be the "classic successes." Projects he and his colleagues considered "successful" that did not include a signed agreement were often designed for achieving other purposes. These other purposes included clarifying and developing general consensus on public policies or scientific opinion, training parties in negotiation skills, identifying issues and developing conflict assessments, developing a common agenda, developing a common sense of needs and opportunities, or reaching better mutual understanding and improving relationships. He draws parallels with Carpenter and Kennedy's[18] non-hierarchical list of goals for conflict management strategies that includes exchanging information, identifying issues and interests, developing acceptable options, and developing recommendations, in addition to reaching agreements.

Of those cases Dotson examined that sought agreement, slightly over half failed to reach it. However, other areas of achievement fell into three categories: improved understanding, improved relationships, and improved process. Improved understanding included scrutinizing existing frameworks, creating new solutions, stimulating later change in parties, and establishing a new "base camp" for the next round of negotiations. Improved relationships included restoring community where it was lacking (and thereby possibly "dampening the ardor for litigation"), bringing parties together without their attorneys, and rebuilding trust through honest communication. Finally, an improved process included establishing new mechanisms for ongoing communication and dialogue, fostering intercultural understanding so that future rounds (e.g., appeals) are confined to narrow and specific areas of the dispute, and creating momentum and interest in future phases of joint problem-solving.

Dotson wonders if "conclusive resolution" is the appropriate goal. He concludes with a concern that if written agreements were the only standard of success, some might (erroneously) take on only those cases that are likely to reach an agreement. This is a difficult but important point. Dotson argues that perhaps success of a mediation effort should be evaluated by looking forward: What does it create, rather than what does it resolve? Even if an agreement were not signed, an effort could be considered successful if it were seen as moving parties and institutions toward a sustainable society.

In her seminal contribution on the evaluation of consensus-building processes, Innes[19] considers both the challenges and opportunities presented by doing evaluation. She considers criteria of process to be the ultimate bellweather of consensus-building success, in that "consensus-building stands or falls on the acceptability of its process."[20] She considers the process of evaluation to be difficult, requiring many different methods. However, evaluation can prove useful to various people at all stages of the consensus-building process – before, during, and after.

Before Innes ventures into her discussion of criteria, she first considers the possible outcomes of consensus-building in addition to, or possibly instead of, an agreement. She suggests that these outcomes fall into four categories: (1) new relationships (increased trust, decreased future conflicts), (2) new partnerships, organizations, and processes, (3) knowledge and learning (about issues, about others' concerns and constraints, about new possibilities and problem frames), and (4) social, intellectual, and political capital.

Innes' most revolutionary contribution is her application of the theories of self-organizing systems, or "complexity theory,"[21] to understanding the possible contributions of consensus-building. She considers the typical context for consensus-building as one on the "edge of chaos" where rapid and ongoing adaptation by the system is required. Processes for resolving conflict in such contexts must allow for such self-organization, feedback, and information sharing. Thus additional criteria for successful conflict resolution should address a system's new or improved capacity to successfully undergo such ongoing adaptation and respond to an ever-changing context.

In addition to those just mentioned, other process criteria she ventures include building civil discourse (respectful, face-to-face, equal access to information), inclusiveness, purpose-driven, engaging, incorporating high quality information, challenging of assumptions, and thorough in exploration of issues and creative responses. Innes argues that if process criteria are met, then the outcome criteria outlined above will also likely be met.

Using the New Community Meeting (NCM) model and content analysis in their study of a community consensus building process, Gwartney, Fessenden and Landt[22] looked for success by asking three questions: First, did the process increase new types of positive interactions among previously alienated individuals and groups; second, did the process encourage core participants to promote collaboration beyond their group to peers and other previously alienated individuals and groups; and finally, did the process encourage core participants, their peers and their constituents to collaborate in a way that benefited the larger community? Gwartney and colleagues found through content analysis of four community newsletters that following consensus building processes, positive interactions improved in type and tone, previously alienated constituencies referred to one another in more positive and less alienated tones, and all parties increasingly collaborated to benefit the larger community.

While attempts to outline criteria for evaluating successful environmental conflict resolution have been relatively rare, common themes can be seen across the varying works reviewed above. For example, there seems to be almost universal agreement that 'agreement' is only one of many possible criteria. Fairness, "soundness," and flexibility of the agreement are important, as are other outcomes such as new options, improved relationships among the parties, and increases in parties' skills and understanding. Parties' satisfaction with both outcomes and process is also important. The process should also be perceived as fair, inclusive, relationship-building, and judicious in its use of time and money. Finally, long-term criteria for success, when discussed, focus on agreement implementation and parties' improved capacity to address future conflicts and to collaborate more broadly. These themes find their analogs in calls for evaluation in conflict resolution more generally.

GENERAL CONFLICT RESOLUTION

As interest in questions of effectiveness and evaluation has increased throughout the field of conflict resolution, writers have struggled to define criteria and frameworks. Some writers confine their analysis to a particular domain, such as family mediation or court-annexed mediation, while others attempt to address general processes. The categories of criteria outlined in these works can supplement our consideration of criteria discussed in the environmental domain.

Ury, Brett, and Goldberg[23] draw on their experience with management/labor disputes in the coal industry to examine the costs of disputes and dispute resolution processes and to develop potential mechanisms for minimizing those costs. They suggest that the following four factors be examined to evaluate dispute resolution approaches:

1) *Transaction Costs* – the costs generated by the dispute itself and of reaching a solution, including direct costs and opportunity costs of time and money, decline of working relationships, damage incurred by the parties to their reputations as well as their assets, loss of their social capital (good will) and limited ability to expend social capital on other matters or on future disputes.
2) *Satisfaction with outcome* – the degree to which affected parties are satisfied with the outcome of the dispute and the process used to achieve that outcome.
3) *Durability of solution* – endurance of the agreement over time. The degree to which the solution discourages recurrence of the same or of similar disputes, and the degree to which it is enforceable and implementable in a manner that discourages further disputes over implementation.

4) *Effect on relationships* – the degree to which the process improves and maintains functional working relationships among the parties.

Ury, Brett, and Goldberg note that the four criteria are interrelated. For example, dissatisfaction with a dispute resolution outcome or process implies a strain on relationships, which can lead to a higher probability of recurrence of the dispute and thus to higher transaction costs. Sometimes the disputants might choose to incur higher transaction costs up front through extended information gathering and negotiating in order to reduce strain or dissatisfaction over the longer run. The four categories and their associated costs can collectively be called the "costs of disputing."

Pruitt, et al.[24] studied long-term success in community mediation through analysis of recorded remarks during mediation sessions and interviews with participants. They defined short-term success as reaching an agreement, being satisfied with it, and reaching mediation goals; they defined long-term success as complying with the agreement in the long-term, improving relationship quality, and avoiding new problems. Surprisingly, they found that clarity and feasibility of the agreement were unrelated to compliance or any other aspect of long-term compliance. In fact, in their study, no relationship was found between short- and long-term success as they defined them. However, the perception that mediation was fair (procedural fairness) and thorough *was* positively related to long-term success, as defined above. The implication of their work is that long-term success is not connected to the agreement itself, but rather to how the parties viewed the process and changes in the quality of their relationship.

Kakalik, et al.[25] looked at the use of mediation and early neutral intervention in six federal district courts that had implemented pilot programs to reduce delays and litigation costs as mandated by the Civil Justice Reform Act of 1990. The six programs differed considerably in their features, including whether use of ADR was voluntary or mandatory, the point in the litigation when referral to ADR occurred, the number and length of ADR sessions, the type of provider, and the costs to the parties. Of the three-hundred litigation cases examined in each of the six districts, half were referred to ADR. The comparison group remained in the traditional court process which included bilateral negotiations and judicially facilitated settlement or adjudication.

Nine criteria were used to compare the ADR and non-ADR cases. These were: (1) time to disposition measured from first filing to case closing; (2) costs to the parties (defined primarily in terms of lawyer work hours per litigant, but also by examining monetary legal costs and time spent by the litigants on the case); (3) costs to the court of start up and administration of the ADR program; (4) likelihood of monetary exchanges between the litigants; (5) likelihood of settlement when cases are referred to ADR; (6) lawyer and litigant perceptions of how fairly the cases were managed; (7) lawyer and litigant satisfaction with case management; (8) lawyer and litigant satisfaction with the ADR process itself; and (9) overall opinions and recommendation of litigants, lawyers, and ADR providers.

The data came from court records including detailed case processing and docket information for the cases studies; mail surveys of ADR providers (66% response rate), lawyers (50% response), and litigants (12% response, no statistical analysis conducted on litigant response data); personal interviews with judges, ADR providers, court staff, and lawyers during site visits to each of the six districts; and each district's cost and delay reduction plans and reports produced to comply with the 1990 Act. Analysis of the data showed no significant difference in time to disposition, lawyer work hours per litigant, lawyers' perceptions of fairness of case management and satisfaction with case management between ADR cases and non-ADR cases. ADR cases in all six districts were more likely to produce a monetary exchange than litigated cases, and settlement of ADR cases ranged from 31 to 72% across the six districts, with a correspondence between higher settlement rates and holding the first ADR session later in the litigation process. Attorneys noted that ADR sessions held "too early" are not useful as the parties are not ready to settle and ADR seemed more likely to produce a settlement when it was initiated after the litigation discovery process had already revealed basic facts and positions on the issues.

By contrast, Wissler's[26] large study of 1,730 general civil cases mediated in Ohio in the period 1992-2000 found that ADR was *most* useful when it occurred early in the case. Mediations were drawn from cases filed in nine different Ohio courts of common pleas. Data was collected from mediation exit questionnaires completed by attorneys, parties, and mediators. Wissler found that "when mediation was held sooner after the case was filed, settlement was more likely, fewer motions were filed, and the case disposition time was shorter, even in cases that did not settle in mediation."[27] Because many of the studies comparing ADR and non-ADR cases use ADR cases with long periods between filing and mediation, this finding may help to clarify some of the mixed results in the literature. In her extensive review of other civil mediation studies comparing ADR and non-ADR cases, Wissler notes these mixed results or no differences between ADR and non-ADR cases in time from filing to resolution/agreement, in amount of discovery, or in compliance rates.

Organizing criteria into clusters or categories can begin to surface the underlying values that guide any evaluation process. Bush[28] reports on a workshop hosted by the University of Wisconsin Law School in 1987, attended by dispute resolution practitioners and researchers, and devoted to identifying and measuring quality in dispute resolution processes. He identifies six clusters of quality components as emerging from the workshop:

1) Individual satisfaction – the parties feel satisfied with the process and the outcome;
2) Individual autonomy – the parties' capacity to resolve their own problems without reliance on external institutions is enhanced;
3) Social control – public and private institutions are strengthened in their ability to control social unrest and to minimize exploitation by special interests;
4) Social justice – existing inequities in distribution of power and wealth are ameliorated, or at least not exacerbated;
5) Social solidarity – common values and reference points are provided and reinforced, increasing solidarity in a pluralistic society; and
6) Personal transformation – disputants have the opportunity to grow and change, becoming less self-centered and more responsive to other interests.

As the workshop participants noted, these are not complementary criteria that can be applied as a set to evaluate any particular process. Rather, they are to some degree mutually inconsistent and competing objectives, among which tradeoffs will have to be made or at least priorities established.

Bush argued that ambiguities in values prevent meaningful application of the criteria to evaluate actual disputes and that criteria for evaluating quality are inextricably linked to one's vision of an ideal society and so no universal definitions of quality can be expected to evolve. In particular, he notes that the quality criteria for an individualistically-oriented evaluator will differ markedly from those of an evaluator with a collective vision of society. He suggests that no single set of criteria for defining success will emerge given diversity in social vision and that perhaps the best we can hope for is careful elucidation of the vision underlying a particular set of criteria.

This diversity in underlying values and priorities is reflected in ongoing debates in the literature about the relative efficacy of various conflict resolution processes. For example, Edwards[29] articulated the concern that informal ADR problem-solving processes must not be allowed to erode the role of the courts in disputes where broad public values are at stake. In particular, he warned that mediation of environmental disputes, such as toxic waste cleanup, may allow private groups to establish standards for cleanup without the checks and balances of a formal process with public scrutiny and the rigorous application of environmental law. Such agreements, he argued, are negotiated and then presented *de facto* to regulatory officials who then must approve them or send the disputants back to protracted litigation. He also argued that matters of national policy cannot be mediated without the hazard of "second class justice," i.e., decisions not accountable to law and public policy. He concluded by noting that ADR has a place in disputes where the law is well defined and disputants cannot delimit public rights in a manner convenient to resolving their dispute. Clearly, he remains wary of ADR applied to policy disputes.

Brunet[30] also questioned whether ADR can provide the same quality of justice as court rulings. He argued that ADR processes undermine the function of substantive law in promoting clear norms, guiding social behavior, and upholding the public values that underlie legal principles. Consequently, society as a whole may be worse off, even if the direct disputants have negotiated a solution that satisfies them. The law may be progressively undermined as "creative solutions" send a pernicious signal that the rights and duties of the law are no longer paramount and are unworthy of enforcement. He posited that widespread use of ADR may increase the number of disputes by causing legal norms to atrophy so that potential disputants no longer have a strong incentive to abide by the letter of the law as they can negotiate away some of the consequences of non-compliance.

Brunet observed that litigation produces public goods, beyond the effects on the litigants themselves. The cumulative stock of court decisions improves and clarifies the law over time, reducing costs and uncertainty for future disputants. ADR, he argues, bypasses the need to develop and publicize substantive law on matters of public importance. Further, ADR has no clear guidelines for gathering and exchanging information as the courts require in discovery procedures, raising the possibility that decisions made through ADR may not be based on full and accurate information.

In addition, Brunet posits that the absence of a judge, an effective and credible public authority, diminishes the quality of ADR outcomes, as does the absence of requirements for adequate legal representation for each party. He concludes by observing that ADR provides a competitive impetus to the attorneys and judges to better address excessive delays and costs, and that court-annexed ADR may be an excellent mechanism for ensuring that substantive law and procedural concerns are incorporated into ADR processes.

Clagett[31] echoes many of Brunet and Edwards' concerns while acknowledging ADR's place in environmental problem-solving. Regarding negotiated rulemaking and policymaking, he urges both EPA and the U.S. Institute for Environmental Conflict Resolution (IECR) to address the following tensions: serving the public interest versus seeking consensus, confidentiality of issues that should be subject to public scrutiny, balancing power and political legitimacy, the need for professionals trained in dispute resolution methods who also understand complex environmental frameworks, and finally, differences between EPA and IECR systems of dispute resolution.

Thus, those critical to ADR provide alternative criteria for success that must be considered when comparing litigation with negotiation, consensus-building or other environmental conflict resolution processes. These include the quality of the outcome from the perspective of 'the public' ("public scrutiny") or someone in whom the public has vested judging powers, the precedent set by the outcome and the clarity it adds to future dispute resolution attempts, and that the decision or agreement be based on full, shared, and accurate information, and address existing environmental law.

In reviewing various frameworks for evaluation of conflict resolution in domains outside environmental conflict resolution, we once again find comparable themes. Agreements can be assessed for clarity, feasibility, durability, implementability, and parties' satisfaction. Processes can be evaluated for their costs, fairness, and thoroughness. In addition to reaching agreement, other outcomes include improved relationships, enhanced good will, and increased social capital. Long-term assessments can be made on parties' compliance and ability to resolve future disputes, and on agreement implementation. Less prominent in these other domains than in ECR were concerns about agreement "soundness" or public scrutiny, perhaps because agreements in some other domains (e.g., community mediation, labor disputes) have less broad implications for public welfare and public policy. However, these criteria were voiced by ADR critics. Also less commonly mentioned was assessment of parties' new skills and understanding. New themes found in this broader review included criteria of social control, social justice, and parties' personal transformation.

Across diverse fields, scholars have addressed various aspects of defining and evaluating success in dispute resolution. While only a relatively small subset of this effort has focused on environmental conflict resolution, common themes for evaluation begin to emerge across these domains. We see common concern for identifying process and outcome criteria that expand traditional definitions of success beyond simply agreement or participant satisfaction. We see recognition of long-term as well as short-term criteria, and system-level criteria as well as individual-level. Yet still lacking is the systematic review, organization, operationalization, and application of some or all of these criteria to specific cases. Particularly those articles that have reviewed research in environmental dispute resolution conclude that such additional efforts are essential to further understanding of what constitutes success, how it should be measured, and the use of evaluation in improving dispute resolution.

The call for comparative analysis is clear. But through what sort of framework? Comparative analysis, at least in its most systematic and reliable form, is possible only if cases each report on similar dimensions. However, most case analyses typically are done through the particular lens of their authors, and thus few case reports have information on comparable variables. As a first step toward true comparative analysis within the field, we have developed and tested a standardized and comprehensive list of criteria to be used for reporting on case studies.

CRITERIA FRAMEWORK

We utilized three main sources to identify useful criteria for evaluating success in ECR: existing literature; interviews conducted with ECR practitioners, parties to environmental conflicts, and researchers; and reasoning by analogy from related fields where evaluation research has a longer history, such as psychotherapy process and outcome research. Applying the criteria to actual cases forced us to clarify and operationalize each criterion (see Appendix A). Feedback from outside practitioners and researchers was garnered through several public presentations of this framework, and criteria were added and adjusted accordingly.

A significant issue considered in identifying criteria for effective ECR intervention was whether the task was to identify (1) criteria that *define* success, or (2) criteria that *predict* success. The former interpretation would be a case of identifying ways of conceptualizing the "dependent variable" – that is, what practitioners, conflicting parties, courts, and the larger society consider to be an "effective resolution." By contrast, the latter interpretation would be a task of identifying those "independent variables" that significantly and reliably predicted, or correlated with, success. Examples of this latter category include Bourdeaux, O'Leary, and Thornburgh's[32] identification of the influence *on* success of participants' control over process, communication opportunities, and presence of key stakeholders, and Lamb, Burkardt and Taylor's[33] discovery that success is improved when the process is staged, with values addressed before technical issues. Throughout our analysis, we struggled with the specific variables' positions in this causal equation. We chose to focus on the first category and task, *defining* success, but acknowledge that many of these criteria may be conceived by some as more appropriate for predicting success, where success is defined yet another way.

CONCEPTUAL CATEGORIES

Through our extensive review, analyses and discussions, we recognized the usefulness of putting the various criteria within a conceptual framework or higher-order organization. Thus, we propose that criteria for success, and hence definitions of success, fit into one of six conceptual categories: (1) Outcome Reached, (2) Process Quality, (3) Outcome Quality, (4) Relationship of Parties to Outcome, (5) Relationship Between Parties/Relationship Quality, and (6) Social Capital (see Figure 2.1). Conceptual categories can be useful tools for framing and considering further questions, such as: What does this categorization tell us about underlying dimensions or assumptions about the goals and practice of conflict resolution? Do categories suggest criteria that may be missing? We use this framework to organize our investigation into the many criteria available for evaluating success.

Category One: Outcome Reached

The first category of criteria is *Outcome Reached.* The notion of reaching a settlement or concluding an agreement as a measure of successful resolution is fairly universal, and often implicit. For example, in their study of successful community mediation, Pruitt, et al.[34] identify **reaching an agreement** to be their first criterion of success. Bingham[35] analyzed 161 cases of mediated environmental disputes and found the parties' objective in 132 of the 161 cases was to reach an agreement. For Bingham, reaching agreement meant that negotiations resulted in a signed agreement, or that "the parties reported verbally that they had reached an agreement and could describe its terms."[36] In Bingham's analysis, both **unanimity** and **ratification** were required in order to consider that an agreement had been reached. More recently, much discussion has centered on the value of substituting **consensus** for unanimity as the decision rule (where a decision is accepted when all parties may not agree with, but "can live with," the result). Whether the decision rule is unanimity, consensus, or some alternative, the associated criterion would be one of reaching the goal of the chosen decision rule.

Mediation training will often focus on reaching an agreement as the goal of mediation sessions. For example, Moore's[37] classic training book defines mediation as an intervention by a neutral third party "to assist disputing parties in voluntarily reaching their own mutually acceptable settlement of issues in dispute,"[38] and focuses mediators ultimately toward "achieving formal settlement." Mediators are taught to gear their efforts toward producing an agreement that itself meets certain criteria (i.e., cost-efficiency, clarity, realism, and ability to withstand public scrutiny for fairness).

EFFECTIVE ENVIRONMENTAL CONFLICT RESOLUTION CRITERIA CATEGORIES

I. **CRITERIA: OUTCOME REACHED**
- Unanimity or Consensus
- Verifiable Terms
- Public Acknowledgement of Outcome
- Ratification

II. **CRITERIA: PROCESS QUALITY**
- Procedurally Just
- Procedurally Accessible and Inclusive
- Reasonable Process Costs

III. **CRITERIA: OUTCOME QUALITY**
- Cost- Effective Implementation
- Perceived Economic Efficiency
- Financial Sustainability/Feasibility
- Cultural Sustainability/Community Self-determination
- Environmental Sustainability
- Clarity of Outcome
- Feasibility/Realism (legal, political, scientific/technical)
- Public Acceptability
- Efficient Problem-Solving

IV. **CRITERIA: RELATIONSHIP OF PARTIES TO OUTCOME**
- Outcome Satisfaction/Fairness as Assessed by Parties
- Compliance with Outcome Over Time
- Flexibility
- Stability/Durability

V. **CRITERIA: RELATIONSHIP BETWEEN PARTIES (RELATIONSHIP QUALITY)**
- Reduction in Conflict and Hostility
- Improved Relations/Trust
- Cognitive/Affective Shift
- Ability to Resolve Subsequent Disputes
- Transformation

VI. **CRITERIA: SOCIAL CAPITAL**
- Enhanced Citizen Capacity to Draw on Collective Potential Resources
- Increased Community Capacity for Environmental/Policy Decision-making
- Social System Transformation

Figure 2.1: Criteria Categories

Controversy exists concerning whether or not reaching an agreement should be the primary criterion for success.[39] It is the quintessential example of a short-term-focused criterion. Although an agreement may be necessary for success, it may not be sufficient. An agreement may not last beyond the ink drying before disagreements about implementation, new environmental contingencies, or changes in the relationship between the parties render the original Outcome impotent at directing parties' future behaviors. By contrast, others may argue that focusing beyond the Outcome ventures into a realm of implementation that may be beyond the mandate and/or responsibility of certain conflict resolution processes (e.g., litigation). What may distinguish these apparently competing perspectives is their focus on either immediate or long-term criteria of success, as well as different conceptualizations of how conflicts are resolved. We will address this argument in later categories.

Category Two: Process Quality

The second criteria category is that of *Process Quality*. Once an Outcome is reached, one of the first considerations in predicting its successful implementation, and in fact determining whether or not a successful Outcome actually has been concluded, is consideration of the process used to reach the Outcome. Such considerations include those labeled "procedural justice," encompassing the procedural components of the commonly cited criteria of "satisfaction" and "fairness." This body of work addresses parties' satisfaction with the process utilized to resolve the dispute, as well as the parties' sense of the fairness of the process utilized, including considerations such as consistency, accuracy, and representation. Also included are concerns about inclusiveness and the costs of the process.

Procedural Satisfaction and Fairness

Although the terms **procedural satisfaction** and **procedural fairness** can be measured separately and can be argued to be different concepts, for the most part researchers have combined these concepts. Though some researchers[40] believe the terms carry different connotations, these same researchers do not find any empirical evidence to support the contention that respondents draw any such distinctions. In fact, factor analyses conducted by Lind and Tyler[41] show that items asking about satisfaction, fairness, and the propriety of a procedure, and trust in the procedure all load onto the same factor.

Lind and Tyler[42] do, however, appear to draw some distinctions between the variable *procedural fairness* and *procedural justice*. Measures of procedural fairness ask direct questions about *the fairness of a specified procedure,* whereas measures of procedural justice ask questions about *features of a pro-*

cedure (i.e., does decision-maker consider all sides before making decisions, does decision-maker take enough time to consider policy decisions, etc.).

Various researchers break down the criterion of *procedural fairness* into components. The following seven procedural elements have been outlined by various researchers:[43] consistency, decision control (control over final decision), process control (control over presentation of evidence before decision is made), neutrality (honest and lack of bias), competence (making factual decisions), politeness, and respect for rights.

Different elements of procedural justice have been explored in greater detail. Researchers have also found that the meaning of justice varies depending on the nature of the dispute or allocation involved. In formal settings, emphasis is placed on bias suppression, decision quality, consistency and representation. On the other hand, in cooperative situations, focus is placed on consistency, decision quality, and ethicality.[44] Roehl[45] designed a measure of procedural justice in the context of mediation and court procedures. Her scaling process found subscales for presenting evidence, coercion, third party fairness, time/cost, clarity of rules, overall fairness, and respect for the disputants. In a laboratory setting, she found that although the outcome of hearings were identical in fair and unfair hearings, outcomes were perceived as unfair when the process was unfair.

Reasonable Process Costs

Questions about the cost of a conflict resolution process are often raised. Raab[46] suggests "process-related resource savings" as one criterion to examine whether some dispute resolution processes are less expensive in terms of time, money, and other resources, as compared to a baseline case of litigating the dispute. Examination of costs encompasses both long- and short-term aspects. Short-term considerations include up-front benefits and costs to disputants, and benefits and costs to others who are not direct participants – such as taxpayers, the environment, resource users not at the table, and the "public interest." Long-term cumulative costs are considered under Outcome Quality and relate to the stability and flexibility of the Outcome, the frequency with which the parties need to revisit its provisions and negotiate adjustments, and the effectiveness of the initial process in building capacity among the parties to resolve future related disputes.[47]

While there is a general sense that ADR processes are less costly than litigation[48] very little discussion exists regarding what types of costs and whose costs need to be considered (i.e., taxpayer and "public" costs in addition to costs borne by disputants themselves). Better understanding of the costs of ADR processes can help us more systematically compare different ways of resolving disputes. Moreover, private firms and public agencies accustomed to taking a litigation approach to disputes want to know whether to taking a litigation approach to disputes want to know whether there are real economic advantages to ADR processes.[49]

Little empirical evidence is available beyond perceptional evidence to document actual cost differences. Alternative dispute resolution may not be less expensive in terms of up-front costs as both litigation and mediated negotiations require information to be gathered on the environmental issues, the parties' needs, and on potential solutions. Moreover, in many cases the parties simultaneously prepare for litigation and negotiations, or litigation is ongoing concurrent with alternative dispute resolution. The case for cost savings in ADR processes may lie in comparing long-term cumulative costs over multiple stages of dispute resolution, implementing Outcomes, and revising Outcomes as conditions change.

One characteristic of a cost-effective process is development of a shared information base regarding the resource. The process must provide incentives for the best technical and scientific information to be made available for use by the parties. Information sharing among the disputants implies that some trust among the parties has been established. When disputes are settled in litigation, technical expertise and information is used as a weapon with which to defend one's interests and there is little incentive to seek out objective data and to build a common understanding of technical matters that can lead to better management of the resource. Consequently, voluntary negotiations may be likely to lead to better and lower cost access to information.

Category Three: Outcome Quality

The third criteria category is that of *Outcome Quality*. Beyond the simple fact of reaching an Outcome, "success" usually incorporates evaluation of aspects of the Outcome itself. Presumably, an Outcome can be evaluated by a neutral outside party, or society, on the basis of certain criteria, regardless of the relationship of the principal parties to that Outcome. These are often short-term criteria. Criteria that fall into this category include cost-effective implementation, perceived economic efficiency, financial feasibility/sustainability, cultural and environmental sustainability, outcome clarity, and feasibility (legal, political, and scientific/technical).

Cost-Effective Implementation

Implementation costs include monitoring (e.g., metering water use, evaluating species restoration), enforcement (e.g., penalties included in Outcome), and responding to new conditions (e.g., drought, more scientific knowledge about a river system, species, needs). These costs should be considered up front in order to craft an Outcome that gives the parties incentives to comply (thus reducing monitoring and enforcement costs) and that specifies a process for dealing with the unexpected.

It would be useful to learn more about how process costs (above) and implementation costs interact. If the parties spend more early on, take their time and don't rush to resolution, does this reduce implementation costs and dispute recurrence? The cumulative expenses are a key issue as complex policy disputes can involve hundreds of millions of dollars in land, water, lost income, technical studies, regulatory costs, and legal fees.

Perceived Economic Efficiency

Perceived economic efficiency asks whether the dispute resolution outcome is viewed as creating net benefits ("net benefits" are benefits minus costs) that would not have been available otherwise. In order to rigorously assess economic efficiency for the outcome of an environmental conflict, it would be necessary to describe all of the relevant costs and benefits and to quantify them in dollars. However, less formally, economic efficiency asks "was it worthwhile?" "Are the costs justified by the benefits?" This concept of weighing benefits and costs is central to the "mutual gains" negotiation framework described in *Getting To Yes*[50] and applied to environmental and public policy disputes in Susskind's work.[51] It is sometimes called "creating value" or converting zero sum negotiations to positive sum negotiations.

Outcomes which are voluntary agreements are likely to satisfy this criterion for the direct parties to the agreement. If the agreement fails to provide improvements for those who sign on, compared to their available alternatives to the agreement (BATNAs[52]), they would decline to bind themselves to the agreement. Litigated outcomes, and other outcomes that do not involve voluntary consent of the parties, are unlikely to satisfy this criterion for the immediate parties. However, they still may be perceived as producing net benefits for society as a whole. Net benefits may arise from avoiding the costs of prolonged litigation, from improved natural resource management, from cleaner air or water, and from better sharing of information and technology among the parties. Many analytic challenges arise in documenting and quantifying the various types of benefits and costs that arise from resolving a dispute, thus perceptions regarding benefits and costs are more often assessed. As noted earlier, O'Leary and Husar's[53] survey of environmental and natural resource attorneys found that several benefits to ADR-produced Outcomes were perceived, including fairer cost allocation, mutually beneficial solutions, and the resolution of difficult technical issues.

Sustainability

Sustainability implies practices that allow for preservation of current resources in such a way that future generations will have comparable resources available to them. We have investigated three criteria of sustainability: *environmental sustainability*, given the natural resources involved in the Outcome, *cultural sustainability/community self-determination*, given the "cultures" and lifestyles impacted by the Outcome, and *financial sustainability/feasibility*, considering the distribution of costs and benefits over time. The sustainability criteria apply here to allocation and management arrangements that are produced by a dispute resolution process, e.g., the outcomes of the process. Characteristics of the process itself that may contribute to the long-term durability of the arrangements are discussed above under stability.

Environmental Sustainability

An outcome that is **environmentally sustainable** needs to carefully consider how the resource central to the dispute will be used and managed, and also must consider other linked resources. In the case of water conflicts, other resources might include fish, wildlife, hydropower, other forms of energy associated with water use and conveyance, and alternative land uses related to water allocations. Those concerned with population growth would add that sustainability criteria should include the degree to which an outcome contributes more water for urban development and thus contributes to associated problems such as congestion, reduced open space, increased urban encroachment into wildlife habitat, and poorer air quality. Others argue that a certain level of growth must be taken for granted and that the evaluation of negotiated outcomes should not focus on problems associated with growth but rather on the most sustainable ways to accommodate the growth that is inevitable.

With regard to water resources, sustainable allocation and management arrangements must consider drought and flood cycles and specify how water will be shared in dry years when there is not enough to meet the competing demands that are accommodated in normal years. This implies that negotiated agreements need to carefully spell out how the risks of shortages are to be distributed among water uses and to specify mechanisms for meeting essential water needs under unusual circumstances such as a prolonged drought.

Cultural Sustainability

Cultural sustainability considers the effects of an Outcome on affected communities. These include demographic and economic effects, such as changes in patterns of jobs, income, taxes, etc. Negotiated agreements or other Outcomes may also result in changes in patterns of ownership, changes in decision-making authority or jurisdiction, and changes in the social or cultural "lifeways" of the affected communities or the relative balance of these lifeways (the "cultural mix"). Examples of "lifeways" include irrigated farming, ranching, community gardens, Native American cultural practices, Hispanic ditch associations (acequias), small town life. Concerns about communities are often raised during times of potential change.

This criterion includes concerns over community self-determination and sovereignty. Bargaining processes and Outcomes can shift control and power among the parties. Tribal sovereignty issues are a classic example of this concern. Other examples involve the ability of local communities to control their future by influencing management of the public lands and water supplies that affect their livelihoods and the quality of their environment. Negotiated Outcomes need to consider who has the power and the voice to make or to influence choices, as authorized by the Outcome, versus who will bear the consequence of those choices. This is likely to be linked to perceived fairness. A match between a party's influence over a decision and that party bearing the consequences of that decision is more likely to be regarded as fair than a mismatch between decision power and bearing consequences. For instance, when a federal agency decision in Washington, DC, causes closure of timber operations in rural Oregon, part of the local outrage stems from having no voice in a decision that affects jobs and livelihoods in their community.

Concern with local self-determination is at the heart of much of the dissatisfaction with federal land and water management in the West. Components of community self-determination that may be affected by some Outcomes (such as legal rulings) include: (1) jurisdiction (which level of government has the legal authority to make decisions concerning water allocation and management); (2) ownership (who holds title to the resources and has secure access to their continued use); and (3) control over economic future (who influences the purposes for which resources are available, the types of jobs generated, and economic stability and vitality of local communities).

Financial Sustainability/Feasibility

Economic or **financial sustainability** considers the ability of the parties, including federal taxpayers, to bear costs arising from the Outcome over time. Outcomes sometimes defer tough questions such as whether or not the parties actually have the financial ability to pay for new water supply development or for construction of additional power plants to provide electricity to move water to new locations. In order to reach agreement, costs may be postponed through loans and creative financing arrangements for a decade or more into the future and the impact of these future repayment obligations is not realistically considered when the Outcome is negotiated. This deferral of realistically assessing parties' ability to pay is a concern in a number of settlements of tribal water claims in which new supplies are to be developed. Often the brunt of the costs is borne by the federal government and non-Indian water users, but tribes also obligate themselves to pay millions of dollars per year in water delivery costs and on-reservation infrastructure development without adequate revenue stream to cover these obligations. Evaluation of financial sustainability entails assessing the ability and willingness of those who will bear the costs in the future to meet their financial obligations.

Clarity and Feasibility

Other criteria applied to the Outcome itself include **clarity** and **feasibility/realism.** In addition to cost-efficiency, Moore[54] includes clarity, realism, and ability to withstand public scrutiny for fairness in his criteria for successful agreement. In their mediation study, Pruitt, et al.[55] operationalized their measurement of the "soundness of an agreement" as its rating of clarity and feasibility as assessed by coders.

In addition to political feasibility, Outcomes must conform to existing law (legal feasibility) or involve changing it. They must also be based on scientific and technical assumptions that are considered sound (scientific/technical feasibility).

Many of these criteria of Outcome quality can be evaluated immediately upon conclusion of an Outcome, at least in a predictive sense, and thus constitute short-term criteria of success. However, because the Outcome is implemented in a context that is dynamic and changing, these criteria can and should also be examined as long-term criteria. For example, as the environmental context changes, an Outcome dubbed initially as sustainable may turn out to be untenable under new resource realities. An Outcome judged as financially feasible under certain economic conditions may no longer be cost effective or successful if economic conditions change significantly. Thus, although criteria of Outcome quality are assessable in a short-term time frame (as opposed to other criteria discussed below that can only be assessed *over* time), such assessments should be considered as having low reliability (in a methodological sense), given their probabilistic nature. These criteria should be reassessed in a long-term frame as well.

Public Acceptability

The quality of the outcome may be judged by the general public. Disputants, or the general public, may deem that certain disputes are best addressed in a public forum (see Moore's[56] private-public continuum of processes), or at least that certain Outcomes should be subjected to "public scrutiny." In fact, conflict resolution processes occurring in public, do so in part because they are assumed to be addressing larger public issues. One of the common criticisms of ADR leveled by advocates of more traditional legal processes is that public scrutiny, either by the general public or by someone the public has vested with such powers, is compromised.[57] However, clear methods for assessing public acceptability have not been well articulated.

Efficient Problem-Solving

Negotiation theory suggests that efficiencies can be created when the conflict resolution process allows for collaborative problem-solving. If parties can work together, they can recognize opportunities for mutual gain, and collaborate to "expand the pie." Exchanges can be made that benefit everyone without anyone losing anything ("elegant trades"[58]). An efficient agreement is one where parties have not missed opportunities for "elegant trades,"[59] and where parties have "created value" by problem-solving together.

Category Four: Relationship of Parties to Outcome

The fourth criteria category is that of the *Relationship of Parties to Outcome*. In addition to a neutral, objective evaluation of the Outcome itself, an important measure of resolution success includes how the parties feel about and behave toward the solution. Some argue that in fact this is the primary measurement of success. Are the parties themselves satisfied with the Outcome reached, regardless of outsiders' assessments of the Outcome? Do they feel that terms of the Outcome are "fair" and represent "justice"? Will the parties abide by the Outcome reached? This category of criteria contains both criteria that can be immediately assessed upon reaching an Outcome as well as criteria that can only be assessed over time, in the longer term. Criteria discussed in this category include: outcome satisfaction and perceived fairness, economic components of perceived fairness, compliance, flexibility, and stability/durability.

Outcome Satisfaction/Fairness

In the short term, one can assess parties' sense of **satisfaction** with and **fairness** of the Outcome. As with procedural fairness and satisfaction discussed above, Lind and Tyler[60] found that items asking about satisfaction with outcomes, fairness of outcomes, and the extent to which outcomes reflect the true situation, all loaded strongly on one factor they called distributive fairness.

Such distributive fairness or 'distributive justice' can be assessed in several ways. Pruitt, et al.[61] asked mediation participants to rate their satisfaction level immediately after mediation on a scale from very dissatisfied to very satisfied. Tyler and Griffin[62] found differences in distributive justice principles used in different settings. In short-term instrumental exchanges among strangers, where effectiveness is a more important goal than positive interpersonal relations, equity is typically the predominate distributive justice principle. In exchange relationships among friends and long-term relationships, where maintaining positive interpersonal relations is as important as short-term effectiveness, equality is more important.

Disputants have different beliefs regarding their baseline entitlements and their "fair share" of the costs of resolving the problem being disputed. These "baselines" determine their perceptions of the fairness of a particular outcome and their internal starting point in negotiations. In environmental disputes, often there are key disagreements about the baseline rights and the validity of each party's starting point. Power contests such as litigation, media or public opinion wars, and political power battles frequently are used to convince other parties that their starting points (their baseline assumptions about their rights) are weak.

Many different types of baseline entitlements are invoked in negotiating over rights to use water, including equity, geography, historic water use, or need. In general, each party will select an approach to define its baseline that results in the biggest claim on the disputed resource. There are no objective criteria for choosing *among* the different baseline positions in multi-party bargaining as each baseline can be justified from the particular perspective of the party advocating that baseline. Rights-based claims lead to legal wrangling and development of competing legal positions, while needs-based claims generate differing assessments of the parties' "legitimate" environmental needs and also incite behaviors that amplify needs-based claims, such as high population growth and development of new uses for environmental resources.

Fairness may be perceived partly as reciprocity – in terms of what I give up versus what you give up. However, recognition of reciprocity also depends on the parties' internal baselines. To use an example of water disputes again, agricultural interests may receive no "credit" from environmentalists for implementing new conservation practices if environmentalists take a view that farmers historically "wasted water."

Another component of fairness involves allocating dispute-related costs. In policy disputes, Outcomes can involve new infrastructure, revised regulations, and other elements with large price tags. The distribution of these costs among the parties (and taxpayers) often is a key factor in the overall context of the dispute resolution process. To what extent should differing financial capabilities play a role in allocating costs? For instance, should both wealthy and poorer disputants bear equal shares of the costs of obtaining additional water to resolve their conflict? Should differences in historic access to and benefits from the disputed resource affect how benefits are distributed under the new Outcome? If shares of past damage to the resource can be attributed among the parties (relative pollutant loads, for instance), how should this affect apportioning of costs?

As noted above, perceptions of outcome fairness and satisfaction also are linked to perceptions of process fairness. Analysis of the fairness of a specific dispute outcome involving changes in resource allocation must focus on both the process employed and the outcome achieved. Voluntary negotiations in which all disputants have a voice and the parties build up trust in one another's integrity and ability to abide by Outcomes may be more likely to be perceived as fair than solutions imposed by courts and bureaucrats.

Compliance

Some criteria involving the relationship between the parties and the Outcome can only be assessed over time. Probably the most common indicator of long-term success we have reviewed is **compliance.** Compliance is often assessed as present or absent, and often through self-report, as in "have you followed all the terms of your agreement?"[63] or with a scale running "no violation" to "major violation."[64] Compliance may be assessed after as short a period as four months[65] or after several months or even years.

Another common measure of compliance is relitigation, i.e., these parties returning to court on an issue related to this dispute or to dispute the terms of their agreement. Quite a large literature exists using relitigation rates as measurements of successful resolution in divorce mediation.[66] The literature is mixed on the virtues of litigation versus ADR in reducing relitigation. However, one potentially serious confound in the way both non-compliance and relitigation are measured is that they both typically encompass cases where parties have a further dispute but also cases where parties have returned to court to make official some mutually agreed upon alteration to the original agreement. Thus "flexibility" may be coded as "noncompliance."

Flexibility

In much of this same literature, **flexibility** is seen to be a positive measurement of success, and in fact linked to durability. An Outcome is flexible if parties feel they can modify their Outcome in a mutually agreeable way as circumstances change. In fact, divorce mediators often encourage parents to return to court to modify agreements in light of new circumstances, explaining what may look like higher relitigation rates among mediation samples over litigation samples.[67] Flexibility, whether formally documented with the court or informally agreed to between the parties, is positive in that it is a measure of the parties' ability to resolve new issues on their own.

Stability/Durability

The stability of an Outcome can only be assessed as implementation of an Outcome proceeds and the parties face new challenges that test the strength of the original agreement. However, it is possible to consider the factors that contribute to stability, as an Outcome is being formulated, so that appropriate adjustments to enhance stability can be built in during negotiations. Features to enhance stability include incentives built into the Outcome to reinforce the parties' commitment to its provisions, such as penalties for non-compliance.

Other types of incentives to enhance stability include carefully staging the satisfaction of each party's key needs over the course of implementation. So long as each of the parties still needs something essential to fulfill their interests from the Outcome, they have an incentive to uphold the Outcome in the face of internal and external challenges. If one of the parties has all of their needs satisfied early on, they have little reason to contribute political and economic influence if the Outcome later is in jeopardy.

Stable Outcomes need to withstand new environmental constraints and changing values, to be able to adapt to unexpected events such as droughts, floods, or new court rulings. One way to provide this adaptability is through specifying in the Outcome a process for negotiating modifications as needed (see "flexibility" above). Such a provision needs to address who sits at the table when modifications are discussed, the power of the various parties to influence modifications (including veto power), and principles for sharing the costs of adopting new water management and water allocation strategies. For Outcomes that will require years to fully implement, some mutually agreeable ongoing forum is essential to enhance stability.

The quality of the relationships that evolve among disputants over the course of the conflict also affects stability. Development of mutual assurance, reciprocity in "give and take" and trust that other disputants will not use their influence to undermine implementation of the Outcome all contribute to stability.[68]

Meierding[69] defines stability or durability as short- and long-term compliance with agreements. Pointing to studies that indicate that in divorce mediation, one-third of successfully negotiated agreements are no longer working within a year, Saposnek defines success in conciliation in terms of "workable, written agreements that hold up over time."[70] After Davis and Roberts (1988) found that 42% of agreements in their family conciliation sample had broken down within 18 months, they comment that it "is somewhat artificial to assess the impact of mediation simply in terms of the durability of an access agreement arrived at in the course of one negotiating session. A better measure would be to try and assess whether... these parents are better equipped to negotiate together."[71] Unless the relationship has changed to allow for further conflict resolution, quality Outcomes may not be adequate.

Category Five: Relationship Between Parties (Relationship Quality)

The fifth criteria category is that of the *Relationship between Parties* (or *Relationship Quality*). In addition to assessing compliance with the Outcome, most discussions of success include some indicator of the quality of the relationship between the parties. The ECR practitioners we interviewed in particular felt this category of criteria to be important, as did evaluation research in other social interventions such as psychotherapy.[72] Various methods are outlined, including direct measurements of relationship quality, ability to resolve future disputes, reduction in conflict and hostility, cognitive/affective shift, and transformation. These criteria are usually assessed over time, and in the long term.

General Relationship Quality

Pruitt, et al.[73] assessed improvement in long-term **quality of the relationship** directly with two scales: current relationship with the other party (very unpleasant to very pleasant) and a measure of whether the relationship worsened, remained the same, or improved. They also coded for any development of new problems. They found that problem-solving training contributed to improved relations between the parties. As has been shown in marital therapy, this result suggests a critical role of rehearsal in problem-solving in conflict resolution sessions. Long-term relations between distressed couples in marital therapy improve with problem-solving training. Complex skills, including problem-solving skills, only improve with practice.

The criterion of **improved relations** seeks to capture changes in the way parties see and relate to one another. To note change, one must first note the nature of the original relationship as a baseline for comparison. In addition to rating the relationship, changes can be noted in how the relationship is discussed, the tone of communication used (hostile, conciliatory), the effort parties expend to protect themselves, and their sense of trust.

Ability to Resolve Future Disputes

Another relationship quality criterion is the **ability of parties to resolve inevitable future disputes.** As discussed above, a good measure of success would be parties ability to negotiate together in the future.[74] In their study of relitigation, Keilitz, et al.[75] point to the importance of parties' ability to maintain amicable and cooperative relationships, an ability they say is unlikely to be permanently improved by only a limited time in mediation. Changes in entrenched patterns of relating take time, a willingness and desire to change, and practice.

Reduction in Conflict and Hostility

Johnston, et al.[76] considered **reduction in conflict and hostility** to be an indicator of success in their studies of conflictual couples. The researchers measured conflict using the Straus Conflict Tactics Scale,[77] which is comprised of 18 behavioral questions on how disagreements have been managed during the previous year. Other mediation studies[78] measured reductions in hostility using scales such as the O'Leary-Porter Scale (OPS) on which parents self-report the frequency of hostility displayed in front of a child. Although these measures have been developed primarily for use in marital and family conflict, analog measures could be developed for other domains. In fact, in studies of international conflict, one of the primary indicators of successful "resolution" is an end or reduction in overt hostilities.[79] Gwartney, Fessenden, and Landt[80] applied this idea in their study using the New Community Meeting (NCM) model to evaluate how participants related to each other over the course of the NCM process.

Cognitive and Affective Shift

Interviews with both local and regional ECR practitioners produced repeated references to a "shifting" in the way parties saw each other as a result of an effective process, in essence a **"cognitive shift."**[81] Although this indicator of successful resolution was commonly noted by practitioners, both within the environmental domain and outside, little research has been done on it directly within conflict resolution research.[82] Other areas of research, however, have sought to capture or measure such changes or shifts in relationships. Primary among these is psychotherapy research.

Authors in the therapy literature that discuss "cognitive shifts" fall into three basic theoretical camps: behavioral, cognitive, and family systems. The first school, the behaviorists, sees cognitive shift in terms of acceptance. Change refers to compromising with and accommodating to a partner, whereas the authors argue 'acceptance' refers to a letting go of the struggle to change and in some cases embracing those aspects of the partner which have traditionally been precipitant of conflict. Previously offensive behaviors, including the partner's pursuit of his or her own interests, are accepted; the value of accepting the partner's point of view is underscored.

In this first school, the third party practitioner (in this case, the therapist) fosters such a shift with four types of intervention, some of which parallel common mediation interventions. First, the therapist listens and rephrases the problem without the blame and accusation that usually surrounds problems. This intervention is generally referred to as reframing. Second, the therapist promotes a style of communication that involves having the parties learn to discuss the problem as an "it," a common enemy rather than something that one does to the other, or as a problem to be solved. Third, the therapist attempts to increase each person's tolerance of the other's negative behavior. Methods for shifting a person's view of the other's problem behavior include pointing out the positive features of the negative behavior. Fourth, the therapist encourages each party to become less dependent on the other, and therefore more accepting of their partner's imperfections. With more independence, the other's imperfections are less critical and the focus shifts from other to self. The actual occurrence of a cognitive shift is assessed through an interview procedure, where changes in feelings are noted.

The second school of therapy that explicitly aims for cognitive shifts as a goal is the cognitive-behavioral school. These cognitive theorists speak directly about the concept of cognitive shifts. They also criticize the behaviorists, above, believing that "acceptance" as a goal is too vague and suggests lack of change.[83] They argue that the goal is a balance of changes in cognition, behavior, and affect. Conflict is in part the result of information processing errors. Cognitive variables commonly causing conflict among parties that thus need to shift are in five areas: selective attention, attributions, expectancies, assumptions, and standards.

Specific interventions have been designed to modify cognitions in each of these five areas of conflict.[84] For example, some parties may selectively attend to the negative aspects of the other parties and/or the relationship. An intervention goal would be to shift attention to a more representative balance of both positive and negative aspects of the other parties or relationship. An intervenor can use specific interventions in each of the five areas described above to achieve cognitive shift. Shifts are assessed through self-report questionnaires, interviews, and scoring of observational data taken during conflict resolution sessions.

Finally, the family systems school suggests that in families, organizations, and communities, shared narrative stories evolve that define "reality" within a given group. The presence of one narrative reality effectively excludes certain other interpretations of reality. The goal of this therapeutic approach is to facilitate or promote change in specific stories or the relationship between stories, thus changing the parties' experience of the existing reality.

Sluzki[85] actually bases what he calls a "transformative process" in therapy in part on earlier work on mediation. Interventions are designed through addressing various dimensions of the story and attempting shifts in specific ones. Dimensions for shifts include *time*, *space*, *causality*, *interactions*, *values*, and *telling-style*.[86] For example, if a story is ahistoric in content, the intervenor might request a history of the characters to provide an avenue for change by setting the story in context. Cognitive shifts are assessed through the therapist noting shifts on these dimensions of the stories told.

Finally, it may be more accurate to conceive of these shifts as both affective and cognitive, that is, not only do parties change their stored representation of the others' characteristics and behaviors, but also change their basic evaluation or emotion associated with the other from more negative to more positive. As such, changes should be able to be noted not only in how the other is conceived, but also in how the other is evaluated. The therapeutic approaches discussed above all explicitly or implicitly note such changes as part of "cognitive shifts."

Transformation

Fundamental changes in how one thinks and feels about another are typically irrevocable, i.e., once a new explanation for another's behavior is understood, the old ways of explaining and perceiving become useless and are discarded. Such changes have the feel of changes in physical states, where matter takes on new appearance and new form. A criterion for success that currently is broadly discussed throughout the conflict resolution field is the notion of **"transformation."**[87]

Some argue that conflict presents an opportunity for individual and collective moral growth.[88] More specifically, this moral growth is toward a social vision that integrates individual freedom and social conscience, and integrates concerns over justice and rights with concerns about care and relationships.[89] This moral growth can occur if conflict resolution processes help people to change their old ways of operating and to achieve new understanding and new relationships through conflict. Others argue that transformation occurs at the level of social institutions and social relations.[90]

Transformation may appear as the parties' renewed sense of their own capacity to handle challenges, as empathy for and acknowledgement of the others' circumstances,[91] and as evidence of other major shifts in perception beyond just shifts in perceptions of the other, e.g., of relationship context, of paradigm, of social and political context, of "the problem" or of tools and solutions. Though transformation has only recently begun to be researched empirically, results suggest parties recognize its benefits, and mediators can be trained to work with third parties in this direction.[92]

Category Six: Social Capital

Though not in our original framework, the sixth and final criterion category was added after numerous discussions with researchers and practitioners. We distinguish this criterion category from others in that this criterion category includes criteria that address positive changes that occur in the *larger system* in which this conflict is embedded: changes that go beyond the relationships between these particular stakeholders and/or beyond the particular issues in this conflict. These changes are grouped loosely together as Social Capital.

Social capital, like other forms of capital, is a potential resource that must be drawn upon to realize its value. It is the capacity for individuals to command resources that comes from having social connections. However, social capital does not inhere in individuals; rather, it is a characteristic or possession of relationships and communities.[93] It has been defined as potential assistance relationships between people,[94] "generalized reciprocity,"[95] the capacity for individuals to command scarce resources by virtue of their membership in networks or broader social structures,[96] the aggregate of these actual or potential resources linked to a durable network,[97] or even the capability for trusting strangers.[98] In Innes'[99] discussion of consensus-building, reviewed above, she identifies the importance of a system's (i.e., a community's) increased capacity for responding that comes from cooperation and coordination. Allen's[100] work on community conflict resolution considers social capital as stemming from the transformation of social interaction to collective action. This occurs through increased acquaintance and personal knowledge, as trust and reciprocity develop between individuals and organizations. Gwartney, Fessenden, and Landt[101] note evidence of collaborating in new ways that benefit the larger community.

Of the many ways that social capital has been operationalized in theory and research and has been predicted to appear, three themes appear that could be considered as criteria. The first, **enhanced citizen capacity to draw on collective potential resources**, builds on group members' ability to rely on potential assistance from others in their network. Processes that build relationships and weave networks among people increase their collective potential resources. The increased potential for assistance and greater collective resources allows for greater risk-taking and creativity. Fukuyama[102] argued that such 'trust,' even among strangers, allows for people to spontaneously work together for common purposes.

Innes[103] argued that the most important result of a consensus-building process may not be an agreement per se, but the increased ability for a community to handle future challenges. Two additional criteria speak to this new ability for responding. First is the **increased capacity of the community for environmental and public policy decision-making**. Many practitioners have argued that the true test of success is whether or not parties can translate their new way of resolving conflict to new issues, new relationships, and new domains.[104] New skills learned and new patterns of behavior provide an increased capacity in the community for cooperation. As a result, activities are better coordinated and duplication of efforts is reduced. Information is shared rather than hoarded, and in fact information gathering becomes a joint activity. As a system, the community is more efficient, and it is more able to act proactively rather than reactively to new challenges.

Finally, successful conflict resolution produces a community that has truly evolved into an integrated, adaptive, learning system – one that has undergone

social system transformation. The community's or system's integration results in coordinated responses to new crises and challenges. The system is more resilient.[105] The positive changes produced through the conflict resolution process create the capacity for continued learning and improved action,[106] and fosters "double-loop learning"[107] where ways of problem-framing are themselves reexamined, and creative new responses are considered. Perceptions of responsibility shift to encompass the entire community system.

As one assesses increases in social capital, it is important also to assess the presence of the two types of conditions that foster its formation and use: networks and perceptions. While networks include such "horizontal associations" as volunteer groups, community associations, and social clubs, other networks that are vital to social capital are communication channels between diverse groups and community residential stability. Perceptual conditions fostering social capital include the perception that one has relationships based on mutual reciprocity and assistance, that one is interdependent with others in the community, and that one can trust the community's members and institutions. While these networks and perceptions form a foundation for the development and strengthening of social capital, they, in turn, are further strengthened as social capital is used. Networks and positive perceptions are reinforced as those relationships are drawn upon for mutual assistance, engendering still more social capital.

Across multiple literatures, and through feedback from many practitioners and scholars, we have assembled a comprehensive list of ways to conceive of "success" in environmental conflict resolution. Our organization into conceptual categories provides a structure for thinking about the varying goals contained within conflict resolution processes. Rather than providing simple answers to evaluation questions, systematic consideration of the criteria for success raises further important issues. We turn to these issues now.

DISCUSSION

Defining Success

Our examination of the many possible criteria for success highlights that there are inherent tradeoffs in working for "successful" environmental conflict resolution. Increasing stability of an Outcome may reduce flexibility. Increasing the inclusiveness of the process may reduce the likelihood of unanimity or consensus. Environmental and cultural sustainability may be at odds. Therefore, it is not possible to simply develop a "scorecard" or "report card" that checks for high performance on all criteria. Ultimately, criteria must be selected and prioritized, a task that requires the application of values. This task moves from the realm of science or research into the realm of policy, where the community values are articulated, negotiated, integrated, and applied. Once criteria are chosen and prioritized, assessments can be made on the merits of a given conflict resolution process, always keeping in mind that the assessment is ultimately linked to questions of value.

A Framework for Ongoing and Future Work

Before engaging in comparative analysis, one must have a common framework: a 'yardstick' to allow for comparison on the same dimension. One of the main goals of this chapter was not only to explore the varying criteria for success that researchers and practitioners discuss and utilize, but also to develop a framework that would allow for comparative analysis. The articulation and organization of criteria begins this process; the criteria are further developed and operationalized in Appendix A's Guidebook. The following chapter will discuss the application of this framework to cases for both individual case analysis and comparative work – tasks demonstrated in subsequent chapters.

We advance this framework with hopes that future case studies also will use this framework to be thorough in their case reporting, so that future comparisons can continue to be made across cases using a consistent set of variables. Such standardization of a reporting framework will allow researchers to begin accumulating information about successful practices across cases. In addition, it will allow for more thorough, systematic, and unbiased comparisons of different conflict resolution *methods*, such as we begin to do in Chapter 8. Many have begun to argue, in both environmental conflict resolution and in other conflict resolution domains[108] for contingency planning: for the thoughtful selection of methods appropriate to a given conflict context. Knowing more about the differing strengths and weaknesses of various methods can go far toward eliminating the zero-sum, competitive approach that often exists between, for example, advocates of litigation, legislation, and negotiation. Such knowledge also will serve the best interests of those whose lives are most affected: those who struggle with how best to resolve their disputes.

NOTES

[1] P.A. Gwartney, F. Fessenden, and Gayle Landt, "Measuring the Long-term Impact of a Community Conflict Resolution Process: A Case Study Using Content Analysis of Public Documents," *Negotiation Journal* 18(1) (2002): 51-74; Marc Howard Ross and Jay Rothman, *Theory and Practice in Ethnic Conflict Management: Theorizing Success and Failure* (New York: St. Martin's Press, Inc. 1999).

[2] J.P. McCrory, "Environmental Mediation—Another Piece For the Puzzle," *Vermont Law Review* 49 (1981): 77-79.

[3] Lawrence E. Susskind and Connie Ozawa, "Mediated Negotiations in the Public Sector," *American Behavioral Scientist* 27, no. 2 (1983): 255–79.

[4] Leonard G. Buckle, and Susann R. Thomas-Buckle, "Placing Environmental Mediation in Context: Lessons From Failed Mediations," *Environmental Impact Assessment Review* 6 (1986): 55–70.

[5] Gail Bingham, *Resolving Environmental Disputes: A Decade of Experience* (Washington, DC: The Conservation Foundation, 1986).

[6] Rosemary O'Leary and Maja Husar, "That Environmental and Natural Resource Attorneys Really Think About Alternative Dispute Resolution: A National Survey," *Natural Resources and Environment,* 16, no.4 (2002): 262-264.

[7] Lisa A. Kloppenberg, "Implementation of court-Annexed Environmental Mediation: The District of Oregon Pilot Project," *Ohio State Journal on Dispute Resolution,* 17, no.3, (2002): 559-596.

[8] Lawrence E. Susskind, Mieke van der Wansem, and Armand Ciccarelli, "An Analysis of Recent Experience with Land Use Mediation – Overview of the Consensus Building Institute's Study," in *Mediation Land Use Disputes Pros and Cons* (Lincoln Institute of Land Policy, 2000).

[9] Rosemary O'Leary and Susan Summers Raines, "Lessons Learned from Two Decades of Alternative Dispute Resolution Programs and Processes at the U.S. Environmental Protection Agency," *Public Administration Review*, 61, no.6, (November/December 2001): 682-692.

[10] Oregon Department of Justice, "Collaborative Dispute Resolution Pilot Project," A report submitted January 30, 2001, to the Honorable Gene Derfler, Senate President, The Honorable Mark Simmons, House Speaker, and The Honorable Members of the Legislature.

[11] Patricia Orr, "ECR Cost Effectiveness: Evidence From the Field," Briefing (Tucson, AZ: U.S. Institute for Environmental Conflict Resolution, April 16, 2003).

[12] Rosemary O'Leary and Maja Husar, "That Environmental and Natural Resource Attorneys Really Think About Alternative Dispute Resolution: A National Survey," 2002.

[13] Western Governors Association, *Park City Principles* (document produced following a series of three workshops developed by Western Governors Association and Western States Water Council), 1991.

[14] These principles have recently been revisited in a 1998 Western Governor's Association document. See Chapter 10 for more details.

[15] J.C. Neuman, "Run River Run: Mediation of a Water Rights Dispute Keeps Fish and Farmers Happy for a Time," *University of Colorado Law Review* 67 (Spring 1996): 259–339.

[16] Resolve, *What is Success in Public Policy Dispute Resolution? Building Bridges Between Theory and Practice,* A roundtable sponsored by Resolve and the National Institute for Dis-

pute Resolution, Washington, DC, 1997; National Institute for Dispute Resolution, *Final Report: Fund for Research on Dispute Resolution* (Washington DC: Author, 1996).

[17] A.B. Dotson, "Defining Success in Environmental and Public Policy Negotiations," (Unpublished manuscript, no date).

[18] Susan Carpenter and W.J.D. Kennedy, *Managing Public Disputes* (San Francisco: Jossey-Bass, 1988).

[19] Judith E. Innes, "Evaluating Consensus Building," in *Consensus Building Handbook,* eds. Lawrence Susskind, Sarah McKearnon, and Jennifer Thomas-Larmer, 631-675 (Thousand Oaks, CA: Sage Publications, 1999).

[20] Ibid.

[21] e.g. Fritjof Capra, *The Web of Life: A New Scientific Understanding of Living Systems* (New York: Anchor Books, 1996), and others.

[22] P.A. Gwartney, F. Fessenden, and Gayle Landt, "Measuring the Long-term Impact of a Community Conflict Resolution Process: A Case Study Using Content Analysis of Public Documents," 2002.

[23] William Ury, J.M. Brett, and Stephen B. Goldberg, *Getting Disputes Resolved: Designing Systems to Cut the Costs of Conflict* (San Francisco: Jossey-Bass Publishers, 1988).

[24] Dean G. Pruitt, et al, "Long-term Success in Mediation," *Law and Human Behavior* 17 (1993): 313–330.

[25] James S. Kakalik, et al, *An Evaluation of Mediation and Early Neutral Evaluation Under the Civil Justice Reform Act,* Report produced by the Institute for Civil Justice (Santa Monica, CA: Rand, 1996).

[26] Roselle L. Wissler, "Court-connected Mediation in General Civil Cases: What We Know From Empirical Research," *Ohio State Journal on Dispute Resolution* 17 (3), (2002): 641-704.

[27] Ibid., 697-698.

[28] Robert A.Baruch Bush, "Defining Quality in Dispute Resolution: Taxonomies and Anti-taxonomies of Quality Agreements," *Denver University Law Review* 66 (1989): 335–380.

[29] Harry T. Edwards, "Commentary: Alternative Dispute Resolution: Panacea or Anathema?" *Harvard Law Review* 99 (1986): 668-684.

[30] Edward Brunet, "Questioning the Quality of Alternative Dispute Resolution," *Tulane Law Review* 1, no. 62 (1987): 1-56.

[31] M. P. Clagett, "Environmental ADR and Negotiated Rule and Policy Making: Criticisms of The Institute For Environmental Conflict Resolution and The Environmental Protection Agency," *Tulane Environmental Law Review* 15 (2002): 409-417.

[32] C. Bourdeaux, Rosemary O'Leary and R. Thornburgh, "Control, Communication, and Power: A Study of the Use of Alternative Dispute Resolution of Enforcement Actions at the U.S. Environmental Protection Agency, *Negotiation Journal* 17(2) (2001): 175-191.

[33] B.L. Lamb, N. Burkardt and J.G. Taylor, "the Importance of Defining Technical Issues in Interagency Environmental Negotiations," *Public Works Management & Policy* 5(3) (2001): 218-223.

[34] Dean G. Pruitt, et al., "Long-Term Success in Mediation," *Law and Human Behavior* 17 (1993).

[35] Gail Bingham, *Resolving Environmental Disputes: A Decade of Experience,* 1986.

[36] Ibid., 73.

[37] Christopher W. Moore, *The Mediation Process: Practical Strategies for Resolving Conflict* (San Francisco, CA: Jossey-Bass, 1987).

[38] Ibid., 6, 14.

[39] A.B. Dotson, "Defining Success in Environmental and Public Policy Negotiations," no date; Judith E. Innes, "Evaluating Consensus Building," 1999.

[40] E. Allen Lind, and Tom R. Tyler, *The Social Psychology of Procedural Justice* (New York: Plenum Press, 1988).

[41] Ibid.

[42] Ibid.

[43] G.S. Leventhal, "What Should Be Done With Equity Theory?" in *Social Exchange: Advances in Theory and Research*, eds. K. J. Gergen, M. S. Greenberg, and R. H. Weiss (New York: Plenum Press, 1980); E. Allen Lind and Tom R. Tyler, *The Social Psychology of Procedural Justice,* 1988; Tom R. Tyler, "The Psychology of Procedural Justice: A Test of the Group-value Model," *Journal of Personality and Social Psychology* 57 (1989): 830–838; C. Bourdeaux, Rosemary O'Leary and R. Thornburgh, "Control, Communication, and Power: A Study of the Use of Alternative Dispute Resolution of Enforcement Actions at the U.S. Environmental Protection Agency," 2001.

[44] E. Barrett-Howard and Tom R. Tyler, "Procedural Justice as a Criterion in Allocation Decisions," *Journal of Personality and Social Psychology* 50 (1986): 296–304.

[45] Janice A. Roehl, "Measuring Perceptions of Procedural Justice," (doctoral dissertation, George Washington University 1988).

[46] Jonathan Raab, *Using Consensus Building to Improve Utility Regulation* (American Council for an Energy Efficient Economy: Washington, D.C., 1994); also Judith E. Innes, "Evaluating Consensus Building," 1999.

[47] Bonnie Colby and Gail Bingham, "Economic Components of Success in Resolving Environment Policy Disputes," *Resolve Newsletter* 28 (1997).

[48] Lawrence E. Susskind, Mieke van der Wansem, and Armand Ciccarelli, "An Analysis of Recent Experience with Land Use Mediation – Overview of the Consensus Building Institute's Study," 2000; Oregon Department of Justice, "Collaborative Dispute Resolution Pilot Project," A report submitted January 30, 2001; Rosemary O'Leary and Susan Summers Raines, "Lessons Learned from Two Decades of Alternative Dispute Resolution Programs and Processes at the U.S. Environmental Protection Agency," 2001; Rosemary O'Leary and Maja Husar, "That Environmental and Natural Resource Attorneys Really Think About Alternative Dispute Resolution: A National Survey," 2002; Lisa A. Kloppenberg, "Implementation of Court-Annexed Environmental Mediation: The District of Oregon Pilot Project," 2002.

[49] Kirk Emerson, "A Critique of Environmental Dispute Resolution Research," (presentation to the Conflict Analysis and Resolution Working Group Seminar, University of Arizona, 1996); Judith E. Innes, "Evaluating Consensus Building," 1999.

[50] Fisher, William Ury, and Bruce Patton, *Getting to Yes,* 1991.

[51] Lawrence E. Susskind, Ravi K. Jain, and Andrew O. Martyniuk, *Better Environmental Policy Studies: How to Design and Conduct More Effective Analyses* (Washington, DC: Island Press, 2001) 236-9 and 273-6; Lawrence E. Susskind, P. Levy, and Jennifer Thomas-Lerner, *Negotiating Environmental Agreements: How to Avoid Escalating Confrontation, Needless Costs, and Unnecessary Litigation* (Washington, DC: Island Press, 2000).

[52] Best Alternative To a Negotiated Agreement (BATNA), a strategy notion from Roger Fisher, William Ury, and Bruce Patton, *Getting to Yes,* 1991.

[53] Rosemary O'Leary and Maja Husar, "That Environmental and Natural Resource Attorneys Really Think About Alternative Dispute Resolution: A National Survey," 2002.

[54] Christopher W. Moore, *The Mediation Process: Practical Strategies for Resolving Conflict* (San Francisco, CA: Jossey-Bass, 1987).

[55] Dean G. Pruitt, et al., "Long-term Success in Mediation," 1993.

[56] Christopher W. Moore, *The Mediation Process: Practical Strategies for Resolving Conflict,* 1987.

[57] M. P. Clagett, "Environmental ADR and Negotiated Rule and Policy Making: Criticisms of The Institute For Environmental Conflict Resolution and The Environmental Protection Agency," 2002.

[58] Howard Raiffa, *The Art and Science of Negotiation* (Cambridge, MA: Harvard University Press, 1982).

[59] Ibid; see also Lawrence Susskind and Jeremy Cruikshank, *Breaking the Impasse: Consensual Approaches to Resolving Public Disputes* (New York: Basic Books, 1987).

[60] E. Allen Lind and Tom R. Tyler, *The Social Psychology of Procedural Justice,* 1988.

[61] Dean G. Pruitt, et al., "Long-term Success in Mediation," 1993.

[62] Tom R. Tyler and E Griffin, "The Influence of Decision-Maker Goals on Resource Allocation Decisions," *Journal of Applied Social Psychology* 66 (1991): 1629-1658.

[63] N. Meierding, "Does Mediation Work? A Survey of Long-Term Satisfaction and Durability Rates for Privately Mediated Agreements," *Mediation Quarterly* 11 (1993): 157–170.

[64] Dean G. Pruitt, et al., "Long-term Success in Mediation," 1993.

[65] Ibid.

[66] P.A. Dillon, and Robert E. Emery, "Divorce Mediation and Resolution of Child Custody Disputes: Long-Term Effects," *American Journal of Orthopsychiatry* 66 (1996):131–140; Joseph R. Johnston, Linda E. G. Campbell, and Mary C Tall, "Impasses to the Resolution of Custody and Visitation Disputes," *American Journal of Orthopsychiatry* 55 (1985): 112-119; Susan L Keilitz, Harry W.K. Daley, and Roger A. Hanson, *Multi-State Assessment of Divorce Mediation and Traditional Court Processing*, project report for the State Justice Institute (Williamsburg, VA, 1992); Joan B. Kelly, *"Mediated and Adversarial Divorce Resolution Processes: An Analysis of Post-Divorce Outcomes,"* final report prepared for the Fund for Research in Dispute Resolution, (1990); Jessica Pearson, "Family Mediation," in *A Report on Current Research Findings – Implications for Courts and Future Research Needs,* ed. Susan L. Keilitz, National Symposium on Court-connected Dispute Resolution Research, State Justice Institute, 1993 (1994); Jessica Pearson and Nancy Thoennes, "Divorce Mediation: Reflections on a Decade of Research," *Mediation Research,* in eds. K. Kressel and Dean Pruitt, 9-30 (San Francisco, CA: Jossey-Bass, 1989).

[67] Susan L Keilitz, Harry W.K. Daley, and Roger A. Hanson, *Multi-State Assessment of Divorce Mediation and Traditional Court Processing*, 1992.

[68] K. William Easter, "Economic Failure Plagues Public Irrigation: An Assurance Problem," *Water Resource and Research* 29 (1993):1913-1222; C.F. Runge, "Institutions and the Free Rider: The Assurance Problem in Collective Action," *Journal of Politics* 46 (1984): 154-181.

[69] N. Meierding, "Does Mediation Work? A Survey of Long-Term Satisfaction and Durability Rates for Privately Mediated Agreements," 1993.

[70] Donald T. Saposnek, *Mediating Child Custody Disputes: A Systematic Guide for Family Therapists, Court Counselors, Attorneys, and Judges* (San Francisco, CA: Jossey-Bass, 1983), as cited in J.A. Walker, "Family Conciliation in Great Britain: From Research to Practice to Research," *Mediation Quarterly* 24 (1989): 34.

[71] G. Davis, and M. Roberts, *Access to Agreement* (Milton Keyes, UK: Open University Press, 1988), as cited in J.A. Walker, "Family Conciliation in Great Britain: From Research to Practice to Research," *Mediation Quarterly* 24 (1989): 34.

[72] Further documentation of its importance in literature, as well as additional measurement approaches, can be found in Tamra Pearson d'Estrée, "Achievement of Relationship Change," in *The Promise and Performance of Environmental Conflict Resolution,* eds. Rosemary O'Leary and Lisa B. Bingham, 111-128 (Washington, DC: Resources for the Future, 2003).

[73] Dean G. Pruitt, et al., "Long-term Success in Mediation," 1993.

[74] G. Davis, and M. Roberts, *Access to Agreement*, 1988.

[75] Susan L Keilitz, Harry W.K. Daley, and Roger A. Hanson, *Multi-State Assessment of Divorce Mediation and Traditional Court Processing*, 1992.

[76] Joseph R. Johnston, Linda E. G. Campbell, and Mary C Tall, "Impasses to the Resolution of Custody and Visitation Disputes," 1985.

[77] Murray A. Straus, "Measuring Intrafamily Conflict and Violence: The Conflict Tactics (CT) Scales," *Journal of Marriage and the Family* 41 (1979):75–86.

[78] e.g. Christopher W. Camplair, and Arnold L. Stolberg, *Benefit of Court-Sponsored Divorce Mediation: A Study of Outcomes and Influences on Success* (Publisher Unknown, 1990).

[79] Patrick Regan, "Conditions of Successful Third-Party Intervention in Interstate Conflicts," *Journal of Conflict Resolution* 40 (1996): 336-359; Duane Bratt, "Assessing the Success of UN Peacekeeping Operations," *International Peacekeeping* 3 (1997): 64-81.

[80] P.A. Gwartney, F. Fessenden, and Gayle Landt, "Measuring the Long-term Impact of a Community Conflict Resolution Process: A Case Study Using Content Analysis of Public Documents," 2002.

[81] One practitioner described this shift as one noticeable in "the way they held their arms" and in the pronouns parties used to referred to others and to themselves ("we" vs. "us" and "them").

[82] For a new effort to examine shifts, see Berenike Carstarphen, "O.H.M. Shift Happens: Transformations During Small Group Interventions in Protracted Social Conflicts," (Unpublished doctoral dissertation, George Mason University, 2000).

[83] Donald H. Baucom and Norman Epstein, "Will the Real Cognitive-Behavioral Marital Therapy Please Stand Up?" *Journal of Family Psychology* 4 (1991): 394-401.

[84] Ibid.

[85] Carlos E. Sluzki, "Transformations: A Blueprint for Narrative Changes in Therapy," *Family Process* 31 (1992): 217-230.

[86] For more details on these dimensions, consult Carlos E. Sluzki, "Transformations: A Blueprint for Narrative Changes in Therapy," or our research Guidebook in Appendix A.

[87] Robert A. Baruch Bush, and Joseph P. Folger, *The Promise of Mediation: Responding to Conflict Through Empowerment and Recognition* (San Francisco: Jossey-Bass, 1994); Franklin Dukes, "Public Conflict Resolution: A Transformative Approach," *Negotiation Journal* 9(1), (1993): 45-57; Carrie Menkel-Meadow, "Pursuing Settlement in an Adversary Culture: A Tale of Innovation Co-opted or 'the Law of ADR,'" *Florida State University Law Review* 19(1), (1991): 1-46; Carrie Menkel-Meadow, "The Many Ways of Mediation: the Transformation of Traditions, Ideologies, Paradigms, and Practices," *Negotiation Journal* 11 (1995): 217-242.

[88] Robert A. Baruch Bush, and Joseph P. Folger, *The Promise of Mediation: Responding to Conflict Through Empowerment and Recognition*, 1994.

[89] Cf. V. Held, *Justice and Care: Essential Readings in Feminist Ethics* (Boulder, CO: Westview, 1995)

[90]. Carrie Menkel-Meadow, "Pursuing Settlement in an Adversary Culture: A Tale of Innovation Co-opted or 'the Law of ADR,'" 1991; Franklin Dukes, "Public Conflict Resolution: A Transformative Approach," 1993; John Paul Lederach, *Preparing for Peace: Conflict Transformation Across Cultures* (Syracuse, NY: Syracuse University Press, 1995); John Paul Lederach, *Building Peace: Sustainable Reconciliation in Divided Societies* (Washington D.C.: U.S. Institute of Peace Press, 1997).

[91] Robert A. Bush, and Joseph P. Folger, *The Promise of Mediation: Responding to Conflict Through Empowerment and Recognition*, 1994.

[92] Tina Nabatchi and Lisa B. Bingham, "Expanding Our Models of Justice in Dispute Resolution: A Field Test of the Contribution of Interactional Justice" (paper presented at the annual meeting of the International Association of Conflict Management, 2002).

[93] Shawn MacDonald, "Social Capital and Its Measurement" (Unpublished manuscript, George Mason University, 1999).

[94] James S. Coleman, "Social Capital in the Creation of Human Capital," *American Journal of Sociology* 94 (supplement), (1988): S95-S120; James S. Coleman, *Foundations of Social Theory* (Cambridge, MA: Harvard University Press, 1990).

[95] Robert D. Putnam, "The Prosperous Community: Social Capital and Public Life," *American Prospect* 13, (1993): 35-42; Robert D. Putnam, Bowling Alone: America's Declining Social Capital," *Journal of Democracy* 6 (1), (1995): 65-78.

[96] A. Portes, "Social Capital: Its Origins and Applications in Modern Sociology," *Annual Review of Sociology* 24 (1998).

[97] Pierre Bourdieu, "The Forms of Capital," in *Handbook of Theory and Research for the Sociology of Education*, ed. J. Richardson (Westport, CT: Greenwood Press, 1986).

[98] Francis Fukuyama, *Trust: Social Virtues and the Creation of Prosperity* (New York: Simon & Schuster, 1995).

[99] Judith E. Innes, "Evaluating Consensus Building," 1999.

[100] J. C. Allen, "Community Conflict Resolution: The Development of Social Capital Within An Interactional Field," *Journal of Socio-Economics* 30 (2001): 119-120.

[101] P.A. Gwartney, F. Fessenden, and Gayle Landt, "Measuring the Long-term Impact of a Community Conflict Resolution Process: A Case Study Using Content Analysis of Public Documents," 2002.

[102] Francis Fukuyama, *Trust: Social Virtues and the Creation of Prosperity*, 1995.

[103] Judith E. Innes, "Evaluating Consensus Building," 1999.

[104] We define "ability to resolve future disputes," above, as specifically applied to the same relationships rather than new relationships.

[105] Judith E. Innes, "Evaluating Consensus Building," 1999.

[106] Ibid.

[107] Chris R. Argyris, Robert D. Putnam, and D.M. Smith, *Action Science* (San Francisco: Jossey-Bass, 1985).

[108] Ronald J. Fisher, *Interactive Conflict Resolution* (New York: Syracuse University Press, 1997).

Chapter Three

COMPARATIVE CASE ANALYSIS

Developing and Applying Methods

METHODOLOGY OVERVIEW

Goals for Evaluating Criteria and Cases

In her important contribution to the topic of evaluating conflict resolution outlined in the previous chapter, Innes[1] presented both the challenges and the opportunities of performing evaluations. She and her colleagues have found in their separate evaluations of consensus-building processes that what seem like failures may actually be successes, while apparent successes can actually be failures. For example, processes that 'failed' to reach an agreement may have 'succeeded' in establishing an ongoing working relationship. Imbedded in this proposition is the central role of the criteria that are chosen to represent and evaluate success. Her observation also implies that it is important to identify the comparison standard being used. In other words, a given conflict resolution process is always evaluated by how it compares to something, whether it be to an alternative, to an ideal, or to the situation at an earlier time.

Earlier, we outlined the many components that comprise people's varying conceptions of 'success' (see Figure 2.1). In addition to examining the definition of success, we had two additional goals. First, we sought to make case study reporting more thorough, both in terms of the information that is needed to analyze success and to make the reporting more standardized. Second, building on this standardized reporting, we wanted to begin the task of comparatively analyzing cases to gain a new understanding of the links between various practices and processes and their results. This involves not only identifying the appropriate criteria for success, but also detailing how those criteria should be measured, and how the resulting information ultimately might be analyzed and compared.

So that we might begin to use these criteria to understand what was achieved in various cases of environmental conflict resolution, we had to put the criteria into a form that would allow them to be systematically and regularly applied to cases. This measurement framework, or Guidebook, includes all the criteria defined and operationalized through multiple indicators (see Appendix A). Once this common framework for reporting and assessment was developed, it was then applied in a common fashion to several cases, and case studies were generated using similar formats. This allowed the cases to be compared according to each criterion.

This type of analysis using a common framework has generated richly detailed reports of the various case studies themselves, in addition to allowing an important analysis across the cases. Prior to this study, a systematic case comparison across multiple variables had not been done in the environmental conflict resolution arena. Such comparative case analysis allows us to accumulate insights from across cases in order to gain a better understanding of the way different processes are used *to accomplish varying goals* in resolving environmental disputes.

The Value of a Case Study Approach

In order to accomplish the overall objective of developing a framework for evaluating success in environmental conflict resolution, eight case studies of western U.S. water conflicts were analyzed. The case studies were an essential element in the research design. It was through analysis of actual cases that we examined the feasibility of taking abstract success concepts (criteria) and crafting questions designed to allow researchers to gather information needed for their assessment. The process entails transforming abstract ideals of success into concrete applicable measures, an exercise that can only be tested by being applied to actual conflict cases.

The Importance of Comparative Analysis

As outlined in the previous chapter, several calls, most prominently from Innes and Emerson,[2] have been made for a comparative analysis of conflict resolution processes. Comparative claims have been made for increased efficiency, reduced costs, improved relations, and a better quality of justice. However, little if any comparative work has relied on a common analytic framework to substantiate these claims. As Emerson notes, there is an important gap in comparing various 'alternative dispute resolution' (ADR) processes to other more traditional and institutionalized processes, which represent the common alternatives. We begin this comparative analysis process here, with the hope that it will become more common as a universe of similarly documented cases grows.

A FRAMEWORK FOR CONSISTENT CASE DOCUMENTATION: THE CASE ANALYSIS GUIDEBOOK

Development of the Guidebook

In order to examine systematically the selected cases within our criteria conceptual framework, each criterion needed to be made operational. How does one prevent a case review from becoming merely an impressionistic snapshot, framed by the particular "lens" of the reviewer? The answer lies in setting out common conceptual categories in advance, and specifying the way each concept (or, in this case, each criterion) will be measured.

Information gathered in this way will likely be more "objective" in the sense that it can be replicated by various researchers (i.e., other researchers will find the same information), thereby satisfying the "intersubjectivity" criterion considered standard in empirical research.[3] Once the information is gathered in this standard way, it can be compared across cases easier, since each reviewer is using the same "measuring stick." Because we attempted to comprehensively include all criteria considered relevant to success, anyone interested in making comparisons should be able to find their chosen "measuring stick" of interest.

The Case Analysis Guidebook (initially circulated as a "working guide") was formulated based on the criteria conceptual framework outlined in Chapter 2. It is included in full in Appendix A. The guide identifies the information needed to assess the case studies on each of the success criteria.

The Case Analysis Guidebook was first drafted by the principal investigators, and then modified over a period of several months through a series of interactive dialogues (via meetings and email) between the principal investigators and the first case researchers who were primarily graduate students at George Mason University.[4] These discussions served to sharpen the definitions of the success criteria and to generate more specific ideas about the information needed to operationalize them. The Guidebook was then used by the case researchers as a framework for approaching and framing each case, looking for sources, and analyzing the case itself. This process is reviewed below. As a result of this "testing" or "piloting" of the Guidebook as a tool, several revisions were incorporated. These included a further clarification of the criteria, the creation of additional indicators to operationalize these criteria, a further discrimination of initially overlapping constructs, and the inclusion of information to aid researchers, including sources and the best times to assess the variables.

Our piloting and subsequent case research revealed that criteria ideally should be assessed at several points in time over the course of a conflict: (1) baseline, before the resolution process, (2) during the resolution process, (3) immediately upon completion/signing, (4) short-term after the agreement or settlement, and finally, (5) long-term after the agreement or settlement. Most criteria are not capable of being assessed at every stage; we discuss difficulties in assessment in Chapter 9. However, knowing which criteria to assess at which stage can make evaluation easier (see figure 3.1, and individual criteria in the Guidebook Appendix). Any criteria that measure change will require a baseline assessment against which to gauge any movement. Most criteria are best assessed at one or two different points in time, and are more difficult to assess at other points. Such knowledge can assist case researchers in gathering information most effectively.

			IMPLEMENTATION	
BASELINE before resolution process	DURING the resolution process	Immediately upon COMPLETION or signing	SHORT-TERM after Outcome	LONG-TERM after Outcome
1	**2**	**3**	**4**	**5**

Figure 3.1 Stages for Criteria Assessment

A CONTEXT FOR EVALUATION: CASES OF WESTERN WATER DISPUTES IN THE AMERICAN WEST

Selection of Cases

Eight cases were selected for analysis after reviewing a larger set of western U.S. water conflicts. The cases were chosen based on geographic diversity across the southwest, the diversity of issues contributing to each conflict, the types of resolution mechanisms attempted and preliminary evidence that sources would be available for researchers to analyze the case.

Four of the cases are described in great detail in Part II, Chapters 4-7, but each of the eight cases is reviewed briefly here.

The Big Horn case involves a dispute between two tribes and their neighboring non-Indian irrigators over water for tribal purposes (including instream flows) on the Wind River Reservation in western Wyoming. The conflict has been active since the 1970s, though its roots date to non-Indian settlement in the 1800s. Litigation has been the primary conflict resolution mechanism employed in the conflict.

The Edwards Aquifer case in Texas involves a conflict over dividing limited regional water supplies between agricultural and urban groundwater pumpers and the endangered species and recreation that depend on spring flows that are linked to groundwater levels. The dispute began in the 1950s and has been addressed through litigation, political bargaining and state legislation.

The Lower Colorado River case involves the states of Nevada, California and Arizona, along with agricultural, urban, tribal and environmental interests in those states. Disputes over dividing the waters of the Colorado River date back to the early 1900s and have been addressed through litigation, multiparty negotiations and state and federal legislation or administrative actions.

The Mono Lake case involves a dispute over the effects of water diversions in the Sierra Nevada mountains among rural, urban, tribal and environmental interests in southern California. The dispute dates back to the 1960s and has been addressed by litigation, multi-party negotiations, state legislative and administrative actions, federal agency actions, and federal legislation.

The Pecos River case is a dispute over the interstate allocation of surface water between New Mexico and Texas. The conflict erupted in the 1940s and appeared to have been resolved by an interstate compact. It re-emerged in the 1970s and has been addressed through litigation and state legislation.

The Pyramid Lake case involves a dispute over water supplies, and water quality in two interconnected river basins in western Nevada. The parties include two tribes, several cities and counties, the states of Nevada and California, and multiple federal agencies and environmental interests concerned with endangered species of fish and wetland preservation. The dispute has been active in various forms since the early 1900s. In the past two decades, it has been addressed through litigation, multi-party negotiations, state and federal legislation and administrative actions, and water rights acquisitions by environmental organizations.

The Salt River case involves a dispute over the allocation of water among rapidly growing cities, a Native American tribe and agricultural interests in central Arizona. The conflict has been active in various forms since the early 1900s. It has been addressed through litigation, multi-party negotiations, and federal legislation.

The Snowmass Creek case is a dispute over water diversions and instream flow needs between a ski resort, local residents and environmental interests in the Rocky Mountains of Colorado. The dispute emerged in the 1970s and has been addressed through litigation, multi-party negotiations and state legislation.

The eight case studies share the common thread that they involve multi-party conflicts over water resources in the American West. The cases can be viewed as a representative, stratified sample selected to represent several different classes of "classic" conflicts over western water. Two cases are disputes between states over the allocation of interstate rivers (Lower Colorado River, Pecos River). Three cases are disputes between Native American tribes and neighboring non-Indian water users (Big Horn, Pyramid Lake and Salt River). Five cases (including two of the tribal cases previously mentioned) involve conflicts between consumptive water uses (farming, urban development, snow making) and environmental needs for water in streams and wetlands (Big Horn, Edwards Aquifer, Mono Lake, Pyramid Lake and Snowmass).

In addition to representing the classic configurations of parties and issues found in western water disputes, the cases also provide a representative cross section of attempted dispute resolution processes. Two cases were eventually resolved after lengthy negotiations among the parties, with the resulting agreements formalized (and altered somewhat) as acts of Congress (Pyramid Lake, Salt River). In another two cases, the "resolutions" came through court rulings after lengthy litigation (Big Horn, Pecos River). The "agreements" analyzed for two other cases were administrative actions–the issuing of a ruling by the California Water Resources Control Board in the Mono Lake case and the promulgation of federal agency rules in the Lower Colorado River case. In the remaining two cases, the agreement analyzed was state legislation (Edwards Aquifer and Snowmass Creek). While it was necessary for the researchers to select a specific form of resolution (i.e., court ruling, legislation, etc.) and time period in the course of the dispute in order to focus their efforts, the histories of each case reveal a rich mixture of different conflict resolution processes. Litigation was a vital motivating factor at some point in every case and court rulings specifically prompted the two legislative agreements, the Mono Lake administrative ruling and the Pyramid Lake and Snowmass negotiated agreements.

By examining quite different types of "resolutions," this study went beyond prior research in environmental conflict resolution, which only compared across negotiated agreements. We expanded beyond negotiated agreements because we wanted to develop a means of evaluating "success" across the variety of resolution mechanisms that are actually utilized by parties in attempting to bring closure to environmental conflicts. This proved quite challenging as some success criterion that apply quite clearly to negotiated agreements (process quality, reaching agreement) are difficult to apply to litigation and court rulings or legislation. However, it seemed essential to the purposes of this project to go beyond negotiated agreements in order to understand whether and how "success" could be measured and compared across different types of "resolutions."

TERMINOLOGY AND THE FOCUS OF ANALYSIS

One of the goals of this book is to provide a framework that allows for comparison not only across cases, but also across conflict resolution methods. The research methodology has been developed and tested on cases involving various combinations of negotiation, administrative rulemaking, legislation, and litigation. This allows for comparative analysis and a preliminary discussion of the strengths and weaknesses of various methods of conflict resolution.

However, it is difficult to develop a general terminology for evaluation because different methods of conflict resolution have different goals and outcomes. For example, certain processes aim to produce "agreement," while others do not. Therefore, we adopted the general term "Outcome" (capitalized to signify its special use) to encompass the output of whichever process is being considered.

Many of the analyzed cases extend over several decades, with the conflict moving through stages of litigation, negotiation, agency actions and legislative responses, producing multiple Outcomes. For the purposes of applying this research methodology, case study analysts were instructed to select a particular Outcome and time period on which to focus. In several cases, two or more very closely related Outcomes were analyzed together. This occurred particularly in the Pyramid Lake case (a negotiated agreement was modified by Congress and enacted as federal legislation), the Snowmass Creek case (a local negotiated agreement required state legislation to legitimize provisions of the agreement), and the Pecos River case (a court ruling was followed by a negotiated agreement that was then formalized by the court as a stipulated ruling).

Whether selecting a single Outcome or several that are closely related, it is essential to choose a well-defined timeframe for this type of case analysis. Failure to define the process and the Outcome would result in criteria being applied to multiple objects simultaneously, and would yield no clear measurement of any specific process or Outcome. This methodology requires carefully defining the boundaries of which process and Outcome is being assessed, even though this may require a bit of artificial separation.

A full case analysis covering the entire history of an environmental conflict and its multiple outcomes is possible, but it would require sequential analyses for different Outcomes and time periods (phases in a complex conflict resolution process). Each phase of litigation, legislative actions, and other processes (along with their respective Outcomes) would need to be treated with distinct sequential applications of this framework in order to ensure that the criteria are meaningfully applied. Thus the framework allows for either form of analysis: single outcome and timeframe, or multiple-stage and outcomes analyzed sequentially.

PROCEDURE

Analyzing the Individual Cases: Applying Criteria

As noted above, a summative evaluation of individual cases, per se, was not our goal. To produce summative judgments of a case's 'success' would require that we choose from among the criteria and weight the criteria to express a 'policy' of what constitutes success. This would then be applied to the chosen case to provide an overall assessment. Instead, one of our goals was a more thorough analysis of individual cases, as well as an analysis that used a common framework to allow for comparative analysis (see below).

Once our methodology was established, researchers could then approach and analyze the water disputes outlined above. One principal investigator (PI) and four of the graduate students involved in the dialogues also served as case study researchers, giving them an opportunity to use the tool they had helped to develop. The researcher on the Snowmass Creek case study was a former graduate student of one of the PI's, who now operates her own economic research business. The researchers varied a good deal in their disciplinary and professional backgrounds, however they all had some combination of expertise in economics, environmental law, environmental policy, conflict resolution, and journalism.

One or two case studies were assigned to each case researcher, with instructions to follow the guide in choosing what information to collect and in organizing the format for reporting the information. The researchers typically began by familiarizing themselves with their cases, and proceeding to track down sources. Some were easily available, while other sources required substantial effort and follow-up for the researchers to locate them in archives or to obtain them from public documents. The availability of the sources varied with each case. The process of gathering information and writing the cases required a substantial investment of time, and many details needed to be addressed midstream. These included instructions on how much the analysts could solicit information directly from parties and the ideal length of the submitted material.

Sources suggested by the PI's fell into two categories, "insiders" (negotiation participants and mediators), and the perceptions of "outsiders" (public opinion, etc.). "Insider" sources included the following: Media interviews/quotes from the parties; internal documents from the parties, such as newsletters, memos, reports to members (environmental groups), stockholders (corporations, utilities), public relations documents, advertisements, videos, web sites; paid media spots commenting on issues; protests, picketing or other public activities; memos (or other documents) circulated among parties after the agreement was developed; and reports by any mediators/facilitators. "Outsider" sources included the following: Newspaper articles (community, region, *High Country News*); editorials or Op Eds; TV and radio coverage; paid media spots commenting on issues; and any protests, picketing, or other public activities. Some researchers made significant use of the internet, revealing the power and richness of this source of information.

In addition to the variability of the available sources, the diverse levels and types of experience among the researchers themselves proved to be both assets and challenges. We include what we learned were helpful characteristics for those who are considering hiring researchers or supervising student researchers to look for:

1) good writing skills
2) persevering (sleuthing) researchers: library, legal docs and Internet skills, good at "digging," but able to distinguish research writing from journalism
3) maturity, refrain from personal judgments or editorializing
4) careful referencing, quoting, meticulous on details, accuracy
5) knowledge of laws and policies and institutional structure, for instance – federal agencies and their structure and knowing whether a settlement needs court approval or not (Guidebook appendices were developed to assist in this)
6) good people skills to solicit information or good "connections"
7) ability to translate technical, scientific legal terms into everyday language.
8) ability to discern the interests of diverse parties and to articulate them; not blinded by own point of view.

It was also found that personal links to the case could prove to be both an asset and a liability. Personal involvement and prior experience with a case could provide context, access to additional sources of information, and a familiarity useful for knowing what to look for. However, for the sake of this analysis researchers were asked to only use publicly accessible information so that we might test accessibility (see Chapter 9), so researchers with personal links had to avoid private sources of information that they knew existed.

It was determined that being "local" to the area of the dispute would be a significant advantage. Certain disputes did not garner much national attention, which made it more difficult to gain access to media sources. Being locally present also would allow researchers to easily gather agency reports, public-meeting minutes, and other documents that are publicly available in theory but not easily gathered from busy staff people at a distance. Also, certain criteria themselves were difficult to assess without being locally present during the process (see criteria under "Process Quality" in Guidebook Appendix A). It is interesting to consider what the impact might be of having a mediator do a case analysis, or at least contribute to the case study.

While our researchers would have found it very useful, we explicitly excluded interviewing as a method of gathering information. The Guidebook was designed to provide for case analysis by graduate students using accessible public information, without personal interviews of mediators and stakeholders. We chose to exclude interviewing for several reasons: the potential for a substantial increase in research costs, the difficulty of independently verifying the information from an interview if it is not published or publicly available (intersubjectivity), and finally, our desire to assess what information would be available without the use of interviews (see Chapter 9). Interviewing also can be an intrusion on the system being studied and can have its own effects. We wanted our framework to be non-intrusive and low cost in collecting and analyzing data, so it would not create burdens for process facilitators and any parties who might wish to utilize it. In practice, we found it was often difficult for researchers to separate the process of requesting (written and publicly available, and therefore replicable) information and asking substantive questions.

Researchers with adequate means and/or access may want to consider the use of interviewing. For gathering certain information, interviews have no substitute, and researchers may decide that it is worth the costs for their purposes.

Once the information was gathered and presented in detailed case study format, we were able to evaluate it in two distinctive ways, with separate objectives. First, the *criteria* themselves in our five categories were evaluated across all eight cases. Second, the information was gathered in order to evaluate the *cases* themselves using each of the five success criteria categories.[5] This would allow for a comparison across different cases to learn how "success" (as measured by the various criteria) varied among the cases and to further test the success evaluation process developed as part of this research project.

Evaluating the Criteria

Once the many criteria had been applied to several actual cases of water disputes, we could begin to consider the usefulness of the criteria themselves. To undertake this epistemological analysis, an analysis of our research tools themselves, we had to first set out our higher-order criteria for judging the applied criteria *qua* criteria. The three measures used to evaluate the performance of the success criteria, as operationalized and applied to the case studies, consisted of accessibility, reliability and validity.

Under *accessibility*, we asked whether information was able to be gathered that answered the questions posed in the guide for each success criterion and if not, why not. We noted the sources of information used, the triangulation of information using independent sources, and the costs and difficulties of obtaining information. We also considered whether the criterion and information requested in the guide might be recast to better match the available information sources.

For *reliability*, we considered whether other researchers would have uncovered the same information and developed the same impressions, and if the researchers sought diverse viewpoints when investigating different questions (and if not, why not). We also evaluated the degree to which collected information was influenced by spurious factors, such as personal connections and the disciplinary expertise of the researcher, the political and legal sensitivities of the case, unorganized government offices, etc.

For *validity* we first considered whether each criterion was a conceptually valid indicator of the success "concept" being addressed. For instance, is *cultural sustainability* a valid indicator of "Outcome Quality" or is *public acknowledgement of outcome* a valid indicator of "Outcome Reached"? Then we asked whether the indicators used to operationalize each criterion were valid for the success concept we sought to measure. If not, we reflected on whether a different kind of question should have been asked for that criterion. We also assessed whether the criterion was applicable, in its current conception, to the different types of cases and Outcomes analyzed and, if not, how it could be refined for broader applicability. For details on the indicators used to operationalize each criterion, refer to the Guidebook in Appendix A.

As we evaluated the success criteria for accessibility, reliability and validity, we also addressed some methodological questions. We examined the information the researchers were able to gather and then considered the timing in the development of a case when information to assess a particular aspect of success was most readily available. We found that, for many success criteria, information becomes available at different points in the development of an agreement and its implementation. For instance, under "Outcome reached" we found that media sources and reports from parties were often available right away to verify that an Outcome, such as an agreement, had been achieved.

However, another aspect of "Outcome reached" was the inclusion of all key parties, and we noted that the failure to include a key player might not become evident until later, when the exclusion caused a problem in implementation. Similar variations in the timing of ability to measure a specific aspect of a success criterion were found for many criteria. Results of this criteria analysis are reported in Chapter 9.

Comparing Cases and Processes

Once the criteria had been operationalized through indicators and collected data, comparisons across cases could be made. Also, because we had chosen cases that were representative of several dimensions of the larger population of water conflicts, such as their issues and conflict resolution processes, tentative comparisons could be ventured across these dimensions as well. These comparisons and conjectures have to remain tentative, given the small number of cases in any given comparison. Results of these are discussed in Chapter 8.

REFLECTIONS ON THE EVALUATION PROCESS

We sought to begin the process of considering successful resolution by identifying criteria and evaluating the process of applying them to real cases. With the process we have developed, we can note how cases may rate according to various criteria. However, we cannot comment on overall judgments of success for these cases because to do so requires an additional step, of prioritizing (and weighting) the criteria, thereby incorporating values. This is ultimately not a research question but a policy question.[6]

However, once personal or community values were articulated, one could use the Guidebook framework to generate judgments of success. After a user (or researcher) had identified criteria priorities (for example, comparing cases on environmental and cultural sustainability, without regard to cost or justice of process), and also made explicit what would be most and least desirable on each of the indicators for the criteria, s/he could then compare only those criteria across cases and produce a "judgment" of success.

As described earlier, Innes[7] and others have called for more comparative case study analysis. If a body of cases were analyzed using this or a similar framework, and indeed if new cases were recorded and reported according to a common framework, then the cross-comparison of cases would be greatly facilitated. In a similar vein, Lewicki, Gray, and Elliott[8] use a common set of questions to facilitate cross-case comparison of "framing" in intractable environmental conflicts. Similar standardized reporting on evaluation criteria topics would allow for comparison and hypothesis-generation across criteria and related conceptual dimensions.

In sum, the evaluation framework in our Guidebook (Appendix A) provides several useful results. First, it helps to standardize the format of case study reports. Our list of criteria and categories are all-inclusive, attempting to reflect the diversity of variables linked by various scholars and practitioners to understanding effective resolution. Guidebook users may decide to use only a subset of criteria depending on their focus and their assumptions about success. However, the presence of all likely criteria in one framework allows one to ascertain information on all these variables so that comparisons can be made across cases. Not only can cases be comparatively analyzed, but resolution processes (such as litigation, researching strategy, informal bargaining, and mediation) can also be compared through this framework (see Chapter 8).

Second, the Guidebook provides a strategy for researching cases. What are the dimensions that various writers and practitioners have considered are important to assess? Where would one go to find such information? When should it be assessed? The Guidebook is structured to provide guidance in what can be a daunting form of research. Third, the Guidebook provides a framework for storing and organizing the many pieces of information relevant to documenting and analyzing these complex cases. It serves as a "filing system" or series of "file folders" that, once familiar, can make the task of organizing information easier and more thorough. Finally, the Guidebook serves as a way to learn about the multiple issues involved in these cases, and about the methodological issues involved in doing any case analysis and/or comparative research.

We now turn in Part II to four case examples of the use of the Guidebook evaluation framework for individual case studies. We return in Part III to our comparative analysis across all eight cases in our sample.

NOTES

[1] Judith E. Innes, "Evaluating Consensus Building," in *Consensus Building Handbook,* eds. L. Susskind, S. McKearnon, and Jennifer Thomas-Larmer, 631-675 (Thousand Oaks, CA: Sage Publications, 1999).

[2] Judith E. Innes, "Evaluating Consensus Building," 1999; Kirk Emerson, *A Critique of Environmental Dispute Resolution Research,* Presentation to the Conflict Analysis and Resolution Working Group Seminar, University of Arizona, April, 1996.

[3] For a good resource on intersubjectivity, as well as current (and past) epistemological standards in social science, consult Martin Hollis, *The Philosophy of Social Science* (Cambridge: Cambridge University Press, 1994).

[4] Eric Abitbol, Annette Hanada, Kathryn Mazaika, and Erin McCandless.

[5] Initial case analyses used the early five-category framework. The sixth category, *Social Capital,* was added subsequently, and most cases were thus revised and expanded. Comparative analysis and criteria analysis were done on the five-criteria framework results.

---

[6] See related work by Kenneth R. Hammond, Lewis O. Harvey, & Reid Hastie, "Making Better Use of Knowledge, Separating Truth from Justice," *Psychological Science,* 3(2), (1992): 80-87.

[7] Judith E. Innes, "Evaluating Consensus Building," 1999.

[8] Roy Lewicki, Barbara Gray, and Michael L.P. Elliott, eds., *Making Sense of Intractable Environmental Conflicts: Concepts and Cases* (Washington, DC: Island Press, 2003).

Part II

SAMPLE CASES

The four case studies presented here are intended to provide useful examples of the application of a common methodological framework. The cases are not intended to provide exhaustive factual coverage of the details of each case. We provide additional references to books and journal articles that readers can consult for more details on these cases. The four cases each were researched by different authors with varying personal knowledge of the case and access to information. The case researchers were asked to use only publicly available information and were asked not to directly interview parties to the case. These constraints were necessary to assess whether or not this methodology could be implemented in a relatively low cost and unobtrusive manner (that is, without collecting primary data and interviewing parties). In the last section of this book, we examine whether the methodology was successfully applied to various criteria in these case reports and reflect on the difficulties that particular criteria present.

The chapters do not reflect the present status of the case; rather, each case was researched at a specific point in time (generally 1999-2000) and each focused on a specified time period and set of conflict resolution processes. All of these cases remain active, in the sense that many of the parties described in these case studies continue to engage one another in problem solving and conflict resolution efforts.

Prior to reading these cases, we recommend that readers examine Appendix A, the Guidebook, in order to understand how the criteria were defined, what information was sought and how the criteria are operationalized in investigating a case.

Chapter Four

THE MONO LAKE CASE

Shaking Up the Established Powers

Case Researcher: Kathryn Mazaika

> Water links us to our neighbor in a way more profound and complex than any other.
>
> *John Thorson, Special Master in the Arizona General Stream Adjudication*

> All water has a perfect memory and is forever trying to get back to where it was.
>
> *Toni Morrison*

Note: This case report illustrates the use of a particular methodological framework. It is not intended to replicate the legal and historical coverage of this case provided in other sources.[1]

Introduction: The Mono Lake case demonstrates the difficulties of addressing environmental impacts from water diverted long distances from its remote (FYI—this is a valley on the Eastern side of the Sierra) origins to supply water for cities. The areas east of the California Sierra Nevada mountains and coastal cities embroiled in this conflict are shown in Figure 4.1.

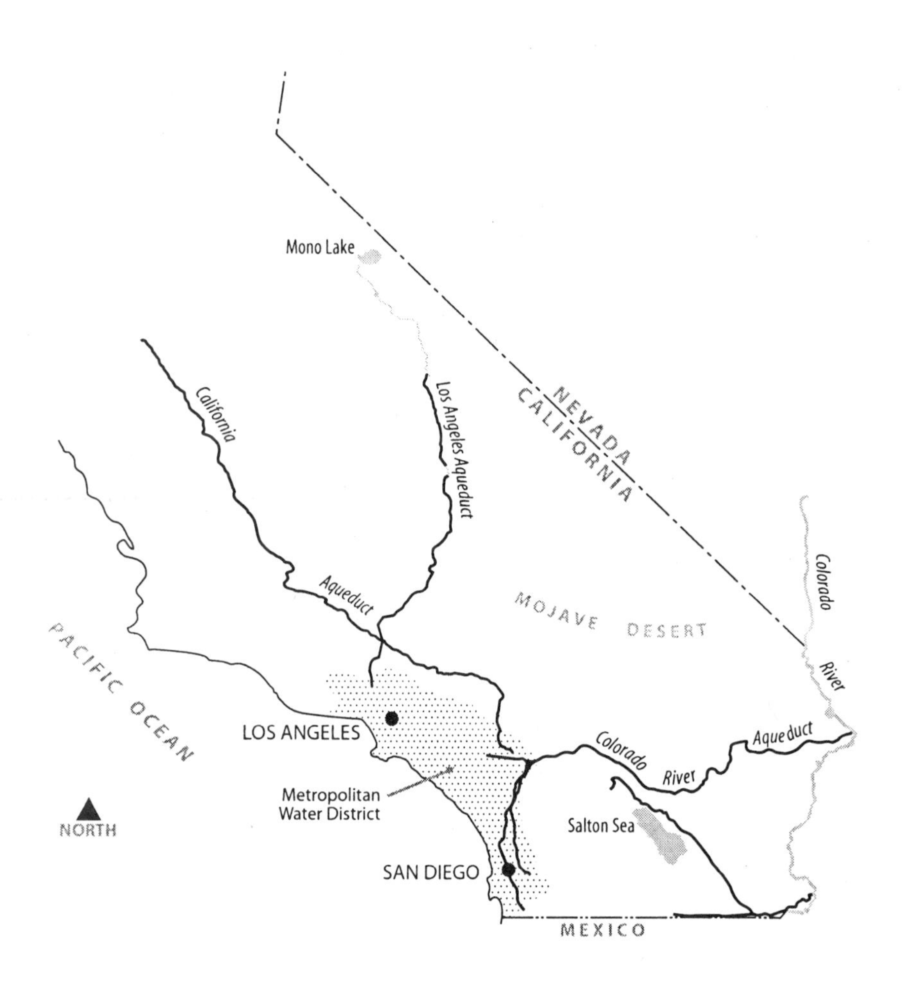

Figure 4.1 Mono Lake Case

Time period: Efforts to restore Mono Lake water levels spanned a period of sixteen years, from 1978 to 1994, when the California State Water Resources Control Board (SWRCB) issued Decision 1631, amending the Los Angeles water rights licenses. This report summarizes the earlier actions that led to Decision 1631, and evaluates the subsequent processes. This report was compiled in 1998 and updated in 2000.

Basic nature of dispute: Concerns about the effects of water diversions by the Los Angeles Department of Water and Power (DWP) on Mono Lake and its ecosystem.

Issues: Mono Lake water level, water diversions pursuant to Water Right License Nos. 10191 and 10192, and instream flows to support fisheries.

Actors and interests:

- **Los Angeles Department of Water and Power**, *Interests:* water rights licenses, aqueduct facilities, water diversions.
- **Mono Lake Committee**, *Interests:* Mono Lake water level, fisheries, streams and habitat.
- **National Audubon Society**, *Interests:* Mono Lake water level, fisheries, streams and habitat.
- **California Trout**, *Interests:* instream flows and fisheries.
- **California Department of Fish and Game**, *Interests:* instream flows and fisheries.
- **U.S. Forest Service, Inyo National Forest**, *Interests:* Mono Basin National Forest Scenic Area.
- **California Department of Parks and Recreation**, *Interests:* Mono Lake Tufa State Reserve.
- **State Lands Commission**, *Interests:* navigable waters (Mono Lake) and the lands beneath them (lakebed and streams).
- **Great Basin Unified Air Pollution Control District**, *Interests:* air quality--particulate matter.
- **U.S. Fish and Wildlife Service**, *Interests:* Mono Lake brine shrimp (*Artemia monica*).
- **Sierra Club Legal Defense Fund**, *Interests:* Mono Lake litigation and precedents.
- **Upper Owens River Landowners**, *Interests:* effects of decreased diversions on the Upper Owens River.

Attempted conflict resolution processes: Litigation, Water Board Evidentiary Hearings, an Environmental Impact Report and the associated California Environmental Quality Act (CEQA) process.

Specific Outcome analyzed: Decision 1631 issued by the California State Water Resources Control board in 1994.

History of the Mono Lake Conflict[2]

1941: Los Angeles Dept. of Water and Power begins diversion of Mono Lake's tributary streams.

1978: The Mono Lake Committee is formed to fight diversions and restore the lake.

1979: A public trust lawsuit is filed by a consortium of environmental groups led by the Mono Lake Committee.

1983: The California Supreme Court rules that the state has an obligation to protect Mono Lake.

1985: The Caltrout I water license challenge lawsuit is filed.

1986: The Lee Vining Creek lawsuit is filed.

1989: Lawsuits are combined into coordinated proceedings and the SWRCB is granted a 4-year stay to conduct Environmental Impact Report (EIR).

1990: The Caltrout II water license challenge lawsuit is filed.

1991: The SWRCB is ordered to keep lake levels at 6,377 feet.

1990-94: The Restoration Technical Committee (consisting of all parties) meets to oversee interim fishery habitat restoration programs.

1994: The SWRCB issues its landmark decision (1631) ordering lake levels to 6,392 feet and allowing Los Angeles to divert up to one-third of its previous amounts once the desired lake level has been achieved.

1997: The SWRCB holds hearings on the Department of Water and Power's (DWP) proposed lake restoration plans. Hearings are halted after one month when all parties agree to most of the plan's measures.

1998: The SWRCB orders the implementation of stream and waterfowl restoration measures in the Mono Basin.

Background

The City of Los Angeles, through the Department of Water and Power, began to divert water from the Mono Basin in 1941 pursuant to Water Rights License Nos. 10191 and 10192. Between 1941 and the early 1980s, Mono Lake water levels declined over 40 feet and water levels decreased in the four streams emptying into Mono Lake. These changes in the availability of water caused changes in the landscape and in the species that relied on the lake, streams and surrounding environment.

The Mono Lake case is marked by a long history of lawsuits between environmental groups and the City of Los Angeles over the issue of restoring Mono Lake and its habitat. Legal precedents were set through a series of lawsuits brought by the Audubon Society and California Trout that established protections for public trust resources and to restore fisheries. Mono Lake water rights cases were consolidated in the El Dorado Superior Court. In Decision 1631, the El Dorado Superior Court addressed the task it was delegated by the Court of Appeals in the California Trout case by setting and overseeing interim instream flows until the State Water Resources Control Board (SWRCB) completed the process.

The Audubon Society, Mono Lake Committee, Friends of the Earth and four Mono Basin landowners filed an action against the City of Los Angeles in 1979 to force it to allow more water to flow into Mono Lake. These groups argued that the diversions were damaging the public trust values of the waters that had previously flown into Mono Lake. The California Supreme Court in 1983 followed Marks v. Whitney[3] by stating that the ecological and recreational values the parties sought to protect were among the purposes of the public trust doctrine.[4]

Two separate actions were filed against the City of Los Angeles in the mid 1980s by Dahlgren and the Mono Lake Committee after heavy rains caused dams to spill over into Rush and lower Lee Vining Creeks. These actions sought to protect the fish and water that spilled into these creeks. In 1985 and 1987, the Mono County Superior Court ordered the DWP to release flows for fish in the creeks.[5]

California Trout, Inc., the Audubon Society, and the Mono Lake Committee filed suit against the SWRCB in 1985 (CalTrout I). These plaintiffs asserted that the SWRCB violated Section 5946 of the California Fish and Game Code when it issued the 1974 water right licenses that permitted the City of Los Angeles to appropriate all water in the four streams flowing into Mono Lake. They argued that the licenses should be rescinded because they did not include provisions for bypasses to protect fisheries in the four diverted streams. In its decision, the California Court of Appeals outlined the legislative intent of the Fish and Game Code to prevent streams from drying up, as had happened in the Owens River. They directed the SWRCB to amend the city's licenses to require sufficient releases, or a bypass, that would protect fisheries.[6]

Those same plaintiffs filed a second action in 1989 to challenge the modifications in the DWP's water licenses. They sought permanent instream flows for the four streams, as well as interim instream flows until the permanent flows could be set. The Court of Appeals directed the trial court to set interim flows while the SWRCB began its process for a long-term solution.[7] This decision required the DWP to "release sufficient water into the streams from its dams to reestablish and maintain the fisheries which existed in them prior to its diversion."[8]

These five cases were consolidated in 1989 in the El Dorado County Superior Court as the *Mono Lake Water Rights Cases*. In June 1990, Judge Terrence Finney entered a preliminary injunction establishing interim flows in the four Mono Lake streams (*Cal Trout I* and *Cal Trout II*). In April 1991, Judge Finney issued another preliminary injunction that required the DWP to release sufficient flows to maintain the Mono Lake water level at 6,377 feet.[9] These injunctions were renewed and all further judicial proceedings were stayed while the SWRCB began the process that culminated in Decision 1631.[10]

I. CRITERIA: OUTCOME REACHED

A. Unanimity or Consensus

Decision Unanimous. The State Water Resources Control Board (SWRCB) issued Decision 1631 on September 28, 1994. It used 43 days of testimony from the evidentiary hearings held in 1993-1994 and the Mono Basin water rights Environmental Impact Report (EIR) as the basis for its decision.[11] Though DWP's manager of Mono Basin studies expressed disappointment with the decision, DWP's Commission President promised they would not appeal.[12] DWP's General Manager expressed concern about replacement water, but also a willingness to move forward after years of litigation.[13]

In a joint press conference with the Mono Lake Committee and the other parties, Tito stated, "The time has come to accept the state's judgment and move on, to work constructively to establish reliable supplies to replace the water that is being dedicated to preserve the Mono Basin environment."[14] Other Los Angeles officials viewed the decision as, "a conservation-minded step toward peaceful resolution of water conflicts."[15]

B. Verifiable Terms

The written agreement is available publicly on a website[16] and in numerous libraries.[17] All five SWRCB board members voted to adopt the decision.

The SWRCB order sets a goal for the level of Mono Lake at 6,392 feet above sea level with varying diversions permitted as the lake level rises. The order also includes a provision for the state to re-examine the situation should the lake level not reach this goal by 2014. It sets minimum flows for all four streams emptying into Mono Lake during wet/normal/dry years, and it requires the DWP to prepare plans to restore stream and waterfowl habitat in the Mono Basin.[18] It requires continuous monitoring above and below diversion points as well as self-reporting when flows are not met and it states that the Mono Lake water level will be measured annually on April 1. It also regulates flows in the Owens River and coordinates these discharges with Grant Lake operations and management. The order requires the DWP to conduct a cultural resources inventory and to prepare a Cultural Resources Treatment Plan to assess and mitigate impacts to resources as the streams and the lake are rewatered. The decision also suspends grazing for a minimum of ten years, with future grazing subject to SWRCB approval.[19]

Terms of the proposed SWRCB order appeared in California newspapers.[20] Major newspapers around the United States and throughout California publicized the agreement.[21] Parties to the agreement were satisfied with the terms of the water diversions, though the DWP still needed to work out the details of plans to restore Mono Basin fisheries and waterfowl habitat.[22] An editorial appearing in the *Los Angeles Times* suggested that the DWP thought the state plan went too far, but would not dispute the decision. It would, however, dispute the restoration plans as "overkill."[23]

C. Public Acknowledgement

The Mono Lake Committee, National Audubon Society, California Trout, State Lands Commission, Department of Parks and Recreation, U.S. Forest Service, and U.S. Bureau of Reclamation joined the DWP's Commission President and General Manager in a Sacramento Press Conference to announce the no-appeal agreement.[24]

D. Ratification

The California State Water Resources Control Board issued the decision. The DWP, Mono Lake Committee and other interested parties participated in both the evidentiary hearings and the California Environmental Quality Act (CEQA) process that produced the EIR and they were active participants in the SWRCB process that produced the decision.

The El Dorado Superior Court exercised jurisdiction over the *Mono Lake Water Rights Cases* while the SWRCB conducted its process to amend Los Angeles' water rights licenses.[25]

II. CRITERIA: PROCESS QUALITY

A. Procedurally Just

Martha Davis, the former Executive Director of the Mono Lake Committee, described the SWRCB process as very thorough and thoughtful. She noted that three years were spent preparing the EIR and public hearings around the state provided opportunities for public testimony.[26]

B. Procedurally Accessible and Inclusive

The public had numerous formal opportunities to provide input into the development and to comment on the environmental impact report and the evidentiary hearings.

1. Public Notice

The SWRCB issued a Notice of Preparation of an EIR in January 1990 and sent it to over 500 parties.[27] It was also widely published in newspapers. The SWRCB and City of Los Angeles, along with representatives from Mono County, California Trout, Inc., the U.S. Forest Service, Department of Fish & Game, National Audubon Society, and the Mono Lake Committee, reviewed proposals to prepare the EIR and selected the consultant who prepared the document.[28]

The SWRCB circulated a Draft EIR for public comment in May 1993 that proposed seven alternatives, including a no-project and no-diversion alternative and five alternatives for varying Mono Lake levels.[29] In addition to 74 letters submitted during the public comment period, it received over 4,000 letters expressing support for the protection of Mono Lake.

2. Public Participation

Opportunities to provide input into the preparation of the EIR were numerous.[30] A September 1989 public hearing provided the public with opportunities to comment on the scope of the SWRCB review of the water rights licenses, public trust uses of Mono Basin water and other beneficial uses of water diverted from the basin.

Five technical advisory groups were formed in October 1989 and included participants from federal, state, and local governments, environmental groups, colleges and universities, private consultants, and members of the public. These parties helped identify issues the EIR should address as well as sources of information.[31]

Evidentiary hearings were held between October 20, 1993 and February 18, 1994 in Sacramento, California. One day was set aside for hearings in Lee Vining to take the testimony of Mono Basin residents. There were over 40 days of hearings, 125 witnesses and more than 1000 exhibits entered into evidence.[32]

Parties that participated in the evidentiary hearings included federal, state, and local agencies, environmental groups, and consultants representing local residents.[33]

In each of the three hearings held in Los Angeles,[34] Mammoth Lakes[35] and Sacramento[36] to accept non-evidentiary policy statements, the public, government agencies, environmental organizations, legislators, and Governor Pete Wilson expressed support for protecting Mono Lake and restoring it to a lake level of 6390 feet or higher.[37] A few comments in Los Angeles expressed concern about the DWP's ability to provide a reliable water supply. The Mono Lake Committee Associate Director Ilene Mandelbaum commented following the Sacramento hearing that, "never before has the meaning of 'public trust' been so eloquently stated than at these public hearings. While the evidentiary hearings are extremely important to the outcome of the Board's decision, there is nothing more compelling than the voices of the public speaking out for Mono Lake's protection."[38]

The Draft Environmental Impact Report (DEIR) included several survey instruments, which covered two broad categories: recreation and economics. The DEIR evaluated the regional economic importance of recreation and tourism in a number of ways. It surveyed visitors and their spending patterns at five locations: Mono Lake, Lower Reaches of the Mono Lake Tributaries, Grant Lake Reservoir, Lake Crowley Reservoir and the Upper Owens River. It then used this data to evaluate the economic impacts of the proposed Mono Lake level alternatives over a 20-year period.[39] It also used contingent valuation methods in surveys at Mono Lake, Lower Reaches of the Mono Lake Tributaries, Grant Lake Reservoir and Lake Crowley Reservoir to estimate the willingness of users to pay over and above the current costs for visiting these areas.

Another survey of 600 California households used contingent valuation methods to estimate the public's willingness to pay for the benefits of three different Mono Lake levels.[40] Commentators on the DEIR raised several concerns over the method of the household survey. First, the lake levels in the survey did not correspond with the alternative lake levels and suggested inconsistent resource conditions. Secondly, black and Hispanic households were under-represented in the survey. Third, statistical confidence levels were lacking for estimates of preservation values and differences in those values. And finally, the estimates of willingness to pay for future years should have been discounted. The SWRCB responded to each of these comments in turn.[41]

3. Public Access to Technical and Substantive Information on Issues

The SWRCB-Division of Water Rights provided a copy of the agency's mailing list for the Mono Lake project. It contains names and addresses of approximately 500 organizations, agencies, universities, elected officials, newspapers and private citizens. The newspapers included, the *Mammoth Lake Review Herald*, the *Mono Herald*, *Bridgeport Chronicle-Union*, and the *Tahoe Daily Tribune*, as well as United Press International.[42] SWRCB staff also noted that notices and documents were available in public libraries and that a public relations group (that has since disbanded) prepared various media spots. Copies of this group's files were not available.[43]

4. Public Education on Scientific and Technical Issues

The Mono Lake Committee has extensive teaching and training materials designed to inform the public about watershed issues and other preservation issues linked to the Mono Lake Basin. Some of the MLC's programs include walking tours, seminars, classroom excursions for children, and sponsored or hosted research programs on Mono Lake's wildlife and environment.[44] The MLC newsletter has been a steady source of technical and political information regarding the lake basin and efforts to restore it to its pre-diversion state. The MLC also staffs a Los Angeles office that provides direct support to the community, and participates in the Los Angeles Water Conservation Council (LAWCC). This author did not find any information regarding educational activities by other stakeholders.

C. Reasonable Process Costs

A *Los Angeles Times* article in 1991 estimated DWP litigation costs for Mono Lake at $7 million dollars.[45] In 1992, the estimates for litigation had risen to $12 million dollars.[46] A more recent article placed the DWP's costs at $15 million, with about half going to scientific studies, and the rest toward legal fees and award fees to its opponents. Kenneth Downey, Assistant City Attorney for Los Angeles who had been with the case since 1979, found the costs "exorbitant." He viewed paying opponent fees as, "equivalent to building your own scaffold for the hanging." This article noted that the attorneys for both sides agreed that mounting legal fees influenced the city's decision to "retreat."[47]

Morrison & Foerster provided $250,000 in pro bono legal fees to the Mono Lake Committee and National Audubon Society. Their legal team included three attorneys: Patrick Flinn, Bruce Dodge and Bryan Wilson. This team estimated that the pro bono legal fees were used up before 1980. The firm agreed to take 50 percent of its regular fees since then (up to 1991) in order to stay with the case. All together, the Mono Lake Committee estimated that the DWP outspent it on legal fees, five-to-one. Moreover, technical experts on Mono Lake provided testimony for free and the Mono Lake Committee membership provided donations in response to emergency legal appeals.[48]

The SWRCB funded three persons per year of the Mono Lake process, according to staff. This meant that two environmental specialists and one engineer worked full-time on the project. A fourth person provided legal support, though not full-time. This information was summarized in correspondence rather than through provision of department records.[49]

III. CRITERIA: OUTCOME QUALITY

A. Cost Effectiveness

1. Summary of Cost-Sharing Arrangements

The costs of implementing the agreement are likely to be borne primarily by the City of Los Angeles as it develops alternative sources of water. These costs have been supported through public monies made available from the state and federal government, and through grant-seeking assistance from the MLC. While there does not appear to be any explicit intent or principles that guided how the parties planned to share costs of implementation, the record contains numerous instances of group efforts to obtain funding to implement the agreement. MLC participated in the efforts to raise state and federal funds for Los Angeles to develop water supply programs. Among these funds are: $36 million from the state legislature to develop water reclamation and conservation facilities, monies from the federal government to create 120,000 ac-ft per year of reclaimed water through project development, and $10 million to support water reclamation and conservation. New facilities and conservation supported through these state and federal funds are expected to replace an estimated 141,250 ac-ft. Other efforts include developing water markets through conservation-pricing strategies, and developing and implementing best management practices expected to save 700,000 ac-ft of water annually.[50] Since the decision, DWP has committed additional staff and funds for conservation, water recycling, and groundwater management.[51] Among DWP's recently adopted resolutions were approximately $1 million in additional funds to continue monitoring waterfowl, fish populations, and limnology of the lake.[52]

2. Costs to Parties Who Participated in Process

The DWP's costs have steadily risen in its efforts to maintain its water rights licenses. In 1989, the DWP estimated it would cost about $15 million dollars to replace water and energy supplies that were lost when El Dorado County Judge Terrence Finney ordered it to halt all diversions from the Mono Basin.[53] The DWP has increased funding to monitor waterfowl, fish populations, and the limnology of the lake, and has committed additional staff to oversee and support these activities.

3. Costs Borne by Taxpayers

According to a 1994 article in the *Sacramento Bee*, Los Angeles residents had been paying $38 million a year to replace water lost to the 1989 court injunction, with expectations for these costs to rise over time. The State of California is providing $36 million for reclamation and conservation projects, and federal taxpayers in 1994 provided another $5 million for these facilities.[54]

4. Costs Borne by Others

In its closing brief in the SWRCB process, the Mono Lake Committee used figures derived from its and Cal Trout's model to estimate costs to Los Angeles ratepayers. This model suggested that fulfilling a 6,390-foot lake level would cost ratepayers less than one percent, or about $0.16, per month. Setting a lake level goal of 6,405 feet would raise that cost by a penny.[55]

B. Perceived Economic Efficiency

Economic efficiency was a consideration for both DWP and MLC in terms of evaluating the costs of replacement water versus the benefits gained by restoring Mono Lake and its surrounding ecosystem. State and federal funding of up to $48 million for LA projects and local water rebates for customers who install water conservation devices were integral to crafting an agreement that was acceptable to both parties.[56] MLC's primary goals in its efforts to restore Mono Lake to its pre-diversion levels (environmental restoration and conservation) were evident, however, in its closing briefs when it emphasized the overriding benefits of restoring Mono Lake versus a projected increase of less than one percent per month to Los Angeles ratepayers.[57]

C. Financial Feasibility/Sustainability

The costs of implementing Decision 1631 are addressed in a number of ways. The decision evaluates costs in terms of the beneficial uses provided by the water the DWP diverts. The decision focuses on impacts in quantity and quality to municipal water supplies. Efforts have been made to meet the shortfalls imposed by the agreement through funds provided by the State of California and the federal government, conservation, and increasing supplies from other sources.

Legislation at the state and federal level provided the DWP with flexibility in developing alternative water supplies. The California Legislature (Assembly Bill 444) provided $60 million in funds to develop new sources of water for Los Angeles.[58] The legislation included an important incentive: in order to access the funds, the DWP had to reach an out-of-court agreement with environmental groups on ways to save the lake. The Environmental Defense Fund lent its assistance in developing the legislation and potential alternatives.[59] After four years of tough negotiations that characterized the parties as "warring factions with symmetrical paranoia,"[60] the DWP agreed to credit Mono Lake diversions with water supplies developed through AB 444 funds.[61] The funds, dwindling to $36 million over the four-year period, would go to four projects: a Los Angeles-based water conservation program, the Sepulveda Reclamation Project, the East Valley Reclamation Project and the West Basin Reclamation Project.[62] H.R. 429, the Western Water Bill, provided Central Valley farmers with opportunities to sell their water rights as well as funds to develop water reclamation projects.[63] The Bureau of Reclamation announced funds of $5.3 million (1994) and $8.3 million (1995) to support Los Angeles water reclamation projects.[64] Taken together, these funds provided the DWP with opportunities to develop replacement options for water that would stay in the Mono Basin.

The DWP obtains water from three sources: groundwater, the Los Angeles Aqueduct (Mono and Owens Basin) and the Metropolitan Water District (MWD). It relies on the MWD most in dry years when supplies are limited in the Los Angeles Aqueduct. On average, the DWP purchased 13 percent of its water from the MWD, which obtains its water from the State Water Project and the Colorado River. Since 1989, the DWP had been prohibited from withdrawing water from the Mono Basin in order to maintain a lake level of 6,377 feet.[65] Thus, Los Angeles Aqueduct supplies have not been available. The DWP also addresses water shortages through 22 water conservation programs (including tiered water pricing) which save up to 30 percent of water use.[66]

The effects of the agreement will be felt most by the City of Los Angeles early on, as diversions are restricted in order to achieve specified lake levels over time. Los Angeles has facilitated permanent changes in water use through conservation, reducing water needs by about 15 percent. The city also seeks to address future demands through increased groundwater use, continued water conservation, reclamation and recycling, and the gathering of additional supplies from the MWD.[67] Because the DWP perceives supplies from the MWD to be uncertain (given Colorado River uncertainties and Endangered Species Act impacts on the State Water Project), it expects to develop more expensive alternatives.[68] While water quality from the Mono Basin is high and provides good dilution of minerals in Owens River water, it was estimated that changes in water quality standards might necessitate construction of a water treatment facility regardless of the Mono Basin diversions.[69]

The implementation of Decision 1631 is expected to increase water supply and power costs to the DWP and its customers. While these costs may appear to be unfair to Los Angeles and its residents, they tend to offset the years of internalized costs to fish, wildlife, and residents of Mono Basin in the form of lost water supplies.

Costs will vary depending on the individual costs of conservation programs that reduce demand, costs of replacement water, water shortage costs, and costs of replacement power. Estimates range from $300 per acre-foot for water conservation programs to $700 per acre-foot for reclamation programs.[70]

The true costs of replacement water were debated based on the two models and their inherent assumptions. Some of the assumptions were considered unrealistic (high water shortage costs will be passed onto Los Angeles residents), while others were unverifiable (effects of water conservation and pricing).[71] Given these differences, the SWRCB estimated costs based upon evidence in the record. These cost estimates seemed rather high, adopting the high end of the water cost range and adding an additional twenty percent for an extra-conservative estimate.[72] Thus, the current prices (1994) of $230 per acre-foot were estimated to cost $520 per acre-foot in a dry year, $400 per acre-foot in a normal year, and $370 per acre-foot in a wet year for replacement water.[73]

Rather than reiterate its decision, the SWRCB referenced the economic benefits provided by fishery and public trust resources that a full economic analysis of the decision would include. It concluded that sufficient water would be available to meet the municipal needs of Los Angeles while restoring lake levels and fisheries in the Mono Basin, and that neither water supply nor power costs would make implementation impossible.[74]

The City of Los Angeles, the Mono Lake Committee and the Audubon Society agreed that the costs of replacing power generated by plants on the Los Angeles Aqueduct would be around $125 per acre-foot. Combined costs for fisheries and public trust resources were estimated at $8.5 million until the lake level reached 6,391 feet and $5.6 million thereafter.[75] The SWRCB considered these costs reasonable, given that Southern California Edison's rates adjacent to the DWP service area are twenty percent higher.

D. Cultural Sustainability/Community Self-Determination

Cultural resources can include sites, features, and locations of archeological, historical, architectural and ethnohistorical origins. They may also include ceremonial locations and traditional food gathering areas presently in use. The DEIR considered inventories of archeological records and conducted a literature search and field assessment of recorded resources.[76] Additional information on cultural resources in the Mono Basin is available in the Draft Environmental Impact Statement prepared by the Inyo National Forest in 1988.[77]

Before contact with European settlers, the Mono Lake Paiute lived in Mono Basin while the Owens River Paiute lived in the lower Owens River. Mono Lake Paiute were known as Kuzedika or "fly larvae eaters." They tended to live more informally and move about, as opposed to Owens Paiute who settled in villages. The Owens Paiute hunted, cultivated two irrigated crops and gathered other foodstuffs. The Mono Paiute hunted, gathered seeds, berries, bulbs and grasses, and in the summer collected alkali fly larvae and Pandora moth larvae. Both groups also fished, collected seeds and hunted game in the Upper Owens River. In the fall, they collected pine nuts from Jeffrey pine. The first transitions in their lifestyle occurred when European settlers came to Mono Basin in the late 1800s to mine gold found in the area. They soon assimilated into the European culture as laborers and traders of goods.[78]

The DEIR notes that Native Americans continue to live and work in the Mono Basin and have initiated an effort for federal recognition. Today, they are known as the Mono Lake Indian Community.[79] They continue to use traditional resources and practice cultural activities, and they raised concerns with the U.S. Forest Service about maintaining these practices when it prepared its DEIS for the Mono Basin National Forest Scenic Area. Among the concerns were access to public lands to hunt deer, conduct rabbit drives, fish, and to gather traditional foods (kutsuvi, buck berries, willows, wild onions and waterfowl), raw materials for crafts, and herbs for medicinal and ritual purposes.[80] The Mono Lake Indian Community reiterated these concerns in its comments on the DEIR in addition to raising concerns about water rights claims that have not been satisfied.[81]

Given the richness of cultural resources found in the Mono Basin, particularly in riparian areas, Decision 1631 requires the DWP to develop a Cultural Resources Treatment Plan. The plan should protect and provide access to resources important to the Mono Basin Native American community, including areas for traditional uses (if requested), and include provisions for unexpected discoveries of cultural resources. Moreover, the SWRCB required that a monitoring plan be included to ensure effectiveness of the treatments.[82] The SWRCB pointed out that this was particularly important because of the restoration work in the streams anticipated by the decision. Because these impacts were difficult to anticipate, the SWRCB decision instead created provisions for the DWP to act as the resources and physical activities became known.

In many ways, the entire State of California was affected by the decision because of the interconnected nature of the state "plumbing." The key communities that were affected included the residents of Mono and Inyo Counties, California and the Los Angeles DWP service area. Other communities included recreational users of the Mono Basin and Inyo National Forest.

Towns within Mono County include Lee Vining, Mono City, June Lake and Mammoth Lakes. Towns within Inyo County include Bishop, Lone Pine and Big Pine. Major employment sectors include services, trade and government. Trades and services support tourism, the main source of employment. Mining and agriculture also provide employment, though their contribution to the local economy is small.[83]

Most (79 percent) of the land in Mono County is in publicly owned.[84] The Mono Basin National Forest Scenic Area includes land around the lake, portions of the four diverted streams and irrigated DWP pastures.[85] Lands used for agriculture support livestock production and most of this occurs by lease from the DWP, the U.S. Forest Service or the Bureau of Land Management. The Mono Sheep Company and Inyo Sheep Company lease land along the tributaries owned by the DWP.

Mono County regulates land use on private and DWP lands in the Mono Basin and Upper Owens River. The DEIR noted three pending developments. In Mono Basin, the Conway Ranch is an approved recreational-residential development on 880 acres. Another Mono Basin development is the Tioga Inn, a 120-unit inn with restaurant, gas station, mini-mart, and residential dwellings.[86] Since the preparation of the DEIR, the Trust for Public Land has purchased Conway Ranch. It will eventually transfer the land to Mono County to use for open space, wetlands mitigation banking, fish rearing facilities and other preservation.[87]

The DWP, the federal government and private parties own land in the Upper Owens River. There are four private ranches within the area considered in the DEIR.[88] These ranches include the Owens River Ranch, John Arcularius Ranch (including a proposed expansion), Howard Arcularius Ranch, and Inaja Land Company.[89]

The DEIR predicted reductions in forage production, and benefits to vegetation and wildlife along diverted tributary streams for all target lake level alternatives. Insignificant countywide economic impacts were expected. However, the potential development of rural properties raised questions of significant growth-inducing impacts.[90]

The DEIR noted that the DWP would terminate irrigation releases from Gibbs, Lee Vining, Walker and Parker Creeks to achieve a 6,390-foot lake level.[91] This change was predicted to affect the operations of the Mono and Inyo Sheep Companies. It would require the companies to either reduce their sheep herds by half or find summer forage elsewhere.[92] Under the same 6,390-foot lake level alternative, overall recreation use and benefits would increase, with opportunities increasing around Mono Lake and decreasing around Lake Crowley and Grant Lake. The DEIR predicted increased annual benefits of $2.7 million.[93] While achieving the 6,390-foot lake level would result in decreases in agricultural production, it would be offset by increases in recreational spending. The overall regional economy would increase.[94]

Information on the differences in socioeconomic indicators was not readily available, although the DEIR did reflect anticipated changes in employment and personal income upon implementing each proposed alternative. Personal income was expected to increase an average of $130,500 annually, while employment would decrease by 7.3 FTE jobs.[95]

E. Environmental Sustainability

All the various factors noted under this criterion were considered in the legal framework that guided the development of Decision 1631. . The CEQA process, for instance, required considering the environmental impacts of modifying Los Angeles' permitted water rights. Among the review factors included in the CEQA process and in developing modified permits were the effects of drought, natural resources committed (water for streams and Mono Lake), projected resource use over time (water and power), and environmental impacts including endangered species. Moreover, environmental agencies and interest groups participated in developing the agreement that produced the decision. Details of specific issues considered in developing Decision 1631 follow below. Decision 1631 considers instream flows necessary to re-establish and maintain fisheries to pre-diversion conditions, measures to provide periodic channel maintenance and flushing and other measures that would facilitate restoration of the fisheries.[96] Each of the four creeks (Lee Vining, Walker, Parker and Rush) were considered in turn and the decision generally incorporated Department of Fish and Game recommendations for streamflows during dry, normal and wet hydrologic years.[97] This was an area where there were differing views on the technical approaches used to establish flows, particularly in the models used.[98]

Other public trust resources and beneficial uses of water considered in Decision 1631 included birds and other wildlife in the Mono Basin, organisms in Mono Lake that provide food sources for birds, riparian vegetation, wetland/meadow habitat, air quality, visual and recreational resources and water quality.[99] For each of these resources the decision considered pre- and post-diversion conditions and the measures that would facilitate restoration to conditions most protective and feasible in relation to diversions, past practices, and evolving systems.

Wildlife considerations took into account the abundant former bird numbers (waterfowl) that used Mono Lake as a major stop-over during migration (eared grebes, red-necked phalaropes, and Wilson's phalaropes) and as a nesting area (California gull, Caspian terns). It also considered the presence of sensitive or listed species (Western snowy plover),[100] as well as the water levels necessary to re-establish wetlands and restore waterfowl habitat.[101] The decision also referenced the EIR for analyses of impacts to 39 special-status species in the Mono Basin and concluded that some may benefit from the decision (osprey and bald eagle) and none are expected to suffer adverse impacts from the lake levels and streamflows the decision establishes.[102]

Air quality, water quality and visual and recreational resource considerations include the fact that the Mono Basin is designated as a PM_{10} (inhalable fine particles < 10 microns) non-attainment area pursuant to the Clean Air Act,[103] and a National Forest Scenic Area pursuant to the California Wilderness Act.[104] In addition, Mono Lake is a California state reserve[105] with its water quality protected by the federal anti-degradation policy pursuant to the federal Clean Water Act.[106]

The decision concluded that the instream flows recommended in the four creeks will cause the water level in Mono Lake to reach 6,390 feet in 29 to 44 years and will sustain a water level somewhere between 6,388 and 6,390 feet upon reaching the desired 6,391 feet in fifty years. Moreover, it concluded that an average lake level of 6,392 feet would protect public trust resources. These include air quality in the Mono Basin, water quality in Mono Lake, Mono Lake brine shrimp and brine fly food sources for migratory birds and secure nesting habitat. It will also provide easily accessible recreation to the Mono Lake Tufa State Reserve.[107]

Depending on the computer modeling approach ("rolling average" versus repeat 1940 to 1989 hydrology), the decision estimated that Mono Lake would reach a water level of 6,390 feet in approximately 18 to 28 years and that in two more years the water level would reach 6,392 feet. Given the limitations in modeling and the differences that may occur from one hydrologic year to the next, the SWRCB reserved the option to adjust diversion criteria should actual conditions vary significantly from those assumed in establishing instream flows.[108]

F. Clarity of Outcome

The Mono Lake Committee generously lent its newspaper clip files for this review. Articles in these files revealed overwhelming public support for the decision. However, despite the decision that brought a long and expensive legal battle to a close, one author speculated that future skirmishes were likely.[109] The DWP's Hasencamp in an interview with the *Review-Herald* saw the decision as a turning point, "the time to end arguing and move on." When asked about his previously expressed disappointment he replied, "I don't feel that that level of protection is necessary, but again the time for arguing is through."[110] Mary Scoonover, the California Deputy Attorney representing the State Lands Commission, viewed the restoration planning as the least specific and therefore subject to a lot of future work. She expressed hope for cooperation among the parties.[111]

The SWRCB's Decision 1631 is available on the Mono Lake Committee's website and is 105 pages.[112] This analysis used that copy extensively. The decision is also available in the University of California library system and is 212 pages in length.

The process of setting a baseline condition was both complex and contentious. The DEIR and the Final Environmental Impact Report (FEIR) explain how the SWRCB set a point-of-reference for analysis. The DEIR used two points-of-reference to compare the impacts of the proposed project alternatives.[113] It compared the impacts of the proposed projects to the Mono Lake water level and streamflows before the August 1989 injunction that prevented diversions. The surface lake level elevation in 1989 was 6,376.3 feet. A point-of-reference scenario was used to approximate water and power supply exports because 1989 was a dry year. The goal of this adjustment was to represent average water supply and power production levels for later impact analysis. A second point-of-reference, the level in 1941 before diversions had commenced, was used to evaluate the cumulative impacts of the proposed project alternatives. The lake level elevation when the DWP began diverting water from the streams was 6,417 feet. The SWRCB in the FEIR stated that the use of these two levels would provide the widest comparison and the fullest disclosure of impacts possible.[114]

All resources discussed in the decision note pre-1941 diversion conditions (as noted in previous sections) and acknowledge the difficulties in finding detailed historical data.

Decision 1631 sets Mono Lake elevation levels and allows for diversions as interim levels are achieved. While there was not wholehearted agreement with the lake levels set,[115] this review did not identify any variance or confusion on these ultimate lake levels.

G. Feasibility/Realism

1. Legal Feasibility

The SWRCB decision is in direct response to litigation establishing the protection of public trust resources and fisheries.

2. Political Feasibility

State and federal legislators as well as California's governor supported the decision. The legislators that provided statements of support for Mono Lake protections included State Senators Patrick Johnston (formerly representing Mono County), Frank Hill, Mike Thompson, Dan McCorquodale, Quentin Kopp, Nick Petris, and Milton Marks and Assemblymembers Richard Katz, Byron Sher, and Jackie Spier. U.S. Representative Norman Mineta also provided a statement of support, as did local politicians. The mayor of Mammoth Lake and Mono County Supervisors and Los Angeles City Council members lent their support for protections as well.[116]

The SWCRB issued Decision 1631 in response to directives from the El Dorado County Superior Court. The order within the decision was not conditioned upon any needed legislation, although funds provided by the California and federal legislature did facilitate reaching an agreement.

The California State legislature specifically appropriated $60 million in funds to protect Mono Lake.[117] AB 444, sponsored by Assemblyman Phil Isenberg, Assemblymen Bill Baker and Assemblyman Richard Katz, made funds available to the City of Los Angeles to resolve the dispute over Mono Lake. The money was available on a matching basis to develop water reclamation and conservation projects. Though it took the City of Los Angeles four years to agree with the Mono Lake Committee on how to use the money, politicians supported developing solutions through the use of AB 444 funds.[118]

3. Scientific and Technical Feasibility

This sub-criterion focuses on parties' behavior and willingness to follow through (based on personnel and resources), supply resource or technology (scientific/technical assumptions) and make financial commitments. None of these factors apply to Decision 1631. The parties announced their willingness to abide by the decision when they issued a No-Appeal statement.[119] The resources and technology are available to implement the solution (see environmental and economic sustainability), and state and federal governments have already allocated funds to the project.

Governor Wilson signed AB 444 on September 19, 1994.[120] The legislature allocated the first $9 million allotment of the $36 million state funds in June of 1994.[121] The Bureau of Reclamation provided the first payment of a grant to the DWP to fund water reclamation projects in 1995. The East Valley Water Reclamation Project, scheduled to begin operating in 1998, will ultimately provide 35,000 acre-feet of recycled water.[122] Despite these financial commitments, Bill Hasencamp, DWP Manager of Mono Basin studies reminded Jason Montiel of the *Review-Herald* that AB 444 funds were matching funds, and the DWP would still have to share the costs of building reclamation facilities.[123] Given these comments, this may be an area to watch over time.

In the wake of the decision, the DWP continued to raise concerns about its ability to replace waters that would have come from the Mono Basin. This issue is one to watch as well. Metropolitan Los Angeles citizens did not view replacement water as a problem and believe there is plenty of water available.[124] Whether the DWP is using this issue for strategic reasons or because of real shortages is not fully clear.

Given that all parties to the agreement announced a "No-Appeal" statement, one might assume they had justified the agreement to their constituencies.

H. Public Acceptability

Environmentalists, the public, resource and regulatory agencies, legislators and the Governor support the SWRCB decision to restore instream flows to Lee Vining, Walker, Parker and Rush Creeks and Mono Lake water levels that were lost when Los Angeles began diverting water from these creeks in 1941. This decision outlines the details of a plan to amend water right licenses, restore public trust resources and satisfies the California Supreme Court's objective of taking "a new and objective look at the water resources in the Mono Basin."[125]

In many ways, there is a sense that wrongs will be righted. Eldon Vestal, whose yellowed field notes from his days as a Fish & Game biologist provided critical evidence in the case, captured this sentiment when he said, "The city of Los Angeles was a tremendous political power over the years, and challenging it seemed like grabbing for a bite out of the moon."[126] David Carle, Park Ranger for Mono Lake Tufa State Reserve, echoed Vestal's jubilation, "Yee-haw! I don't care what the sober-sided alter ego says, it's time to celebrate! Party! Sing songs and dance a jig! Long live Mono Lake!"[127] Marc Del Piero, a member of the Water Resources Control Board said, "Today we are correcting a mistake and putting in place an order that protects the public trust."[128]

Support for the decision was clear among members of the Mono Basin community when asked what they thought about it at Mark Twain Day. Bud Stickles, Lee Vining, said, "It's about time they did something. L.A. has raped this area long enough." Betty Clayton, June Lake, added, "I think it's great. I saw it on T.V. It was the greatest thing." Michael Wells, Lee Vining, recognized Mother Nature's hand in the decision; "I think it's a long time coming and I hope that Mother Nature is good to us this year so it will raise the level to the prescribed amount. It's up to Mother Nature now."[129]

Dennis Martin, supervisor of the Inyo National Forest, had the sense that everyone, except perhaps the DWP, "felt good about [the decision]."[130] Dennis Tito, DWP Commission President, promised, "We are here to accept the state's decision and to work with the Mono Lake Committee and other environmental groups to replace the lost water that will be used for the Mono Basin. We're happy, actually, that the conclusion is now behind us. We can now move ahead to the future. We have worked not only on the reclamation projects but also on the restoration of the streams in the Mono Basin."[131]

California Senator Tim Leslie's support through AB 444 funds for reclamation and conservation projects observed, "I think we can declare the battle of Mono Lake over. . . I think that this reclamation project was a catalyst that helped it all come together. That wasn't the whole agreement. But things weren't happening until that reclamation project worked out."[132]

California Governor Wilson said, "The plan approved today assures protection of Mono Lake's extraordinary environmental values, while also providing flexibility for the renewed diversions to Los Angeles as the lake is restored to healthier levels."[133]

IV. CRITERIA: RELATIONSHIP OF PARTIES TO OUTCOME

A. Satisfaction/Fairness

While there was no evidence of statements of unfairness, several DWP officials expressed personal disappointment. The DWP's manager of Mono Basin studies expressed some disappointment with the decision, but conceded it was time to focus on what is best for Mono Basin and Los Angeles.[134] Counsel for the DWP also expressed disappointment, "On a personal level I'm a little disappointed, but the department has stated that it accepts the decision and we will implement the decision."[135] One writer detected a trace of bitterness in a statement by an Assistant City Attorney for Los Angeles when he said, "We appropriated (water) in accordance with the law of California, and we appropriated water in accordance with the Constitution, but somehow it wasn't good enough. It's just amazing to me."[136] There was also occasional evidence of differences of opinion on how to set baselines and model impacts.

There is no evidence that parties at the table refused to sign-on to the decision. On the contrary, the joint press conference announcing that no parties intended to appeal the decision conveys agreement among the parties.

A continuing theme in the DWP's reaction to Decision 1631 is its concerns for replacing Mono Basin waters. Since 1989, when El Dorado Superior Court Judge Finney issued a preliminary injunction preventing diversions, the DWP has had to obtain Mono Basin volumes of up to 85,000 acre-feet from other sources. Apparently, as of 1998, it had been able to secure these alternate sources at the cost of $38 million per year.

B. Compliance with Outcome Over Time

The Guidebook indicators in this section focus on subsequent litigation (initiated or threatened), records of regulatory/monitoring organizations, ways and presence of documents to verify compliance, and internal recordkeeping of compliance. There is no evidence of subsequent litigation other than a motion filed by the Mono Lake Committee and the National Audubon Society to allow the Restoration Technical Committee (RTC) to continue its restoration works.[137] The RTC had been carrying out interim court-ordered creek restoration since 1990 until the SWRCB issued its decision.

The SWRCB order requires the DWP to keep records of instream flows above and below points of diversion and to measure the Mono Lake water level each year on April 1. Moreover, it includes provisions to report instances when the specified flows are not met. [138] Records of monitoring and compliance used for this review are noted below.

Measuring compliance with the decision, requires among other things, the monitoring of Mono Lake water levels and the development and implementation of stream and waterfowl restoration plans.

Mono Lake water levels have risen steadily. In the summer of 1995, the lake level was 6,376.2 feet above sea level, an increase from 6,374.4 the previous winter.[139] By fall of 1995, the lake level had risen to 6,377.4 feet above sea level--a three foot rise from the previous winter.[140] One year later, the lake level had risen to 6,380.1 feet above sea level. The Mono Lake Committee, who monitors the lake level, reported a 5.5-foot vertical rise since the SWRCB decision. They attributed this rise to two years of above-normal runoff and limited diversions from the four streams.[141] In the fall of 1997, the Mono Lake Committee reported the Mono Lake level at an elevation of 6,382.4 feet above sea level--7.8 feet higher since the SWRCB decision.[142] At the time of this writing in 1998, Northern California has experienced another exceptionally wet winter and spring runoff, and the Mono Lake water level has no doubt continued to rise. Most recent estimates are that the lake level will reach 6,384 feet above sea level by the end of the 1998 summer.[143]

In Decision 1631, the SWRCB charged the DWP with preparing stream and waterfowl restoration plans as well as a plan for managing Grant Lake operations. These documents were due to the Water Board in November 1995, though requests to extend the deadline were already pending in the previous summer[144] and newsletters from the Mono Lake Committee note an extension to November 1996.[145] It is not clear from this review when the documents were actually submitted nor if copies of the plans were obtained despite requests to the Mono Lake Committee, State Water Resources Control Board, Water Resources Archives Library at the University of California-Berkeley and Los Angeles Department of Water and Power. According to the SWRCB Water Rights staff, the Water Resources Control Board will review and issue a draft order [on the plans] by the end of July 1998. The SWRCB staff offered a copy of this document for future project reviews. The Board was expected to adopt the plans in early September 1998 after taking testimony.[146]

The contents of the plans have already raised concerns in the community.[147] Questions over what values should take precedence, what point in time the habitat should be restored, and who should decide are at the heart of these concerns. The focal point of these questions is a proposal to create waterfowl habitat by restoring Mill Creek with waters from nearby Conway Ranch. Members of the community feel they were left out of plans that will sacrifice one creek (Wilson Creek) to restore another (Mill Creek). A member of a family who has lived in the Mono Basin since the 1870s feels that, "restoring a natural creek at the expense of a century of local history is not only outrageous, it's unrealistic."[148] The SWRCB has responded by leading a group known as the Conway Ranch Evaluation Workgroup (CREW) to consider land use options and alternatives.[149] Questions of implementation, therefore, will hinge in part on how the restoration planning process proceeds.

C. Flexibility

This review did not identify any evidence of modifications in the agreement, perhaps because the decision included provisions for SWCRB oversight and action if, and when, necessary. The decision includes provisions for adjustments if hydrology does not reflect the assumptions used in setting instream flows and Mono Lake water levels.

Questions have developed since the agreement about the implementation of restoration plans. The status of these issues has been discussed in the previous section on compliance.

D. Stability/Durability

Mechanisms exist to track progress, particularly Mono Lake water levels. Los Angeles water diversions are modified as particular lake levels are achieved. The sooner the 6,392-foot level has been achieved; the sooner Los Angeles will be able to divert water. There are threshold lake levels below which future diversions are forbidden,[150] and there are contingencies if the lake level goal has not been achieved before 2014.[151]

The DWP, through Decision 1631, is responsible for implementation. The regulatory, academic, local and environmental community that have participated in the process since its beginning join the DWP in this effort through their continuing interests, while the SWRCB has retained oversight.[152]

This review did not identify alternate conflict management or resolution processes. Standard negotiations and a return to the SWRCB appear to be the backstop processes when problems or conflicts arise.

Shifting alliances have emerged as the details of the waterfowl and riparian restoration plans have become known. See the previous section on compliance for more details.

V. CRITERIA: RELATIONSHIP BETWEEN PARTIES (RELATIONSHIP QUALITY)

A. Reduction in Conflict and Hostility

Public statements from the parties have consistently maintained a professional tone. If the tone of the conflict was rising or falling, it was not apparent in statements to the press. Rather, the press characterized the discussions as contests, or as a forum to right wrongs. Headlines such as, "public trust values win," "Mono Lake, environment saved," "Mono nears water war victory," and "L.A. gives up" support and reinforce an adversarial relationship between the parties. The press characterized the environment as the victor and at times recognized the change in the DWP's strategy. Examples include, "DWP: The time for arguing has ended," "Mono's troubled waters at peace," "City of Angels Makes Peace in Water Wars," "A Welcome Truce," and "Peace at Mono Lake."

B. Improved Relations

The relationship between the parties began as an adversarial one, that is, through legal actions brought by environmental groups to stop DWP actions. An important point to bear in mind is that water rights have always been considered within the framework of common law. A formal, legal process continued to frame discussions of the issues throughout the process. One would expect a written agreement or consent document as the outcome of a formal process. Thus, the written legal document the SWRCB generated in this case does not seem unusual given the traditional framing.

Despite the formality of the process, many of the parties spent as much as 16 years together in negotiations. Trust levels were described as fragile[153] early on, which supported the continued use of a structured process. Though the legal process imposed a more formal framework, it also provided third party oversight and protection to the plaintiffs that they would not have otherwise had in these actions. Following the Board's decision, numerous parties expressed hope for future cooperation.

C. Cognitive and Affective Shift

A Los Angeles City Councilwoman has facilitated and led changes in the DWP. Following the Board decision she said, "Some of what you are seeing is a genuine change in the Department. There has been a change coming. Even in Los Angeles, there is a strong environmental movement." She views herself as responsible for fixing the city's past environmental problems.[154] The Commissioner of the U.S. Bureau of Reclamation praised the DWP's change in attitude; "All parties have come to the realization that many serious environmental mistakes were made when we constructed the water projects on which our urban and agricultural sectors now depend."[155] Others concurred with the assessment that the parties were "old antagonists" that had "declared peace."[156] The former Executive Director of the Mono Lake Committee noted earlier signs of shifting views. She observed, "for years the DWP portrayed the fight as a win-lose – if Mono Lake won, L.A. would lose." She expressed her hope for change and the promise of compromise so that each party would get what it needed.[157] Over the years as the parties met to develop a compromise, shifts had occurred though it is not completely clear what brought them about. Some attribute it to escalating costs,[158] others to changing public values.[159]

D. Ability to Resolve Subsequent Disputes

This review did not identify a lot of information to strongly state whether problems were handled constructively. For instance, following the Board's decision, the DWP, the Mono Lake Committee and other parties announced their intent to work together to implement the SWRCB's order that required developing plans for waterfowl and stream restoration and managing Grant Lake.[160] However, when the SWRCB proposed closing out lawsuits brought by the Mono Lake Committee, the National Audubon Society and California Trout, these groups filed briefs requesting continued court jurisdiction until final restoration plans were approved.[161] While the parties continue to meet to hammer out details, the discussions appear to focus on technical issues.[162] Changes in relationships seem to be an unintended and indirect benefit from spending so much time together.

E. Transformation

A new spirit of cooperation at the DWP was noted by a Los Angeles Councilwoman as a sign of "genuine change in the Department."[163] She added, "We are on our way to a new environmental ethic, a new way of supplying water, [a]nd I hope a new era of California politics."[164] Others recognize the precedents set in the Mono Lake case and see it as a model for solving other water disputes.

VI. CRITERIA: SOCIAL CAPITAL

A. Collective Citizen Capacity

Over the course of the negotiations to reach an agreement on the future of Mono Lake and since the decision in 1994, the MLC has worked with the DWP both to secure funds to replace diversions and to expand water conservation programs. During the Mono Lake negotiations the MLC actively participated in efforts to secure state funds ($36 million) to build new water supply facilities – a key issue in developing replacement water.[165] The MLC and the DWP presently work together in implementing the Ultra-Low Flush Toilet Program through the Los Angeles Water Conservation Council, a collaborative of Los Angeles community organizations who, in addition to their respective community interests, promote water conservation and environmental protection. Among the organizations participating in the LAWCC are Adro Environmental (a community development corporation that trains and employs residents from three LA neighborhoods); Asian American Drug Abuse Program (fights drug abuse and runs a treatment program); Calvary Baptist Homes; Iglesia Poder de Dios (small community church promoting educational youth trips to Mono Lake); Korean Youth & Community Church (sponsoring youth trips to Mono Lake); Mothers of East Los Angeles Santa Isabel (promoting a healthy, united local community and sponsoring youth trips to Mono Lake); and Watts Labor Community Action Committee (providing community services and support for residential and community development).[166]

B. Community Capacity for Decision-Making

The MLC has assisted the DWP in its recent efforts to address community concerns over Los Angeles' water supply, water quality and new programs. The DWP General Manager and the MLC Executive Director were scheduled to co-lead the first of fifteen water education workshops sponsored by DWP. These workshops are intended to provide residents and businesses with opportunities to ask questions and learn more about their water supply.[167]

Numerous instances of symposia and conferences also convey the collective efforts of parties to create forums in which the local and statewide community can learn more about the Mono Lake "story" and its connection with legal and ecological systems. Water policy conferences and panels, legal symposia, and local restoration workshops are examples of efforts that have conveyed the Mono Lake decision's connection to other water policy issues around the state (such as the Bay-Delta/Cal-Fed process), the significance of the Public Trust Doctrine, and the state of the Mono Lake stream restoration.[168]

The Mono Lake process also provided the motivation to develop similar collaborative processes to address the specifics of stream restoration efforts. The Conway Ranch Evaluation Work-groups (CREW) illustrates one such example. CREW provided a constructive forum for county residents, resource agencies, and public interest groups to discuss alternatives for the Mill Creek restoration and describe the creek's contributions to the Mono Basin and the lake ecosystem.[169]

Evidence that the community has connected with MLC's efforts to "save" Mono Lake and developed an understanding of the issues is apparent in the prize-winning essay written by a young resident of Lee Vining. The essay recognizes the magnitude and importance of MLC's efforts by describing Mono Lake's connection to plants and wildlife, and the capacity for restoration to provide a focal point for the community in general, and young people in particular.[170]

C. Social System Transformation

The beginnings of a social transformation stemming from the Mono Lake efforts are suggested in a number of local and wider reaching efforts that focus on water conservation in the West. DWP's sponsorship of the Ultra-Low Flush Toilet Distribution Program, and provision of free low flush toilets provides one example of assistance and support provided to the Los Angeles community in general. More far-reaching instances of transformation are evident in the way Los Angeles and the state of California has approached securing water supplies for its community. The former MLC executive director commented that the Mono Lake process provides important lessons in developing and supporting wiser, more efficient water use in California.[171] The Bay-Delta efforts in Northern California and the restoration of the Owens River and Owens lakebed provide two such illustrations.

NOTES

[1] Craig Anthony Arnold, and Leigh A Jewell, "Litigations Bounded Effectiveness and the Real Public Trust Doctrine: The Aftermath of the Mono Lake Case," *8 Hasting West-Northwest Journal of Environmental Law and Policy 1,* (Fall 2001); Brian E. Gray, "The Property Right in Water," *9 Hasting West-Northwest Journal of Environmental Law and Policy 1,* (Fall 2002); John Hart and Nancy Fouquet, *Storm Over Mono: The Mono Lake Battle and the California Water Future,* (Berkeley, CA: University of California Press, 1996).

---

[2] Drawn from *Political Chronology of Mono Lake* [Website]. Mono Lake Committee, 2000 [cited June 16 2000]. Available from http://www.monolake.org/politicalhistory/potchr.htm.

[3] 6 Cal. 3d 251, 98 Cal Rptr. 790, 491 P.2d 374.

[4] "Mono Lake Decision 1631" http://monolake.org/politicalhistory/d1631test.html: 4, 5; California State Water Resources Control Board, Division of Water Rights, *Draft Environmental Impact Report for the Review of the Mono Basin Water Rights of the City of Los Angeles,* Appendix R: Legal History of Mono Lake Controversy, May 1993.

[5] California State Water Resources Control Board, Division of Water Rights, *Draft Environmental Impact Report for the Review of the Mono Basin Water Rights of the City of Los Angeles,* Appendix R: Legal History of Mono Lake Controversy, May 1993.

[6] "Mono Lake Decision 1631" http://monolake.org/politicalhistory/d1631test.html: 4, 5.

[7] Ibid.

[8] California Trout Inc. *v.* Superior Court *(Cal Trout II)* 218 Cal. App. 187 [266 Cal. Rprtr. 788].

[9] "Mono Lake Decision 1631" http://monolake.org/politicalhistory/d1631test.html: 4, 5.

[10] "Mono Lake Decision 1631" http://monolake.org/politicalhistory/d1631test.html: 6.

[11] "Water Board set lake at 6,392; DWP promises no appeal," *Mono Lake Newsletter* 17 (2A): 4, Special Issue 1994; "Mono Lake Decision 1631" http://monolake.org/politicalhistory/d1631test.html: 8; California State Water Resources Control Board, Division of Water Rights, *Draft and Final Environmental Impact Report for the Review of the Mono Basin Water Rights of the City of Los Angeles,* 1993-1994.

[12] "Mono Lake Decision 1631" http://monolake.org/politicalhistory/d1631test.html: 6.

[13] Virginia Ellis, "State Sets New Safeguards for Mono Lake," *Los Angeles Times,* A3, A15, September 29, 1994.

[14] "Mono Lake Decision 1631" http://monolake.org/politicalhistory/d1631test.html: 6.

[15] Anne Bancroft, "L.A. Gives Up--Mono Lake Finally Saved," *San Francisco Chronicle,* A1, A17, September 29, 1994.

[16] http://monolake.org/politicalhistory/d1631test.html.

[17] Melvyl--University of California Library Catalog (http://www.melvyl.ucop.edu)

[18] "Highlights of the Water Board Order," *Mono Lake Newsletter* 17 (2A): 6-7, Special Issue 1994.

[19] "Mono Lake Decision 1631" http://monolake.org/politicalhistory/d1631test.html: 97-104.

[20] Anne Bancroft, "Big Step for Saving Mono Lake," *San Francisco Chronicle* A1, September 21, 1994; Marla Cone, "Plan Calls for Refilling of Mono Lake," *Los Angeles Times* A3, A27, September 21, 1994.

[21] *New York Times*, October 3, 1994; *The Hartford Courant*, November 9, 1994; *The Christian Science Monitor*, November 1, 1994; *San Francisco Chronicle*, September 29, 1994; *Los Angeles Times*, September 29, 1994; *Santa Rosa Press Democrat*, September 29, 1994; *Sacramento Bee*, September 29, 1994; *San Luis Obispo Telegram Tribune*, September 28, 1994; *Woodland Hills Daily News*, September 29, 1994; *San Pedro News Pilot*, September 29, 1994; *Star News-Pasadena*, September 29, 1994; *The Herald-Monterey*, September 29, 1994; *Valley Times-Pleasanton*, September 29, 1994; *San Jose Mercury News*, September 29, 1994; *Times Herald-Vallejo*, September 29, 1994; *Fresno Bee*, September 29, 1994; *Appeal-Democrat-Marysville*, October 3, 1994; *Plumas County Reporter*, October 5, 1994; *Riverside Press Enterprise*, October 5, 1994; *Arcata Eco News*, October 1994; *Contra Costa Times-Walnut Creek*, October 3, 1994; *Mammoth Times*, October 5, 1994.

[22] Anne Bancroft, "L.A. Gives Up--Mono Lake Finally Saved," *San Francisco Chronicle,* A1, A17, September 29, 1994.

[23] Editorial, "DWP's Terrible Case of Mono," *Los Angeles Times,* B7, September 24, 1994.

[24] *San Francisco Chronicle*, September 29, 1994; *Los Angeles Times*, September 29, 1994; *Santa Rosa Press Democrat*, September 29, 1994; *Sacramento Bee*, September 29, 1994; *San Luis Obispo Telegram Tribune*, September 28, 1994; *Woodland Hills Daily News*, September 29, 1994; *San Pedro News Pilot*, September 29, 1994; *Star News-Pasadena*, September 29, 1994; *The Herald-Monterey*, September 29, 1994; *Valley Times-Pleasanton*, September 29, 1994; *San Jose Mercury News*, September 29, 1994; *Times Herald-Vallejo*, September 29, 1994; *Fresno Bee*, September 29, 1994; *Appeal-Democrat-Marysville*, October 3, 1994; *Plumas County Reporter*, October 5, 1994; *Riverside Press Enterprise*, October 5, 1994; *Arcata Eco News*, October 1994; *Contra Costa Times-Walnut Creek*, October 3, 1994; *Mammoth Times*, October 5, 1994; Photograph, *Mono Lake Newsletter* 17(2A): 2, Special Issue 1994.

[25] "Mono Lake Decision 1631" http://monolake.org/politicalhistory/d1631test.html: 5, 6.

[26] Geoffrey McQuilkin, "Martha Davis on Mono Lake," *Mono Lake Newsletter* 16(2) (Winter 1994): 11.

[27] Walter G. Pettit (State Water Resources Control Board-Division of Water Rights), Notice of Preparation for the Review of the City of Los Angeles' Water Rights Licenses, Revised Water Quality Control Plan and the Public Trust Issues of the Mono Lake Basin, January 4, 1990.

[28] "Mono Lake Decision 1631" http://monolake.org/politicalhistory/d1631test.html: 7.

[29] California State Water Resources Control Board, Division of Water Rights, *Draft Environmental Impact Report for the Review of the Mono Basin Water Rights of the City of Los Angeles* 1 (May, 1993): 2-15-24.

[30] "Mono Lake Decision 1631" http://monolake.org/politicalhistory/d1631test.html: 7.

[31] Jim Canaday (State Water Resources Control Board-Division of Water Rights), Letter Inviting Participation on Technical Work Groups. November 16, 1989; Jim Canaday (State Water Resources Control Board-Division of Water Rights), Letter Inviting Wildlife, Riparian Vegetation/Wetlands and Land Use Technical Advisory Group to December 14, 1989 Meeting, December 7, 1989.

[32] "Mono Lake Decision 1631" http://monolake.org/politicalhistory/d1631test.html: 8.

[33] Parties included California Air Resources Board, California Department of Fish & Game, California State Lands Commission, California Department of Parks and Recreation, California Trout, the City of Los Angeles and the City of Los Angeles Department of Water and Power, the Great Basin Unified Air pollution Control District, Haselton Associates, the National Aududon Society, Mono Lake Committee, the Sierra Club, The Metropolitan Water District of Southern California, the United States Fish and Wildlife Service, and the United States Environmental Protection Agency, "Mono Lake Decision 1631" http://monolake.org/politicalhistory/d1631test.html: 10.

[34] Maria L. LaGanga, "State Backs Effort to Raise Mono Lake Level," *Los Angeles Times* A3 A24, October 3, 1993; "Public Hearings to be held by Water Board" *Mono Lake Newsletter* 16(2) (Fall 1993): 6.

[35] "Public Hearings to be held by Water Board" *Mono Lake Newsletter* 16(2) (Fall 1993): 6.

[36] Ibid.

[37] Sally Miller, "Water Board Overwhelmed By Statements Favoring Protection of Mono Lake," *Mono Lake Newsletter* 16(2) (Winter, 1994): 4-5.

[38] Ibid.

[39] California State Water Resources Control Board, Division of Water Rights, *Draft Environmental Impact Report for the Review of the Mono Basin Water Rights of the City of Los Angeles* 2 (May 1993): 3N8-3N11.

[40] Ibid: 3N18-19, Appendix X, Economics.

[41] California State Water Resources Control Board, Division of Water Rights, *Final Environmental Impact Report for the Review of the Mono Basin Water Rights of the City of Los Angeles* 1 (September 1994): 4-144-146.

[42] Division of Water Rights Mailing List File Copy, January 9, 1990.

[43] Telephone conversation with Jim Canaday, State Water Resource Control Board-Division of Water Rights, July 13, 1998.

[44] See the MLC website at http://www.monolake.org

[45] Robert A. Jones, "Mono Lake: A Test for New Regime," *Los Angeles Times* A3, May 8, 1991.

[46] Bob Schlichting, "DWP Loses in Court, Again, and Again...," *Mono Lake Newsletter* 15(2) (Fall 1992): 5.

[47] Dennis Pfaff, "15-year Court Battle Over Mono Lake Leaves Legacy of Legal Fees," *Recorde,r* October 12, 1994; Dennis Pfaff, "It's Not Over Yet-Controversy Still Swirl Around Mono Lake Litigation," *Daily Journal* 1, 5, October 13, 1994.

[48] "Pro Bono Fees and Mono Lake," *Mono Lake Newsletter* 14(1) (Summer, 1991): 5.

[49] Jim Canaday, E-mail re: Questions, July 20, 1998.

[50] http://www.monolake.org/socalwater/altwater.htm.

[51] Frances Spivy-Weber, "Water Conservation Statewide is Insurance for Mono Lake," (http://www.monolake.org/newletter/98fall/insurance.htm).

[52] http://www5ladwp.com/whatnew/bbs/board/AC050200.htm.

[53] Virginia Ellis, "Judge Halts L.A. Diversion of Water from Mono Basin," *Los Angeles Times,* I1, I32, August 23, 1989.

[54] "The Price for Mono Lake," *Sacramento Bee,* October 5, 1994.

[55] "Committee Files Closing Brief in State Water Board Proceedings," *Mono Lake Committee* 16(4) (Spring, 1994): 7.

[56] Paul Rogers, "Rising Water, Rising Spirits: Mono Lake on the Brink of a Stunning Rebirth," *San Jose Mercury News*, July 2 1995.

[57] "Committee Files Closing Brief in State Water Board Proceedings," *Mono Lake Committee* 16(4) (Spring, 1994): 7.

[58] Elliot Diringer and Greg Lucas, "L.A. Is Offered $60 Million To Use Less Mono Lake Water," *San Francisco Chronicle* 1, September 16, 1989.

[59] "California's Ancient and Unique Mono Lake is Saved," *EDF Letter* 26(19) (January 1995): 1, 3.

[60] Marla Cone, "DWP Agrees to Take Less Mono Lake Water," *Los Angeles Times* A1, December 14, 1993.

[61] Martha Davis, "Reclaimed Water Agreement Underscores Feasibility of Protecting Mono Lake," *Mono Lake Newsletter* 16(4) (Spring, 1994): 10, 11.

[62] Ibid.

[63] "H.R. 429: The Western Water Bill," *Mono Lake Newsletter* 15(3): 6, 7, Winter, 1993.

[64] Daniel B. Wood, "Mono Lake Decision Marks Sea of Change in California Water Wars," *Christian Science Monitor* 3, November 1, 1994.

[65] "Mono Lake Decision 1631" http://monolake.org/politicalhistory/d1631test.html: 5, 80.

[66] "Mono Lake Decision 1631" http://monolake.org/politicalhistory/d1631test.html: 81.

[67] "Mono Lake Decision 1631" http://monolake.org/politicalhistory/d1631test.html: 82.

[68] Ibid.

[69] "Mono Lake Decision 1631" http://monolake.org/politicalhistory/d1631test.html: 83.

[70] "Mono Lake Decision 1631" http://monolake.org/politicalhistory/d1631test.html: 84.

[71] Ibid.

[72] "Mono Lake Decision 1631" http://monolake.org/politicalhistory/d1631test.html: 85.

[73] Ibid.

[74] "Mono Lake Decision 1631" http://monolake.org/politicalhistory/d1631test.html: 87.
[75] "Mono Lake Decision 1631" http://monolake.org/politicalhistory/d1631test.html: 88, 89.
[76] "Mono Lake Decision 1631" http://monolake.org/politicalhistory/d1631test.html: 93.
[77] Inyo National Forest, *Draft Comprehensive Management Plan for Mono Basin National Forest Scenic Area* 103-107, September 19, 1988.
[78] California State Water Resources Control Board, Division of Water Rights, *Draft Environmental Impact Reports for the Review of the Mono Basin Water Rights of the City of Los Angeles* 2 (May 1993): 3K-5-6.
[79] Ibid: 3K-13.
[80] Inyo National Forest, *Draft Comprehensive Management Plan for Mono Basin National Forest Scenic Area* 105-106, September 19, 1988.
[81] William J. Andrews, "Letter #31--Comments on Draft EIR" in California State Water Resources Control Board, Division of Water Rights, *Final Environmental Impact Report for the Review of the Mono Basin Water Rights of the City of Los Angeles* 2, September 1994.
[82] "Mono Lake Decision 1631" http://monolake.org/politicalhistory/d1631test.html: 94.
[83] California State Water Resources Control Board, Division of Water Rights, *Draft Environmental Impact Report for the Review of the Mono Basin Water Rights of the City of Los Angeles,* 2 (May 1993): 3N-4.
[84] Ibid.: 3G-13-14.
[85] Ibid.: 3G-24.
[86] Ibid.: 3G-26.
[87] "Mono Basin Updates" *Mono Basin Newsletter* 20(4), 21(1) (Spring-Summer 1998): 6.
[88] California State Water Resources Control Board, Division of Water Rights, *Draft Environmental Impact Report for the Review of the Mono Basin Water Rights of the City of Los Angeles,* 2 (May 1993): 3G-17.
[89] Ibid.: Figure 3G-4
[90] Ibid.: 3G-33-34.
[91] Ibid.: 3N-37.
[92] Ibid.: 3N-28.
[93] Ibid.: 3N-38.
[94] Ibid.
[95] Ibid. and Table 3N-18.
[96] "Mono Lake Decision 1631" http://monolake.org/politicalhistory/d1631test.html: 10, 11.
[97] "Mono Lake Decision 1631" http://monolake.org/politicalhistory/d1631test.html: 17, 22, 26, 37.
[98] "Mono Lake Decision 1631" http://monolake.org/politicalhistory/d1631test.html: 16, 17, 34-37.
[99] "Mono Lake Decision 1631" http://monolake.org/politicalhistory/d1631test.html: 41.
[100] "Mono Lake Decision 1631" http://monolake.org/politicalhistory/d1631test.html: 49-55.
[101] "Mono Lake Decision 1631" http://monolake.org/politicalhistory/d1631test.html: 59, 60.
[102] "Mono Lake Decision 1631" http://monolake.org/politicalhistory/d1631test.html: 61.
[103] "Mono Lake Decision 1631" http://monolake.org/politicalhistory/d1631test.html: 64.
[104] "Mono Lake Decision 1631" http://monolake.org/politicalhistory/d1631test.html: 66.
[105] "Mono Lake Decision 1631" http://monolake.org/politicalhistory/d1631test.html: 67, 68.
[106] "Mono Lake Decision 1631" http://monolake.org/politicalhistory/d1631test.html: 75.
[107] "Mono Lake Decision 1631" http://monolake.org/politicalhistory/d1631test.html: 77, 78.
[108] "Mono Lake Decision 1631" http://monolake.org/politicalhistory/d1631test.html: 79.
[109] Bob Schlichting, "DWP Loses in Court, Again, and Again...," *Mono Lake Newsletter* 15(2) (Fall 1992): 5; Dennis Pfaff, "15-year Court Battle Over Mono Lake Leaves Legacy of Legal Fees," *Recorder* October 12, 1994.

[110] Jason Montiel, "DWP: The Time For Arguing Has Ended," *The Review Herald* 22 (76): A1, A10, October 6, 1994.
[111] Jason Montiel, "Public Trust Values Win With Mono Order," *The Review Herald* 22 (82): A1, A12, October 30, 1994.
[112] "Mono Lake Decision 1631" http://monolake.org/politicalhistory/d1631test.html: 1-105. This link has changed to http://www.monobasinresearch.org/images/legal/d1631text.htm.
[113] California State Water Resources Control Board, Division of Water Rights, *Draft Environmental Impact Report for the Review of the Mono Basin Water Rights of the City of Los Angeles,* 1 (May, 1993): 2-25-28.
[114] California State Water Resources Control Board, Division of Water Rights, *Final Environmental Impact Report for the Review of the Mono Basin Water Rights of the City of Los Angeles,* 1 (September, 1994): 4-3.
[115] Bob Schlichting, "DWP Loses in Court, Again, and Again...," *Mono Lake Newsletter* 15(2) (Fall 1992): 5; Dennis Pfaff, "15-year Court Battle Over Mono Lake Leaves Legacy of Legal Fees," *Recorder* October 12, 1994.
[116] Sally Miller, "Water Board Overwhelmed By Statements Favoring Protection of Mono Lake," *Mono Lake Newsletter* 16(2) (Winter, 1994): 5.
[117] Martha Davis, "Reclaimed Water Agreement Underscores Feasibility of Protecting Mono Lake," Mono Lake Newsletter 16(4) (Spring 1994): 10.
[118] Ibid.
[119] Jason Montiel, "No Parties Appeal State's Decision on Mono Basin," *Review-Herald,* A1, A8, November 6, 1994.
[120] Marla Cone, "Plan Calls for Refilling of Mono Lake," *Los Angeles Times,* A3, A27, September 21, 1994.
[121] Marla Cone, "Mono Lake Plan Could Slash L.A. Water Supply," *Los Angeles Times,* A32, September 18, 1994; Jason Montiel, "Mono Lake Should Rise to 6,392 Feet, Says State," *Review-Herald,* A1, A10, September 22, 1994.
[122] "Money Flowing for Reclaimed Water," *Mono Lake Newsletter* 18(1) (Summer, 1995): 10.
[123] Jason Montiel, "DWP: The Time for Arguing is Ended," *The Review-Herald* 22 (76): A1, A10, October 6, 1994.
[124] "Letters to the Times," *Los Angeles Times,* October 2, 1994.
[125] "Mono Lake Decision 1631" http://monolake.org/politicalhistory/d1631test.html: 96.
[126] Virginia Ellis, "State Sets New Safeguards for Mono Lake," *Los Angeles Times,* A3, September 29, 1994.
[127] David Carle, "It Ain't Over Till the Fat Bird Sings" *Mammoth Times* 14, 43, September 29, 1994.
[128] Jason Montiel, "Mono Lake, Environment Saved," *Inyo Register* 114: A1, A2, September 30, 1994.
[129] "What Do you Think?" *The Review-Herald,* A10, October 6, 1994.
[130] Jason Montiel, "USFS: Mono Decision Supports Our Goals," *The Review-Herald* 22 (79): A-1, A-12, October 16, 1994.
[131] "Mono Lake Decision 1631" http://monolake.org/politicalhistory/d1631test.html: 59, 60.
[132] Jason Montiel, "Leslie: State Money Helped to Secure Mono Basin Agreement," *The Review-Herald* 22 (78): A1, A10, October 13, 1994.
[133] Anne Bancroft, "L.A. Gives Up--Mono Lake Finally Saved," *San Francisco Chronicle,* A1, September 29, 1994.
[134] "Water Board Set ake at 6,392; DWP Promises No Appeal," *Mono Lake Newsletter* 17 (2A) (Special Issue 1994): 4.

[135] Jason Montiel, "Mono Lake, Environment Saved," *Inyo Register* 114: A1-A3, September 30, 1994.
[136] Dennis Pfaff, "15-year Court Battle Over Mono Lake Leaves Legacy of Legal Fees," *Recorder,* 1, 5, October 5, 1994
[137] Jason Montiel, "Groups Seek to Get Mono Basin Ceek Work Moving," *Inyo Register*, December 21, 1994.
[138] "Mono Lake Decision 1631" http://monolake.org/politicalhistory/d1631test.html: 100.
[139] "Lake to Rise Dramatically! Heavy Runoff Year Forecast," *Mono Lake Newsletter* 18(1) (Summer, 1995): 9.
[140] "Mono Lake Up 3 Feet and Still Rising!" *Mono Lake Newsletter* 18(2) (Fall, 1995): 11.
[141] Greg Reis, "Mono Lake Surpasses 6,380'," *Mono Lake Newsletter* 19(2) (Fall, 1996): 7.
[142] Greg Reis, "El Nino Returns--Another Wet Winter Coming?" *Mono Lake Newsletter* Web version, Fall 1997.
[143] Greg Reis, "El Nino Delivers," *Mono Lake Newsletter* 20 (4), 21(1) (Spring-Summer, 1998): 14.
[144] "Streams, Waterfowl Habitat, Waterfowl Management," *Mono Lake Newsletter* 18(1) (Summer, 1995): 5-7.
[145] "Streams, Waterfowl Habitat, Waterfowl Management," *Mono Lake Newsletter* 18(2) (Fall 1995): 6-8.
[146] Telephone conversation with Jim Canaday, State Water Resources Control Board, Water Rights Division, June 30, 1998.
[147] Jane Braxton Little, "Mono Lake: Victory Over Los Angeles Turns Into Local Controversy," *High County News* 29 (23): 1, 10, December 8, 1997.
[148] Ibid.
[149] Jim Canaday, "Memorandum on Conway Ranch Evaluation Workgroup Mailing List," March 13, 1997.
[150] "Mono Lake Decision 1631" http://monolake.org/politicalhistory/d1631test.html: 101.
[151] "Mono Lake Decision 1631" http://monolake.org/politicalhistory/d1631test.html: 100.
[152] "Water Board Decision Stands Unappealed – Restoration Plans to be Made in 1995," *Mono Lake Newsletter* 17(3,4) (Winter-Spring 1995): 5.
[153] Dan Morain, "Mono Lake Supporters Raise Wineglasses to Toast a Victory Over L.A.," *Los Angeles Times,* A27, May 12, 1991.
[154] "State Votes to Raise Mono Lake Level," *The Review Herald* 22(74): A1, A12, September 29. 1994.
[155] Virginia Ellis, "State Sets New Safeguards for Mono Lake," *Los Angeles Times,* A3, A15, September 29, 1994.
[156] John D. Cox, "Mono's Troubled Waters at Peace," *Sacramento Bee,* A1, A24, September 29, 1994.
[157] Dan Morain, "Mono Lake Supporters Raise Wineglasses to Toast a Victory Over L.A," *Los Angeles Times,* A27, May 12, 1991.
[158] Bob Schlichting, "DWP Loses in Court, Again, and Again...," *Mono Lake Newsletter* 15(2) (Fall 1992): 5; Dennis Pfaff, "15-year Court Battle Over Mono Lake Leaves Legacy of Legal Fees," *Recorder* October 12, 1994.
[159] Seth Mydans, "City of Angels Makes Peace in Water Wars," *The New York Times,* A10, October 4, 1994.
[160] Kimberlee Noll, Kelli Du Fresne and Jason Montiel, "Reflections of 1994," *The Review-Herald* 23 (2), January 5, 1995.
[161] "Litigation Deja Vu: Back to Court for Mono Lake" *Mono Lake Newsletter* 18(1) (Summer, 1995): 3.

[162] See Connie Ozawa, "Science in Environmental Conflicts," *Sociological Perspectives* 39(2), (1996): 219-230.

[163] "Water Board Set Lake at 6,392: DWP Promises No Appeal," *Mono Lake Newsletter* 17 (2A) (Special Issue 1994): 4.

[164] Daniel Wood, "Mono Lake Decision Marks Sea Change in California Water Wars," *Christian Science Monitor,* 3, November 1, 1994.

[165] http://www.monolake.org/socalwater/altwater.htm

[166] http://www.monolake.org/socalwater/lawcc.htm.

[167] http://www5.ladwp.com/whatnew/dwpnews/055000.htm.

[168] http://www.monolake.org/newsletter/97winter/symposium.htm; http://www.monolake.org/newsletter/97fall/waterpols.htm; http://www.monolake.org/newsletter/98winter/forum.htm

[169] http://www.monolake.org/newsletter/97fall/community.htm.

[170] First Person, "Sierra Club Scholarship Winner—The Future of My Hometown," *Mammoth Times,* August 10, 2000.

[171] Frances Spivy-Weber, "Water Conservation Statewide is Insurance for Mono Lake" http://www.monolake.org/newsletter/98fall/insurance.htm.

Chapter Five

THE PYRAMID LAKE CASE

How Many Farms, Towns, Fish and Ducks Can the Rivers Support?

Case Researcher: Erin McCandless

> All things are bound together.
> All things connect…
> Man has not woven the web of life,
> He is but one thread
> Whatever he does to the web, he does to himself.
>
> *Chief Seattle to Governor of Washington Territory, 1855*

> If there's water in the big rivers, the small rivers will be full.
>
> *Chinese Proverb*

Note: This case report illustrates the use of a particular methodological framework. It is not intended to replicate the legal and historical coverage of this case provided in other sources.[1]

Introduction: This case illustrates the complexities of addressing multiple environmental needs for water in a region characterized by rapid urban growth and long established irrigated farms and farming communities. The Truckee and Carson River Basins in western Nevada are the sites of this decades-old conflict and are shown in Figure 5.1.

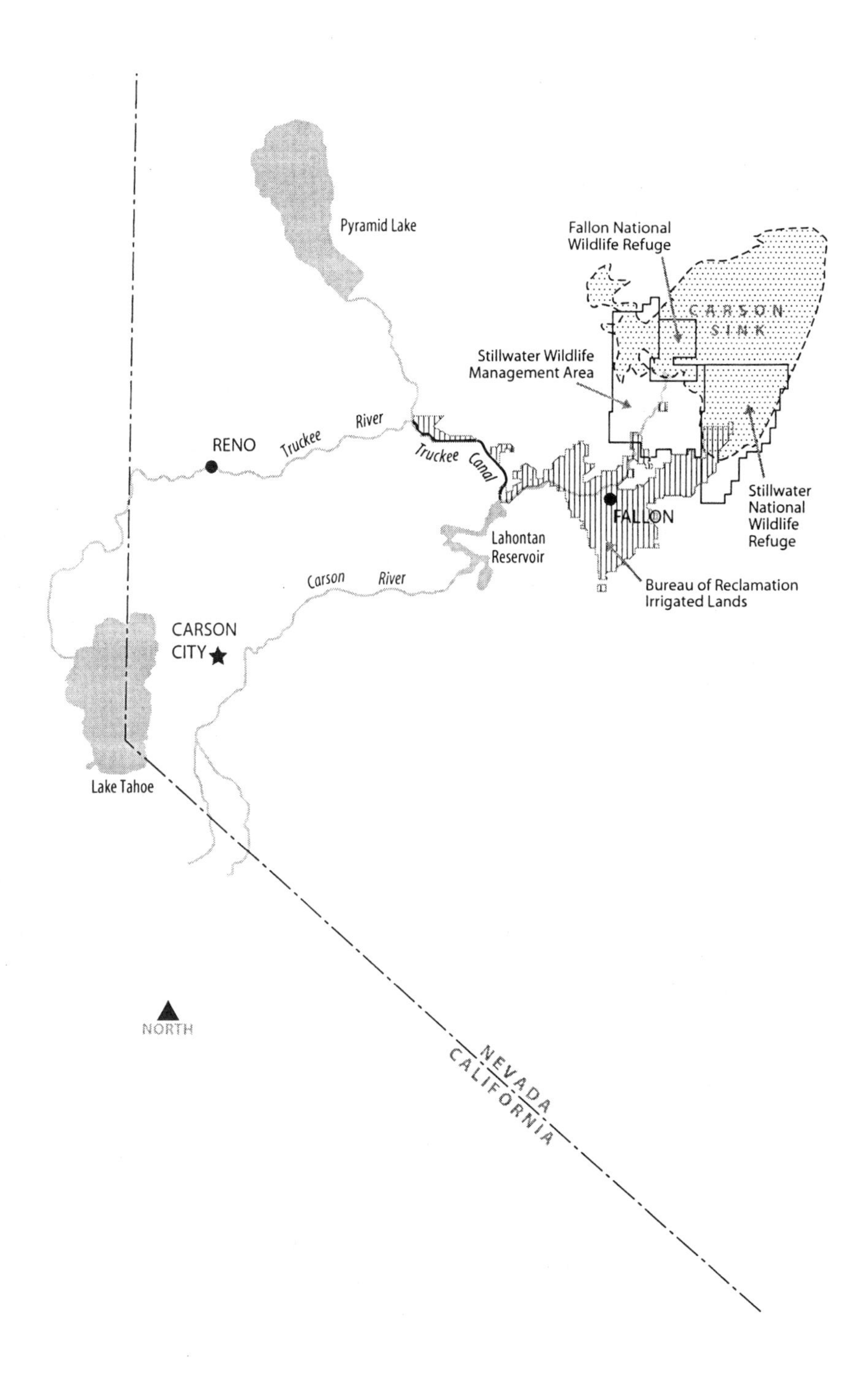

Figure 5.1 Pyramid Lake Case

Time period: The negotiated settlement process began in 1988 and led to Public Law 101-618 (sometimes called the Settlement Act or the Negotiated Settlement). This law had two titles and was originally drafted to provide a financial settlement between the U.S. Government and the Fallon Paiute-Shoshone Indian Tribe. This analysis will focus on the second title of the act:

Title II. The Truckee-Carson Pyramid Lake Water Rights Settlement Act

Through a combination of negotiation, litigation, legislation, voluntary transfers, and consensus-building processes a negotiated settlement was reached, which was confirmed by Congress and written into law in 1990. The Settlement Act attempted to resolve decades of disputes between the states of California and Nevada over the allocation and use of the Truckee and Carson River basins they share. It also settled several conflicts between competing users in the State of Nevada and established a framework for identifying and resolving those conflicts that were not settled by the provisions of the Act.

The Preliminary Settlement Agreement of 1989 and later negotiations and agreements that were mandated by Public Law 101-618 will be incorporated into the analysis to varying degrees, insofar as they add to the explanation of the criteria for analyzing Public Law 101-618. This extends the time period of analysis from 1988-1998, with emphasis on the 1990 Settlement Act process.

Basic nature of dispute: Access and rights to water flowing through the Truckee and Carson River basins and the level of water necessary for Pyramid Lake and the Lahontan Valley.

Issues: Access by all parties to the water rights, the fate of two endangered species of fish (Lahontan cutthroat trout and cui-ui), environmental degradation linked to water diversions (including a large trans-basin diversion by a federally-funded irrigation project), water quality, urbanization, and Native American water rights.

Actors and Interests: Befitting a dispute that crosses state borders and involves both usage and quality issues, a large number of actors were stakeholders in the process. For simplicity, they have been divided into several categories.

Federal Agencies: The Federal Government was a central part of the process, with many agencies involved. The most prominent were:

- **Truckee-Carson Coordination Office**: Established to help coordinate the various federal agencies involved.

- **U.S. Department of the Interior:**
  - **Bureau of Reclamation**, *Interest:* Responsible for Newlands Project.
  - **Fish & Wildlife Services**, *Interest:* Concerned about area wildlife.
  - **Bureau of Indian Affairs** (BIA), *Interest:* Concerned about Native American rights.
  - **Bureau of Land Management (BLM)**
  - **U.S. Geological Survey (USGS)**

- **U.S. Department of Defense:**
  - **Army Corps of Engineers**
  - **Fallon Naval Air Station (NAS)**, *Interests:* Located near Fallon in the Carson River Basin, the Fallon NAS is one of the largest employers in Churchill County. Their water rights have been used to grow crops adjacent to runways, control dust, and to suppress aircraft-caused brush and grass fires.

- **U.S. Department of Agriculture:**
  - **Forest Service**
  - **Natural Resources Conservation Service**

State Agencies: California and Nevada both have interests in the Truckee-Carson River Basins. Their decades-long pursuit of an interstate compact culminated in the Settlement Act. The key state agencies included:

California: In California, the agencies listed below were coordinated under the auspices of the DWR, representing both state and local agencies.[2] California agencies were concerned with maintaining the State's share of Truckee River water and ensuring enough flow on the California side of the river to support existing wildlife.

- **Department of Water Resources (DWR)**
- **State Water Resources Control Board**
- **Department of Fish & Game**
- **Interstate Compact Commission**

Nevada:

- **Department of Conservation & Natural Resources**
- **Division of Environmental Protection**
- **Division of Water Planning**
- **State Engineer**
- **Department of Wildlife**
- **Interstate Compact Commission**
- **Cooperative Extension Service**

Regional Agencies:

- **Regional Water Planning Commission of Reno-Sparks and Washoe County**
- **Regional Planning Governing Board (Reno, Sparks, Washoe County)**
- **Tahoe Regional Planning Agency,** *Interests:* Created by an interstate compact adopted by each state's legislatures and ratified by Congress, this agency acts to control growth and regulate land use planning and development. It also focuses on preserving the pristine water quality of Lake Tahoe.
- **Truckee-Carson Irrigation District (TCID)**, *Interests:* Organized in 1918, it has been operating the Newlands Project for the Bureau of Reclamation under contract since 1926.

Cities and Counties: Five California counties and seven Nevada counties lie within the Truckee-Carson River basins, including about 15 cities and towns. Particularly noteworthy are:

- **Reno**, *Interests:* Both Reno and Sparks are primarily concerned with maintaining an adequate water supply for municipal use and to support continued growth.
- **Sparks:** Although Reno and Sparks are separate chartered cities and were represented separately in the negotiations, they do share a number of agencies, including a tourism board and chamber of commerce.
- **Washoe County**
- **Fallon and Fernley**, *Interests:* These are rural agricultural communities that have served as town centers for the Newlands Project irrigators for decades and which will bear the brunt of the impact if changing water allocations cause declines in irrigated acreage. (Fallon has a municipal water system that sup-

plies drinking water to thousands of Nevadans. This system is affected by Newlands Project water deliveries, and the City believes that its interests are jeopardized by the current implementation of the mandates set in motion by P.L. 101-618.)
- As mentioned above, California communities in the Truckee and Carson River Basins were represented by the California DWR.

Native American Tribes:

- **Pyramid Lake Paiute Tribe**, *Interests:* The Tribe has been, and continues to be, a key actor in the decades of conflicts and proposed solutions. Their interests include the preservation of Pyramid Lake, which provides habitat for the cui-ui, an endangered fish, and of the Lahontan cutthroat trout. The lake and fish are of special cultural and economic/life-sustaining significance to the Tribe. Water is a major component of the Tribe's identity, integrating (the inseparable) people, lake and fish. Newlands Project diversions starting at the beginning of the century deeply affected the Pyramid Lake water levels, and the Tribe's way of life.
- **Fallon Paiute Shoshone Tribe**, *Interests*: The Tribe is interested in the settlement of claims related to the failure of the U.S. to carry out obligations authorized by Congress. This includes a 1978 Act, Public Law 95-337, where the government recognized its failure to meet contractual responsibilities to the Tribe, and increased the size of the reservation due to tribal growth. They wanted more tribal water rights.
- **Washoe Tribe of Nevada and California**

Environmental Organizations:

- **The Nature Conservancy (TNC)**, *Interests:* Spoke on behalf of endangered species and the overall preservation of the environment and resources in the area.
- **Environmental Defense Fund (EDF)**, *Interests:* Improved water management to accommodate environmental needs.
- **Lahontan Valley Environmental Alliance**, *Interests:* A group consisting of citizens from Churchill and Lyon counties, Fallon, Fernley, the TCID, and the Lahontan and Stillwater conservation districts whose aim is to serve as a public forum for water, growth and environmental issues.

- **Sierra Pacific Power Company**, *Interests:* The Sierra Pacific Power Company is a water retailer that supplies directly to users, and a wholesaler, supplying to small purveyors. It supplies water to Reno and Sparks (53,000 customers) and gets unstored water from the Truckee River and its tributaries, while stored water comes from local lakes. It also provides electric service to 235,000 residential and commercial customers in Northern Nevada and northeastern California, and natural gas to 74,000 accounts.[3] Sierra Pacific Power Company holds direct diversion rights to the Truckee River. However, the seniority of these rights vis-à-vis those of the Pyramid Lake Paiute Tribe are a subject of the Truckee River Operating Agreement (TROA) negotiations.

Attempted conflict resolution processes: A combination of litigation, negotiation, voluntary transfers, and consensus-building processes to reach a negotiated settlement, which led to legislation in the form of an Act ratified by Congress. Senator Harry Reid of Nevada, who assumed office in 1987, initiated and led this process. This report was compiled in 1998 and 1999, and edited in 2000.[4]

History of the Pyramid Lake Conflict

Truckee River.

1860s: Alexis von Schmidt forms a company to ship Lake Tahoe water to San Francisco.

1860s-1900: A dam is built for the von Schmidt project. Opposition from Nevada causes cancellation of the project.

1902: The Newlands Irrigation Act is passed to irrigate lands around Fallon, Nevada.

1905: The Newlands Irrigation Project is initiated. Changes in water flows result in increased litigation.

1908: The Truckee River General Electric Company & Floristan Pulp & Paper Company agree on Truckee River flow requirements.

1915: The Truckee River General Electric Decree is issued by a Federal court, granting the Reclamation Service (operators of the Newlands Project) an easement to operate the Lake Tahoe dam. The decree is revised to require the government to adhere to the Floristan flow agreement.

1924-1935: Conflicts arise between lakeshore owners and irrigators over water flows during droughts. One incident involves a steam shovel sent by TCID with an armed guard to dig a diversion trench to restore flows that were curtailed by lowered lake levels. Ensuing negotiations conclude with the 1935 Truckee River Agreement, which still remains the basis for river operations.

1944: The Orr Ditch Decree apportions Truckee water rights between the Pyramid Lake Tribe, Sierra Pacific Power Company, the Newlands Irrigators, TCID, Washoe County Water District, and individual rights holders.

Carson River.

1905: The Anderson-Bassman Decree addresses water rights between upper river users and the Newlands Project users.

1921: The Price Decree adjudicates to California the rights to Carson River water.

1925-1980: Federal litigation covering Carson Basin rights to establish the Newlands water rights culminates in the 1980 Alpine Decree.

Background

Lake Tahoe, on the border of California and Nevada, stretches over an area of 192 square miles with a depth of 990 feet. The lake feeds the Truckee River, which – along with the Carson River – ends in Nevada's Great Basin. The Truckee River flows through Reno and Sparks and empties into Pyramid Lake, which is surrounded by the Pyramid Lake Indian Reservation inhabited by the Pyramid Lake Paiute Tribe. The Carson River flows into the Stillwater Wildlife Management Area and the Carson Sink where any remaining water evaporates. There is a section of Reservation here that is home to the Fallon Paiute-Shoshone Tribe. Conflict over water rights and allocation of the Truckee and Carson Rivers goes back to the early 1900s.

Many different parties have a stake in decisions affecting the use of these rivers. The cities of Reno and Sparks want the water from the Truckee River for municipal use, while the various power companies are concerned with maintaining hydropower for the area. Industry in the area, such as the Floristan Pulp and Paper Company, also rely on water from the rivers for their operations. Irrigation districts, farming dealers, and farming communities represent the agricultural interests in the area. In addition, the Pyramid Lake Indian Tribe is concerned with maintaining the fisheries at Pyramid Lake since they are the primary resources for the Tribe. They also require water for irrigation. Pyramid Lake is home to two endangered species: the cui-ui and the Lahontan cutthroat trout. The lake requires certain water levels for the species' continued existence. Area Tribal leaders and representatives of the Sierra Club, the Nature Conservancy, the U.S. Department of Fish and Wildlife, and the Stillwater Wildlife Management Area have spoken on behalf of the endangered species and the overall preservation of the environment and resources of the area.

In 1859, the Department of the Interior set aside half a million acres as a reservation for the Paiute Indians. In 1874, the President of the United States confirmed the withdrawal of land. The Reclamation Act was passed in 1902, which directed the Secretary of the Interior to reclaim some land and to develop irrigation projects so the land could be homesteaded. In 1903, approximately 200,000 reclaimed acres in Western Nevada became the Newlands Reclamation Project.

Problems over water use began in 1903 with the creation of the Newlands Irrigation Project by the Reclamation Services (later known as the Bureau of Reclamation). The project diverted water from the Truckee River to the Carson River with construction of the Derby Diversion Dam. Once in the Carson River, the water was stored at Lahontan Dam and distributed for irrigation. The Truckee Canal was completed in 1905 as part of the irrigation project. At that point in time there were no established water rights, although there were many concerns about the effects of the diverted water on the area's historical water users. Then, water rights for agricultural landowners and the Pyramid Lake Tribe were established in 1913 through a suit brought by both these landowners and the United States. Eventually, the management of the Newlands Irrigation Project was transferred from the Bureau of Reclamation to the Truckee Irrigation District.

The completion of the irrigation project greatly affected water levels in Pyramid Lake. These lake changes had such a drastic negative impact on the fishery at the lake that, by 1935, the government stepped in on behalf of the reservation to request additional water. The request for additional water was approved in 1944 and issued as the Orr Ditch Decree. Despite this additional allocation of water, the continued diversion of water from the natural system caused further decline in the habitat of the area. In response, the Stillwater Wildlife Refuge was created in 1948 in an attempt to preserve the area. The only problem was that the refuge was located in the Carson River Basin and water levels there did not affect the levels of Pyramid Lake.

With the implementation of the Endangered Species Act of 1967, conflict over water allocation increased. Under the Endangered Species Act, the Lahontan Cutthroat trout was listed as "threatened" and the cui-ui listed as "endangered", with Pyramid Lake declared a critical habitat for both species. The survival of these species was directly related to the water levels at Pyramid Lake. Many court cases were filed to ensure the necessary water levels for the continued survival of these endangered species.

In addition to the problems surrounding wildlife, the population of California and Nevada continued to increase rapidly. This continued growth brought drastic increases in demand for water, leading both states to begin negotiations in 1955 for an interstate compact to allocate their waters. Their negotiations culminated in an agreement 14 years later. The agreement excluded parties such as the Pyramid Lake Reservation and issues such as the environment and endangered species. Because of the limited scope of these negotiations, Congress never approved the interstate agreement.

In 1970, the Stampede Dam was built as part of the Washoe Project Act. The project was established for many purposes, including flood control, irrigation, storage of water in case of drought, power generation, development of fish and wildlife resources, and "other beneficial purposes." The Carson-Truckee Water Conservancy District was created under Nevada law to act as the agency to purchase the water stored by the Washoe Project. The Secretary of the Interior was to control the flow of water of the Truckee River by monitoring the releases from Stampede Dam. Because the United States funded the initial project, there were guidelines set up for the repayment of costs. Repayment would be made possible through the sale of Stampede water.

Several parties raised concerns at various points. First, there were concerns that the project did not protect the Pyramid Lake Indians' interest in preserving the lake and its fishery. However, arrangements were made to assure adequate water supply. Then, just prior to the construction of the dam, the Secretary notified the Carson-Truckee Water Conservancy District that Stampede Dam would only be used for fish and wildlife, recreation, and flood control and that he had no intention of allowing for municipal and irrigation uses. As a result of the Secretary's decision, bargaining power shifted in the regional battles over area water. The allocation of Stampede reservoir fishery needs gave the tribe and environmental organizations a strong position, from which they would negotiate with urban interests who wanted the Stampede water made available for their needs.

Three sets of legal cases defined and attempted to resolve many of the issues. The first set are known as the "Orr Ditch" and "Alpine" cases, which were to decide the Truckee River water rights for the Pyramid Lake Indian Reservation (Reservation) and the Newlands Reclamation Project (Project). The second set were cases involving water controlled through Stampede Dam and Reservoir. Finally, the third set were cases involving the status of the Fallon Paiute Shoshone Indian Tribe. Before describing the cases, the following provides a summary of the litigation.

Summary of Litigation/Negotiations

Court battles occurred throughout the 1980s in which the Carson-Truckee Water Conservancy District sued to approve municipal and irrigation uses of the Stampede Dam Water. The District was concerned with the repayment of dam costs and fulfillment of the area's water needs. In 1982, the U.S. District Court ordered that any remaining water be sold after requirements for the Tribe and endangered species had been fulfilled. By contrast, the U.S. Court of Appeals ruled in 1984 that irrigation and municipal uses were appropriate, but the Secretary was not obligated to sell the water for such purposes.

The court battles did not resolve the area issues, and negotiations outside the courtroom had not been attempted since the agreement between the states of California and Nevada. Interested in resolving some of the conflict, Nevada Senator Reid attempted to bring the water users of the Truckee River together in 1985. A year later, negotiations between the parties began. As a result of these talks, the Fallon Paiute Shoshone Tribal Settlement Act was supported by California, Nevada, local utilities, the Tribes, interested conservation organizations, and the U.S. Department of the Interior. Congress approved the Act in 1990. Some agricultural water rights holders – especially the TCID – and other interested parties opposed some of the provisions of the legislation. Problems surrounding implementation also had yet to be solved. Despite the attempt at inclusive negotiations, it was not until three years after the passing of the Settlement Act that public hearings were scheduled. The discussions at the hearings made it clear that negotiations were once again needed.

A professional facilitator was appointed by Senator Reid to guide new negotiations that proceeded from September 1994 to March 1995. As a result of these discussions, instream flows and water quality were to be strengthened through the development of a framework of operations. Efforts began to improve the planning on the upper Carson and maintain the Lahontan Valley wetlands. A commitment was made to develop tools to achieve minimum levels for recreation and fisheries in the Lahontan Reservoir. It was decided that the Fallon Paiute-Shoshone Tribe would have more independence in their water management. Finally, certain mechanisms were established to protect the Fallon Tribe from water losses, which might result from new operating criteria for the Newlands Irrigation Project.

Attempts to resolve the conflict surrounding the Truckee and Carson rivers began with negotiations outside the courtroom. When the resulting compact was not approved because it excluded many interested parties, the conflict moved to the courts. Upon the failure of the courts to resolve the conflict, group problem solving through formal negotiations was attempted once again. Although this last round of talks produced some partial solutions, unresolved water allocation and management issues remain, as well as questions about the parties' willingness to follow through with the solutions.

Focus of Case Study

This study focuses on the 1990 Settlement Act. The process of negotiating this specific agreement began in 1988 and the Settlement Act was passed into Public Law 101-618 on November 16, 1990. Several key aspects to this agreement are still under negotiation, and many are currently in various stages of implementation.

There have been hundreds of meetings, so the "table" concept as a one-session event is not accurate for this *process*. The whole "agreement" is a multi-layered one, negotiated and implemented in stages.

Public Law 101-618 established a framework to negotiate a resolution of multiple issues and is the "agreement" or outcome for the purposes of this analysis. However, success cannot be evaluated without looking at implementation, and a major component of this process and agreement (the Truckee River Operating Agreement) is still under negotiation at this writing in 1998.

Various related agreements, for example, the 1996 Water Quality Agreement, the Preliminary Settlement Agreement as modified by the Ratification Agreement, and the Contract between the U.S. and the Truckee-Carson Irrigation District (TCID) for the Operation and Maintenance of the Newlands Project, have been successfully negotiated since the Preliminary Settlement Agreement.

Given the interrelated nature of all of these agreements however, looking at any one of them alone would remove it from its context. For example, the Truckee River Operating Agreement (TROA) must fulfill the conditions of the Preliminary Settlement Agreement as modified by the Ratification Agreement (PSA), and this PSA will only go into effect upon the agreement of TROA. To the degree that these other agreements support an understanding of (relative) success of the Settlement Act, they are included in the assessment.

I. CRITERIA: OUTCOME REACHED

A. Unanimity or Consensus

Not Unanimous; there were parties not at the table who left or walked out on negotiations.

The Settlement Act was an incremental process, involving different parties at various phases. Senator Reid, the initiator of the process, considered this the best strategy. In the first round of negotiations, beginning in 1987, the Sierra Pacific Power Company, the Pyramid Lake Paiute Tribe, the State of Nevada, and the Truckee-Carson Irrigation District were brought to the negotiation table. By June 1988, the Truckee-Carson Irrigation District had withdrawn from the process.[5]

A number of new actors were introduced in 1988. One of these was the Fallon Paiute Shoshone Tribe. Senator Reid felt that their legal position was so strong that, combined with the federal government's commitment to see that their situation was remedied, their participation was not necessary earlier. Also, a settlement fund of $43 million had already been agreed upon. In addition, the State of California, the Stillwater National Wildlife Refuge, Fallon Naval Air Station, the Cities of Reno and Sparks, and the US Department of Interior (Bureaus of Reclamation and Indian Affairs)[6] were included in negotiations. Over time, the sessions grew larger as the Lahontan Valley Wetlands Coalition and the Coalition for a Negotiated Settlement were introduced into the process.

The Fallon Paiute Shoshone Tribe, the Pyramid Lake Paiute Tribe, and the Sierra Pacific Power Company supported the 1989 Preliminary Settlement Agreement. In addition, California, Nevada, and the Department of the Interior gave their support. Those in opposition to the Preliminary Settlement included individual agricultural water rights holders and the Truckee-Carson Irrigation District (TCID), with the latter withdrawing from negotiations in mid-1988. It is important to note that this agreement is considered an important building block to the overall process, and was later incorporated into the Truckee River Operating Agreement – TROA – (or final settlement act). The parties who signed the Preliminary Settlement Agreement included the Pyramid Lake Paiute Tribe, Sierra Pacific Power Company and later, the U.S. Government.[7]

<u>Why the TCID – (the irrigators) – left the table</u>

The Newlands Water Protective Association noted in a news release that, "In 1985 the Pyramid Lake Paiute Tribe walked out of a negotiated settlement as it was ready to go to the congress. We don't hear anything about this in the media, yet we hear that the farmers walked out of the 1990 negotiations. We continued to participate in negotiations but were told that other parties could reach agreement without affecting individual water rights. The final bill did cause problems with individual water rights. We tried to make changes that would have protected our rights and not have affected the upstream settlement." They go on to further express their willingness to negotiate.

The Clearwater Consulting Corporation (CCC) Report also notes that a more neutral characterization of events was provided through personal interviews with participants to the process, including state and federal officials as well as irrigators. According to these sources, an impasse had been reached at a particular juncture of the Reid-sponsored negotiations and a mutual decision was reached that it was not productive to go forward. The irrigators were not prepared to accept the demands placed upon them, but they did not express an unwillingness to continue the negotiations. A more accurate description would probably be that the parties were unable to agree. The CCC report notes that this description would also apply to the 1994-5 Second Generation Negotiations.[8]

Several sources say that TCID had little incentive to participate, believing they could do better in court.[9]

Pursuant to the final settlement agreement are ongoing negotiations regarding the TROA. As of mid-1999, 14 parties have been engaged in a discussion of applicable legal principles to be used in developing goals, objectives, and general principles for the TROA.

Parties Participating in Ongoing TROA Negotiations

- US Dept. of the Interior
- Sierra Pacific Power Co.
- City of Sparks, NV
- Washoe County WCD
- City of Reno, NV
- TCID
- Town of Fernley
- Pyramid Lake Paiute Tribe
- Carson-Truckee WCD
- Washoe County
- State of California
- Churchill County
- State of Nevada
- Fallon Shoshone Paiute Tribe

B. Verifiable Terms

The agreement was written and formally signed as an act of Congress in 1990. Its terms were published in the media or posted in public forums. It appears that there was quite a bit of explanatory and educational literature about the act in the general media. There was also a great deal of coverage by the parties themselves to their own constituencies. The Pyramid Lake Paiute

Tribe covered the Settlement Act extensively in their Pyramid Lake Water Resources Newsletter. Sierra Pacific Power Company has covered the issues by Internet, as well as by sponsoring information in the newspapers. There is a wealth of information about water issues generally, and copies of the Settlement Act in local libraries along with newspaper clippings.

C. Public Acknowledgement

The agreement process was well covered in the newspaper, despite being a "private" or "secretive" process, as often described by the public (see letters to the editor below).

Reid distributed his views on the legislation through the newspaper and letters in Aug. 1990.

> Because water is a question of life and death, I said from the beginning that my legislation could never become law unless you, the people of Nevada, were behind it. I stand by that commitment. Some people want to delay this legislation until it is too late for it to become law. MAKE YOUR VIEWS KNOWN, both to me and your state and local officials. We have until early September to decide where to go on this extremely important issue.[10]

The parties themselves appear to have undertaken various media efforts regarding the Settlement Act and their preferred views on the subject. Sierra Pacific Power Company, for example, maintains a web page explaining to people why the agreement is good for the region.

D. Ratification

The 1989 Preliminary Settlement Act negotiated between the Pyramid Lake Paiute Tribe and the Sierra Pacific Power Company required ratification by the federal government due to the involvement of federally-owned reservoirs. It was ratified in 1990, via its incorporation by reference in the Settlement Act. One year later, on November 16, President Bush signed the agreement into law, within Public Law 101-618.

The CCC Report describes the Washington politics that took place to pass the bill, noting that:

> Washington politics proved to be an even more complex negotiation process than that which took place in Nevada, with many Congressmen, Senators, committees, agencies, and the President needing to sign off on the bill to make it law.[11]

The Report explains that while the bill initially died in a package of other water bills in the Senate Water and Power Subcommittee, it later was revived and attached as a rider to S-3084, the Fallon Paiute Shoshone Tribal Settlement Act.

In addition to the ratification process, the TROA will require some degree of judicial approval, particularly by the courts dealing with the Orr Ditch Decree.[12]

II. CRITERIA: PROCESS QUALITY

A. Procedurally Just

The settlement negotiations were private and confidential.

Parties were asked not to reveal information about the negotiation sessions. Furthermore, anything learned in the settlement negotiations could not be used unless it was publicly disclosed. Despite the level of confidentiality, the *Reno Gazette* of May 13, 1998 reported that:

> He [Senator Reid] won the respect of most of the parties because he was seen as an honest broker. He didn't favor any one side and made everybody buckle down and compromise.

Other sources suggest that the parties perceived the fairness of the process differently. The Pyramid Lake Paiute Tribe seems to have been unsatisfied with the conceptualization of the agreement.[13]

> In reviewing decrees and agreements related to our water, they have all been established on the prior appropriation doctrine. It is based on this concept of, "first in time, first in right". This is not understood. If an Indian Tribe at the end of a river system receives the least amount of water than anyone else, plus receiving "discharged" water, it appears that in our position we must work harder to preserve water quality. Presently we are working on establishing our own water quality standards for water reaching the reservation.[14]

The CCC Report notes that one observer saw a difference in atmosphere during the 1988 negotiations. In the previous attempts, an "us" versus "them" attitude, pitting non-Indian against Indian claims often emerged and pushed Indian interests into the background. At the 1988 negotiations, there was a better bargaining environment (with the exception of the TCID), where all parties seemed to not want to return to court battles. After the TCID's departure, the other parties remained and took one issue at a time. Entering the negotiations, "all sides had won some victories but no one felt secure or saw a way to get more of what they wanted through litigation." (CCC Report, fn # 32). While the TCID had extensive legal rights to the use of Carson and Truckee River waters, it found its access to Truckee water reduced by the courts.

The 1988-90 process took place over several years, though there was a deadline placed on reaching agreement with the farmers. Considering that there was no successful agreement during the 90-day negotiations, one could argue that there was not enough time. However, the protracted nature of the conflict was illustrated by the fact that there have been ongoing stalemates in attempting to reach an agreement over the farmer's issues.

The press covered in great detail the perceptions of the irrigators concerning unfair treatment. It seemed to be reported objectively, though occasionally it appeared somewhat alarmist, as if the farmers were being made the scapegoats.[15] The CCC Report postulates that efforts of the irrigators to participate in the Settlement Act have been hampered for two reasons. First, conceptually, because the federal government, Pyramid Lake Paiute Tribe, and Sierra Pacific Power Co. have been moving toward a more risk-based management scheme—trading firm water rights for a physical solution that provides an adequate margin of safety in water-short years. The irrigators believe that they should have rights without risk. The second reason is more political, with power shifting away from the irrigators and toward the Tribe, environmental concerns, and upstream uses, the irrigators' bargaining power has eroded substantially.[16]

B. Procedurally Accessible and Inclusive

1. Public Notice and Public Participation

Although it was difficult to find information on this aspect, it appears that very little time was allotted for public input into the process that was the focus of this analysis. Congressional committees changed the Act and there did not appear to be any comprehensive "education and participation" aspects to the process.

The Draft Environmental Impact Statement (EIS)/TROA describes its "public involvement plan" as a process where interested and affected individuals, organizations, agencies, and governmental entities are consulted and included in the decision-making process. Aiming to solicit public input in defining the public issues surrounding the action, identifying alternatives to be evaluated and to educate the public on the issues, a structured public involvement plan was laid out in August 1992. As required by EIS preparation, scoping meetings were held, and are a "continuing and integral part of the decision process, environmental review, and documentation for the TROA EIR/EIS."[17]

The report describes several other public meetings, including a consultation/education session with the Pyramid Lake Paiute Tribe. It goes on to describe the development of a newsletter that went out to a mailing list of those who had been involved in meetings.[18]

The TCID has made scathing comments regarding the TROA Summary statement: "A public involvement program encouraged the general public and governmental agencies to help identify issues related to the resources in the Truckee River [B]asin."[19]

> This is a vociferous misrepresentation of the facts! There has never, ever been any attempt to make any of those TROA meetings public! And in fact, quite the contrary. On 5 October 1995, this Irrigation District formally protested, in writing, the Department of Interior's conduct of unannounced, clandestine meetings here in Nevada, right in the Truckee Meadows... Federally, there was never any attempt nor encouragement to gain public response(s). And in fact, on those rare occasions when the subject was broached, the Department of Interior position was quite adamant – neither the public nor the press were invited, nor welcome.[20]

To ensure this continued exclusivity, agendas were never publicly noticed in any prescribed or commonly accepted manner. Outside of a sanctioned short "telefax" list to participating agencies, there simply was no notification process. To maintain this subrosa status, the term *meetings* was changed to *negotiations*, and the following caveat was written or stamped on all papers that may have been distributed.

> For settlement negotiations only: Notice: This document is prepared and made available to participants and observers in the negotiation of the Truckee River Operating Agreement referenced in Public Law 101-618 (104 Stat.3289). Receipt and retention of a copy of it by any person or entity constitutes that person's or entity's agreement that it may not be used for any purpose outside of those negotiations, including any ongoing or future litigation or administrative proceedings.[21]

2. Public Access to Information on Issues

The process seemed very well covered in the media, though perhaps more educationally, *after* it was passed into law. December 1993 and April 1994,

Reid convened public hearings on implementation of the Act. Concerned parties (including farmers) had problems with some of the provisions, and implementation issues still needed to be resolved. A second round of negotiations – "The Second Generation Negotiations" aimed at long-term solution was planned by Senator Reid. Other sources suggest that there was not enough access to information:

> Senator Harry Reid has promoted his original water bill as the salvation of northern Nevada. Now it seems the new senate bill is even better: so good in fact that no one should have the right to adequate review time or due process.[22] (Tom Riggins, Fallon)

> ...it is also deeply disappointing that the talks were conducted in secret. Four times over the past decade the farmers and the tribe have been involved in such talks and every time the talks were held behind closed doors. It's time to bring the talks into the open, where everyone can follow the issues and maybe even offer some helpful suggestions to break this impasse. Reportedly, the negotiators have feared that public talks could lead to sensationalization and make matters worse. But it is difficult to see how matters could get much worse since secrecy has accomplished absolutely nothing. Further, this is a matter of grave interest not just to the negotiators but to all water users of the Truckee River water. All users are affected by what other users do. So while urging the talks to resume, we also urge them to be public. More things grow in the light than in the dark of night.[23]

C. Reasonable Process Costs

Bill Bettenberg, chief negotiator for the Department of Interior, indicated he had no budget/accounts for the process. He spoke about the more recent TROA and WQA processes, which he said, "were not cheap," and in most cases the parties have paid their own costs for participating.

Some of the process costs of the secondary negotiations, namely the hiring of a professional mediator, were paid by Senator Reid's office. However, actual figures for this stage of the negotiation are not publicly available.

III. CRITERIA: OUTCOME QUALITY

A. Cost Effectiveness

The Federal Water Masters Office is developing a new computer accounting system to track reservoir operations and water ownership for use on a daily basis to effectively implement the TROA. The system will cost $365,000. The US Department of Interior has requested that the State of California contribute one-fifth of the necessary budget.

1. Costs to Parties at the Table

Each of the mandatory signatories[24] was asked to pay for a part of the implementation costs. For example, each party has been asked to come up with $75,000 for the Water Master. Where the states are concerned, local and metropolitan agencies share the costs. While the Department of Interior has paid for most of the TROA EIS/EIR, the State of California will pick up some of these costs given that California has its own environmental act to comply with, the California Environmental Quality Act (CEQA). Nevada has no equivalent state policy, falls under the National Environmental Policy Act (NEPA) and does not share in EIS/EIR costs.

The Sierra Pacific Power Company calls the Settlement Act Sierra's "foremost water resource project for the future." Not only does it provide the largest yield of any resource identified – sufficient for 50 years or more – but it also secures the community's existing Truckee River supply, and specifically, it more than triples the drought storage available to Sierra. (Sierra's objective in the Negotiated Settlement was to acquire additional storage in upstream reservoirs.)[25]

In describing their estimated yield and costs, they say: "Yield of this project is estimated at 39,000 AF/yr. which reflects the conversion of 42,900 AF (39,000 x 1.1 of irrigation rights). The capital costs associated with this option are surface water treatment, water rights acquisitions and retrofit metering. The Settlement also requires Sierra to use its best efforts to implement Sparks Pit and Donner Lake as water resources. The cost of Donner Lake ($3.8 million) is included at the value of Sierra's last offer to TCID. Sierra has already accomplished the equivalent of the Sparks Pit project through development of other groundwater options, so its costs are not included."[26] This means that, as part of the negotiated settlement, Sierra Pacific is required to find other water sources—to restore flows to Pyramid Lake—and to provide an adequate supply of water for drinking, electricity generation, and the restoration of the environment in the Pyramid Lake region.

In their March 1998 publication, *1995-2015 Water Resource Plan Supplement*, the benefits and "requirements" are listed, with a note that "it remains to be seen whether some of these parties will conclude that the indirect benefits outweigh the perceived costs of the agreement for their interests.

Benefits: Interim drought storage for Truckee Meadows until Settlement Act becomes effective; permanent drought storage for Truckee Meadows; drought storage for Fernley (proposed); certainty regarding interstate allocations of the Truckee and Carson Rivers; improved timing of river flows for threatened and endangered fish; water quality enhancement by flow augmentation; improved instream flows and reservoir levels in CA; wetlands recovery at Stillwater; reduced litigation region-wide; and more water reaching Pyramid Lake.

Costs: Payments to U.S. for drought storage; waiver of Sierra's 'hydroelectric right;' installation of water meters in Reno/Sparks; water conservation in Reno/Sparks; legal/engineering costs to implement settlement; water right purchases for water quality; donation of Sierra's water in non-drought to fishery; 10% over-dedication of water rights for new service by Sierra; and concession of claims to unappropriated water to Tribe.

While the contents of the water conservation plan were not known at the time of the Settlement Act, the July 1996 conservation agreement reached between Sierra, the Tribe, Reno, Sparks, and Washoe County and ratified by the U.S., hold Sierra to the following obligations until 90% meter installations are reached: (1) They must expend up to $100,000 annually for free water-saving devices and field personnel to support lawn-watering limits; (2) expend at least $100,000 annually to encourage projects to retrofit water efficient landscaping and one ongoing requirement; and (3) expend at least $50,000 annually for public education about conservation.[27]

They further discuss the development costs of TROA, and conclude that while it has been higher than predicted, "it is probable that litigation costs would have exceeded the costs of TROA. Most certainly the costs of uncertainty to the community would have grown as the issues in litigation grew." They cite Docket No. 94-6015: "The parties agree that Westpac should continue to pursue the Negotiated Settlement as the leading water supply option in this Water Resource Plan. In addition, Westpac agrees to conduct its negotiations with the parties to the TROA in such a way that costs for the Settlement continue to be cost-effective."[28]

Pyramid Lake Paiute Tribe: "The tribe and the urban users had won a major victory. This act recognized the responsibility to return water to Pyramid Lake and to the wetlands....For the Paiutes, the power has shifted in their favor. The Settlement Act set up a tribal economic development fund of $40 million and a fishery fund of $25 million. The tribe is a force to be reckoned with today."[29]

"In short, the Water Quality Agreement benefits everyone on the TR from the Lake Tahoe to Pyramid Lake."[30]

- for Truckee Meadows urban interests
- improves tourism (TR flows through the heart of downtown Reno)
- settlement of lawsuits
- for Reno/Sparks: dependence heightened during the recent severe drought in the late 1980's – early 1990's when the river frequently dried up during the summer months and wildlife dependent upon the river suffered
- for California: will help meet minimum and preferred flows in the upstream portion of the river; additional water held in storage will improve recreation levels
- will contribute substantially to meeting the water quality standards of the Clean Water Act and inflow stream requirements in CA, NV, and PL Indian Reservations
- will go a long way towards implementing the objectives of the cui-ui recovery plan

2. Costs to the Public/Costs of Not Settling

Senator Reid, in his public letter, noted "[i]f we fail, our way of life will alter drastically because droughts are becoming more frequent. Also, we will lose millions of tax dollars when the federal government, the state, the local governments, the utilities and the Indians go back to court." However, an opposing opinion was offered by OMB, in which they stated that the government had no legal obligation to restore Indian rights considered lost after the Orr Ditch Decree. OMB further opposed the settlement on the basis of taxpayer costs in the form of federal payments for implementation.[31]

3. Costs to Other Parties Not at the Table: Impacts on Ratepayers

Sierra Pacific Power Company, in its promotional literature supporting the agreement, emphasizes that "Water meters guarantee fairness, because they see to it that we're all charged only for what we use. Those who try to save won't subsidize those who don't. Once installed, most of us will pay the same – or less – for water. Those with larger lots and houses will pay more." They also note that customers will not pay for meter installation, nor will Sierra Pacific Power Company. Hence, their water rates will not be raised for this service. This assertion, along with Sierra Pacific's estimate of costs to implement the plan, indicates that the company intends to shoulder implementation plans itself. However, one should note that while Sierra Pacific states that water rates will not rise, it says nothing about electric rates. Therefore, it may be safe to assume that some costs will, as usual, be passed to the consumer; although these may be somewhat less than the actual costs of implementation.

B. Financial Feasibility/Sustainability

The Secretary of the Interior is authorized to enter into an agreement with the State of Nevada for use by the State of not less than $9 million of State funds for water and water rights acquisitions and other protective measures to benefit Lahontan Valley wetlands.[32] He is also authorized to reimburse non-Federal entities for reasonable and customary costs for operation and maintenance of the Newlands Project associated with the delivery of water in following the provisions of this subsection.

The Federal Government (treasury) is authorized to appropriate $25 million for the "Pyramid Lake fisheries fund" (the principle of which is unavailable for withdrawal, with the interest available for the Tribe, for the purposes of operation and maintenance of fisheries).[33] The government is also authorized to appropriate $50 million for the Pyramid Lake Paiute Economic Development fund, where the principle and the interest can be used by the Tribe according to a plan made in consultation with the Secretary of the Interior. [34]

The State of Nevada must provide no less than $4 million for use in implementing water conservation measures pursuant to the settlement described in part 1 of this subsection.[35]

Overall, the cost-sharing mandated by TROA requires the Federal Government to pay for 40% of the administration costs of the Agreement; with California paying 20% and Nevada (with contributions by the Tribe and power company) also paying 40%. Other costs were difficult to determine as they were often subsumed in the operating budgets of the various departments.

C. Cultural Sustainability/Community Self-Determination

According to the National Historic Preservation Act of 1966 (NHPA), as amended, Federal law requires Federal agencies to consider the effects of their undertakings on cultural resources. In preservation of national, State, regional, and local resources of cultural significance, the NHPA (specifically section 106) requires agencies to consider the effects of its actions on "any district, site, building, structure, or object that is included in or eligible for inclusion in the national Register." Other legislation includes the protection of historic and archeological resources by the Federal Government, i.e. the Archeological Resources Protection Act and the Native American Graves Protection and Repatriation Act. Discussions involving the Fallon Shoshone Paiute Tribes, Pyramid Lake Paiute Tribe, the Reno-Sparks Colony and the Washoe Tribe regarding cultural properties began in 1995 and are ongoing (as of June 1998).[36]

Pyramid Lake Tribe members note that the Endangered Species Act is providing mechanisms that allow them to recover cui-ui, a fish species at the core of their cultural and economic lifestyle but that the "Tribe is suffering a sense of loss for cultural preservation by not having the ability to harvest the Cui-ui. One whole generation has not consumed any Cui-ui and there is a strong desire to seek a method of preserving its practice since the Cui-ui are the foundation of our culture."[37]

The irrigators should also be considered, as their lifestyles have and will continue to change as a result of the Settlement Act. "Now farmers instead of Indians are losing water and their livelihood."[38]

One member of the Lahontan Valley negotiating team for the Second Generation Negotiations declared, "If we start removing water from water-righted lands, we will destroy Fallon." He went on to call the changes to the Lahontan Valley "cultural genocide."[39]

D. Environmental Sustainability

Water conservation measures were a major aspect of Public Law 101-618, as well as an overriding goal throughout the negotiated settlement process. This has been an ongoing goal for the Pyramid Lake Paiute Tribe.

Water conservation measures: Naval Air Station (NAS): Under the Settlement Act, the NAS was required to develop a modified land management plan and implement water conservation measures.[40] These saved or conserved NAS waters, thus assisting with conservation of Pyramid Lake resources (fish and wildlife, though primarily cui-ui), and the Lahontan Valley wetlands. A Memorandum of Agreement was signed regarding these purposes, with an expected 2,300 acre-feet per year available.

Reno/Sparks Area residents: Must implement a water conservation program that produces a savings of 10% in drought years. Installation of water meters was required.[41] The Nevada Legislature insisted that the water saved by the installation of meters must be stored for drought protection [and not used for future growth accommodation].[42]

July 1996 Water Conservation Agreement (WQA): Two key concepts of this agreement included: 1) instead of drought-year-only conservation called for in the Preliminary Settlement Agreement as modified by the Ratification Agreement, more emphasis is placed on every-year conservation; 2) local governments agree to continue twice-weekly watering until 90% of water meters required by the Preliminary Settlement Agreement as modified by the Ratification Agreement have been installed.[43]

Water: One objective was to sustain on average 25,000 acres of primary wetlands habitat in the Stillwater National Wildlife Refuge, Stillwater Wildlife Management Area, Carson Lake and pasture, and Fallon Shosone Paiute Indian Reservation wetlands. In order to meet this objective, the U. S. Fish & Wildlife Service determined that an annual average total of up to 125,000 acre-feet of water would be needed. Public Law 101-618 required the Secretary of the Interior to acquire by purchase or other means, enough water and water rights for this purpose.[44]

Endangered Species: A primary purpose of the Settlement Act was to facilitate cui-ui recovery. Public Law 101-618 also has provisions for wetlands protection (section 206). For example, "water rights acquired under this subsection shall, to the maximum extent practicable, be used for direct application to such wetlands and shall not be sold, exchanged, or otherwise disposed of except as provided by the National Wildlife Refuge Administration Act and for the benefit of fish and wildlife within the Lahontan Valley." One objective was to expand the Stillwater national wildlife refuge, with the following aims: maintaining and restoring natural biological diversity within the refuge; providing for the conservation and management of fish and wildlife and their habitats within their refuge; fulfilling treating obligations of the U.S. with respect to fish and wildlife; and providing opportunities for scientific research, environmental education, and fish and wildlife recreation.

Public Law 101-618 also called for several studies. These included requiring the Secretary of the Interior to study and report to Congress by Nov. 9, 1993, on the environmental, economic and social impacts of the water rights purchase program for Lahontan Valley wetlands and the Pyramid Lake fishery; on the feasibility of improving the efficiency of Newlands Project conveyance facilities; and in consultation with Nevada, on administrative operational and structural measures to benefit recreation on the Lahontan Reservoir and the Truckee River downstream from Lahontan Dam. Also required were recommendations to Congress by Nov. 26, 1997, on any revisions of the boundaries of Stillwater NWR to include or exclude adjacent Service and Bureau of Reclamation lands. The Secretary of the Navy, in consultation with the Secretary of the Interior, was required to initiate a study by Nov. 9, 1991, of management actions at Fallon NAS that could reduce the need for water delivery to the Station, with the conserved water to be utilized for fish and wildlife. The Secretary of the Interior, in consultation with the EPA, Nevada and other parties, was charged to study and report to Congress (without a deadline) on the feasibility of using municipal wastewater to improve or create wetlands. And Congress directed the Secretary of the Army, in consultation with the Secretary and other parties, to undertake a study of the rehabilitation of the lower Truckee River for the benefit of the Pyramid Lake fishery.[45]

TROA: As required by Public Law 101-618, the Secretary of the Interior "may not become a party to the operating agreement [i.e. TROA] if the Secretary determines that the effect of such action, together with cumulative effects, are likely to jeopardize the continued existence of any endangered or threatened species or result in the destruction or adverse modification of any designated critical habitat of such species." Section 7 of the Endangered Species Act prohibits Federal agencies from authorizing, funding, or carrying out activities that are likely to jeopardize the continued existence of a listed species or destroy or adversely modify its critical habitat. Agencies consult with the U.S. Fish and Wildlife Service prior to initiating projects to determine the project's compliance.[46] Coordination for TROA under the Fish and Wildlife Coordination Act (FWCA) has been an ongoing process, including Department of Interior, Fish and Wildlife Service, Bureau of Reclamation, California Department of Fish and Game, and Nevada Department of Water Resources.[47]

How the Settlement Act would affect the Endangered Species Act was a major issue; when the OCAP freeze was added to the settlement, the House Merchant Marine and Fisheries Committee members believed that it would diminish the authority of the Endangered Species Act. Don Barry, General Counsel for Fisheries and Wildlife on the Merchant Marine and Fisheries Committee, said that had it been any other bill, they would have killed it. "The only reason the Committee agreed to back off was because of the Tribe's coalition and support. Mr. Barry said that Senator Reid made repeated personal appeals to key members on the Committee and persuaded them "to hold their nose" and agree to the bill because it was good for the Tribe. Senator Reid promised to add language to the bill (210 (b)(9) to protect the Endangered Species Act, and the Pyramid Lake fishery, Anaho Island, and the Lahontan Valley wetlands would benefit."[48]

E. Clarity of Outcome

There was little documentation alluding to the agreement's clarity, or lack thereof, except for a section in the CCC Report that notes a major area of miscommunication concerning the irrigators' (secured) right to water. The Report also notes that the Federal Government does not clearly state to the irrigators what limit exists regarding the amount of water they can use, thus leaving them vulnerable to what they perceive as an open-ended process. This is in addition to their perception that the Tribe's principle message questions the irrigators' fundamental rights to water and articulates the Tribe's objective of eliminating diversions from the Truckee River.

F. Feasibility/Realism

Politically, since the Act was formalized as a Public Law and signed by President Bush, it is eminently feasible. In addition, given the agreement of most – though not all – of the parties, the agreement remains feasible. Legally, the Settlement Act is consistent with the Orr Ditch decree; which supercedes or includes all other orders for the Truckee River: "[n]othing in this section shall be construed as modifying or terminating any court decree, or the jurisdiction of any court."[49]

G. Public Acceptability

An October 1990 newspaper announcement entitled, "Settlement provides future drought protection", noted the following supporters for the settlement: the Building and Construction Trades Council of Northern Nevada; Ducks Unlimited, Inc.; the Nevada Wildlife Federation; Carpenters Local No. 971; the Lahontan Valley Wetlands Coalition; Associated General Contractors; the Pyramid Lake Paiute Tribe; the Nevada Association of Realtors; the Tbiyabe Chapter of the Sierra Club; Coalition for a Negotiated Settlement; Sheet Metal Workers Local No. 26; the Environmental Defense Fund; Economic Development Authority of Western Nevada; Friends of Pyramid Lake; Plumbers and Pipefitters Local No. 350; Northern Nevada Central Labor Council; the Gaming Industry Association of Nevada; Plaster and Cement Masons Local No. 241; Greater Reno/Sparks Chamber of Congress; Builders Association of Northern Nevada; the Lahontan Audubon Society; Nevada Waterfowl Association; Citizens for Private Enterprise; Nevada Landscape Association.[50]

In a July 28, 1990 newspaper article entitled, "Reno to Carson March to Protest Water Bill," an organizer of the march was quoted as saying that "marchers will hand out letters to be signed and sent to Congress protesting the parts of the bill that would dismantle [the] locally controlled TCID in favor of a Federal agency."[51] FACTS (Fair Allocation of Carson and Truckee Systems); a coalition of representatives from TCID, Fallon, Churchill County, Fernly, Citizens for Private Enterprise, the Lahontan Valley Water Users Association and about 900 members of the Fallon and Shoshone Tribes teamed up to fight the bill (or to increase their share of the water). Another article describes their efforts traveling to Washington to lobby Reid for "amendments that would increase storage in the Lahontan Reservoir, secure more water for the Stillwater marshes and obtain an additional 900 acres of water rights that tribal spokesmen say were promised and never delivered."[52]

There is a general perception that the irrigators got a raw deal or, at the very least, that they lost because they didn't negotiate. While sometimes the situation of the farmers was stated in a non-victim-like, less sensationalistic manner, this author didn't come across any articles representing an alternative point of view. The Pyramid Lake Paiute Tribe would not have held this view, given their belief that the farmers were taking what was not theirs to claim.

The CCC Report outlines the thinking on this perspective. "[The D]istrict presents a typical example of the economic pressures confronted by irrigated areas that rely primarily on low-value crops. External market pressures create incentives to leave land fallow or to sell the land and appurtenant water rights. However, few areas are faced with such an unfortunate location as the irrigators of the Newlands Project under post-Settlement conditions. The irrigators are squeezed by two water claims that have been able to mount strong political and legal claims. To the northwest, the Pyramid Lake Paiute Tribe's long-term goal is to eliminate all transbasin diversions from the Truckee to the Carson. The Stillwater National Wildlife Refuge to the east seeks increased high quality flows to support the refuge."[53]

IV. CRITERIA: RELATIONSHIP OF PARTIES TO AGREEMENT

A. Satisfaction/Fairness

Sierra Pacific Power Company: Received the drought protection it wanted – 40 years – provided it develops the storage capacity to meet future urban demands. Mr. Faust, the Sierra Pacific Power Company's lobbyist in Washington, believes that the settlement is fair to the company and looks forward to implementing the Settlement Act with the Tribe.[54]

California: Received the assurance that its water supplies would be protected in the future and saw an end to 100 years of water wars.

Nevada: Also received assurance that its water supplies would be protected in the future (90% of all the water from the Truckee River, no matter how much growth takes place on the California side of the border).

Fallon Tribe: Pratt notes that they wanted justice: "If one defines justice in terms of forthcoming water rights and monetary compensation for past damages, then justice, in this instance, was served."[55] A settlement fund was created for $43 million, to be allocated over a five-year period beginning in 1992 and to be used for economic development.

Environment and Endangered Species: Also benefited from federal money authorized to purchase irrigation rights and to promote agricultural water conservation for cui-ui recovery, as well as wetlands and wildlife protection.

Farmers/irrigators/TCID: The farmers were interested in continuing to utilize Newlands Project water supplies for irrigated agriculture. They withdrew from the process when no agreement could be reached. TCID actively protested the negotiated settlement because they believed it would reduce the amount of water for the Newlands Irrigation Project. Mr. Clinton, a Civil Engineer and consultant to TCID, noted that the loser in the agreement is TCID, which "took it in the shorts."[56]

Office of Management and Budget (OMB): "OMB considers enactment of the Settlement Act a defeat of its policies for controlling the terms and costs of Indian water rights settlements...They consistently took the position that the government had no legal obligation to approve the settlement because the Tribe's water rights had been lost by the federal government in the Orr Ditch Decrees... at the end of the process, OMB officially opposed the settlement on the grounds of cost only."[57]

Pyramid Lake Paiute Tribe: Wanted enough money and water to maintain and enhance Pyramid Lake and its fisheries and it received the promise of both. The tribe is receiving $25 million for the enhancement of its fisheries and $40 million for tribal economic development. The Pyramid Lake Paiute Tribe has expressed dissatisfaction that the money is not enough for the amount of work that is needed, and for the costs of maintaining the lake.

The Tribe commissioned an internal report entitled, "Report with a Review, Opinion, and Recommendations about the "Truckee-Carson-Pyramid Lake Water Settlement" (Public Law 101-618 Title II, 104 Stat. 3294 November 16, 1990) through D.C.-based attorneys. The report briefly concluded that, "the Pyramid Lake Tribe has won a great victory with the Settlement Act. After more than 20 years of litigation and negotiation that began with the Tribe's successful suit against the Secretary of the Interior in the Pyramid Lake Paiute Tribe of Indians v. Morton, the Tribe has now substantial power over the use, development, and management of water in the Truckee, Tahoe, and Carson basins."

Overall, with the exception of the TCID, it appears that the involved parties were at least partially satisfied with the fairness of the outcome. The dissatisfaction of the TCID is covered above in sections I.A. and II.A.

B. Compliance with Outcome over Time

The Settlement Act in Section 210 (a)(3) explicitly states that "On and after the effective date of section 204 of this title, except as otherwise specifically provided herein, no person or entity who has entered into the Preliminary Settlement Agreement as modified by the Ratification Agreement or the Operating Agreement....may assert in judicial or administrative proceeding a claim that is inconsistent with the allocations provided in section 204 of this title, or inconsistent or in conflict with the operational criteria for the Truckee River established pursuant to section 205 of this title. No person or entity who does not become a party to the Preliminary Settlement Agreement as modified by the Ratification Agreement or Operating Agreement may assert in any judicial or administrative proceeding any claim for water rights for the Pyramid Lake Tribe, the Pyramid Lake Indian reservation, or the Pyramid Lake Fishery. Any such claims are hereby barred or extinguished and no court of the United States may hear or consider any such claims by such persons or entities."

It further notes, "Notwithstanding any other provision of law, the operating criteria and procedures for the reclamation project adopted by the Secretary on April 15, 1988 shall remain in effect through December 31, 1997, unless the Secretary decides in his sole discretion that changes are necessary to comply with his obligations, including those under the Endangered Species Act."

Miller, in his Nevada Journal article, notes, "...Reid's law also included a remarkable and explicit abrogation of Newlands property owners' constitutional rights to due process, in the form of a seven year ban on legal redress against any arbitrary and hostile administrative actions taken by the Secretary of the Interior, of which there have been many." He further notes that the Department of Interior has gained great leverage in the deal: "One clear example was the law's suspension of irrigators' rights to litigate against Secretary of the Interior Bruce Babbitt. With no injunctive or legal redress available to irrigators, the Department of Interior's Operating Criteria and Procedures – the rules it imposes on project irrigators – could be institutionalized as a large hammer with which to convert increasingly ravaged farmers into what Public Law 101-618 refers to as 'willing sellers.'"

Litigation involving farmers: TCID filed litigation immediately when the deadline expired. The U.S. Justice Department filed a lawsuit against the TCID, demanding the return of over 1 million acre-feet of water that the farmers had illegally diverted from the Truckee River between 1973 and 1987. The farmers responded with a motion to the U.S. District Court to dismiss the Federal Government's claim for return of the water. Judge Howard McKibben denied the TCID's motion, citing the following reasons: 1) a general public interest for the government to seek recovery of the water; 2) the Supreme Court's acknowledgement of the government's duty to protect the Pyramid Lake Paiute Tribe's interests. The Tribe has continued to litigate with the TCID following the Settlement Act.[58]

Churchill County lawsuit: On December 1, 1995, the county sought an injunction in U.S. District Court to prevent the federal government from acquiring and transferring water rights to the Lahontan Valley Wetlands and/or Pyramid Lake. They called for a more "programmatic Environmental Impact Study" to assess the cumulative impacts of the federal government's/Public Law 101-618's simultaneous actions affecting the Lahontan Valley. In particular, they claimed that water rights transfers would dry up local aquifers and adversely affect homes and businesses.[59] Churchill County and Fallon filed lawsuits in an attempt to block the water buyout program from continuing until a study was conducted on the impacts it would have on the agricultural region. U.S. District Judge Edward Reed, however, dismissed the three lawsuits in March 1997, saying that the city and county did not show that they had been harmed or would be harmed by the programs. He also said that local governments were not the proper parties to sue, but only individuals whose wells would go dry could sue.[60]

Verification of Compliance with the allocations for interstate transfers made in Public Law 101-618 applicable to each State "shall be assured by each State. Within the third quarter following the end of each calendar year, each State shall publish a report of water use providing information necessary to determine compliance with the terms and conditions of this section."[61] A report on the economic, social, and environmental effects of the water rights purchase program [under Section 206 – Wetlands Protection] was called for, to be conducted in coordination with the studies authorized in 207 (c)(5) and subsection 209 (c) of this title. It was required to be reported to the Committees on Energy and Natural Resources, Environment and Public Works, and Appropriations of the Senate, and the Committees on Interior and Insular Affairs, Merchant Marine and Fisheries, and Appropriations of the House of Representatives within three years after enactment.[62]

- 207 (c)(5): water rights acquisition program within the cui-ui and Lahontan cutthroat trout recovery and enhancement program
- 209 (c): the Secretary of the Interior is asked to study the feasibility of improving the conveyance of Newlands Project facilities to the extent that, within twelve years after the date of the enactment of this title, on average not less than 75% of the actual diversions under applicable operating criteria and procedures shall be delivered to satisfy the exercise of water rights within the Newlands Project. The Secretary of the Interior shall consider the effects of the required measures achieve such efficiency on groundwater resources and wetlands in the Newlands Project area.

Assistant Secretary of Water and Science of the US Department of Interior Betsy Reike's testimony on the progress in implementing P.L. 101-618 notes:

> So far, the Act is working largely as advertised. The legislation is not self-implementing but instead establishes a framework of actions, schedules, and incentives designed to lead to agreements and resolution of new problems. Steady progress has been made in implementing most parts of the act.[63]

The existence of the organizations and coordination bodies listed above – as well as the ongoing EIS/EIR – demonstrate that there are funded mechanisms in place to assure and verify compliance with the agreement.

C. Flexibility

The Preliminary Settlement Agreement was modified in Public Law 101-618 to become the Preliminary Settlement Agreement, as modified by the Ratification Agreement. In addition, the Settlement Act itself was amended, with this section concentrating on the latter modification.

The amendments included: a mandate that the Secretary of the Interior recoup illegally-diverted Project waters taken by the TCID; an authorization for Newlands Project water to go to municipal and other uses; an authorization for the federal government to contract with another agency other than TCID; a cut of the requested $75 million to $40 million for the Pyramid Lake Paiute Tribe – $20 million for fishery and $20 million for economic development. The Congressional Committee on Energy and Water, which reviewed Senator Reid's Bill, initiated these changes.[64]

The CCC Report notes that the amendments "were apparently specifically designed to restrict TCID's ability to litigate, and thereby stall implementation of the settlement." (Reid and the Committee on Energy and Water, which reviewed the bill, did not want the farmers to be able to stall the implementation of the settlement.) He also deduces that the actions gave the Secretary of the Interior leverage to force TCID to improve its irrigation systems. A D.C. insider purportedly suggested that Bush was ready to veto the bill if the provisions were not included.[65]

"Reid spokesman Wayne Mehl said Monday he didn't know why the Energy Committee put in the new provision."[66] TCID Manager Lyman McConnell: "My reaction was surprise; it was something I didn't expect...I think in some respects it was vindictive. We were trying to protect our interests and they perceived it as an impediment."[67] Nor was the Tribe pleased, given that their allotted funds were reduced $35 million. They said they would fight the modifications.[68]

Recoupment of water from TCID: On December 8, 1995, the U.S. Justice Department filed a suit in Reno federal district court, on behalf of the Secretary of the Interior, against the TCID. The suit demanded the full return of approximately 1,057,000 acre-feet of waters diverted from the Truckee River between 1973 and 1987 in violation of existing OCAPs. It also sought "in-kind interest."[69] There was a great deal of coverage of this and the ensuing rage from farmers in local papers. The general (media) perception seemed to be that there was foul play and vindictiveness on the part of the Federal Government toward the farmers.

D. Stability/Durability

The Act placed numerous responsibilities on the Department of Interior and other parties. Betsy Reike, in her testimony to the Senate subcommittee, noted that implementation began immediately.

> Mr. Chairman, the Department of Interior has moved aggressively to implement the Act. A substantial number of specific tasks required by the Act have been completed; good progress is being made on most others. In addition, we are taking action on many related matters such as examining means to improve the OCAP. There is much more to do such as developing a creative water right acquisition targeting program and completing a revision of the OCAP. We are however, addressing such issues directly and look forward to continued progress in meeting the goals I identified for you at the beginning of my statement.[70]

In this 19 page statement, she notes the Department of Interior's seven broad water resource management goals, which stem from P.L.101-618 and other underlying responsibilities for the Newlands Project, threatened and endangered fish at Pyramid Lake, and restoration of the Lahontan Valley wetlands. These goals are 1) to promote the enhancement and recovery of the cui-ui and the Lahontan cutthroat trout of Pyramid Lake in compliance with the ESA and the Settlement Act. 2) Protect the Lahontan Valley wetlands from further degradation and improve the habitat of the fish and wildlife that depend on those wetlands. 3) Encourage the development of solutions for water requirements in Washoe, Churchill, and Lyon Counties consistent with recovery objectives for the listed Pyramid Lake fishes. 4) Manage the N.P. to serve water rights efficiently and to meet other authorized purposes consistent with the above goals. 5) Provide a settlement of water issues related to the FPS Indian Tribes. 6) Provide for the settlement of water, fish, and other issues related to the PLP Tribe. 7) Facilitate the apportionment of the waters of the Truckee River, Carson River, and Lake Tahoe between California and Nevada.

Finally, Reike discusses "Other significant issues" that the Department of Interior is working on, including: Newlands Project operating criteria and procedures; assurance of water rights compliance; water rights acquisition targeting; management of the Newlands Project; the pace of water rights acquisition; broader regional approaches; and federal program management.

United States Fish and Wildlife Service: In order to meet the requirement of a long-term average of 25,000 acres of primary wetland habitat in Lahontan Valley, the Fish and Wildlife Service, in coordination with the Nevada Division of Wildlife, has been working to acquire a permanent and reliable supply of water and rights to provide approximately 125,000 acres of water annually.[71]

The U.S. Fish and Wildlife Service released its Draft Environmental Impact Statement, *Water Rights Acquisition for the Lahontan Valley Wetlands,* in July 1995. It began public scoping and planning for this in early 1992 and conducted formal public scoping workshops, bi-monthly public meetings and informal agency meetings. After an extensive comment period, the Final Environmental Impact Statement (FEIS) was due to be issued in September 1996.[72]

Based on the comments received (which were issued on time), the report laid out five alternative actions. It included a required "no action" baseline condition (Alternative 1) and proposed action (Alternative 2). Alternatives 2-5 would meet the Public Law 101-618 objective of sustaining 25,000 acres of primary wetland habitat.[73]

The Service has also been involved in administering part or all of the following provisions of Public Law 101-618: Section 106; Section 206(a)(3); 206 (b); 206 (b)(2); 206(b)(3); 206(e); 206(c), (d), (g). These include such tasks as eliminating the toxic flows of Hunter Drain within the Refuge Boundary and managing the Stillwater NWR wetland habitat. They are also heavily involved in the cui-ui recovery plan.

The Bureau of Indian Affairs has responsibilities to comply with obligations in Public Law 101-618, attachment to Rieke's testimony,

Interior Settlement Strategy Group: Composed of representatives from the Bureau of Indian Affairs, the Fish and Wildlife Service, and the Bureau of Reclamation, this group aimed to coordinate implementation of the many provisions of the Settlement Act. Its first task was to prepare a set of status reports on each of the 31 actions identified and authorized under the Settlement Act. The reports included the Department of Interior's perspective on the legal relationship and implications of each action. The group solicited help from government agencies, the public, and the Pyramid Lake and Fallon Tribes in identifying potential conflicts and ways to resolve these conflicts which would assist in achieving the purposes of the Settlement Act.[74]

Ongoing Dispute Resolution Forums: This author has not come across any "ongoing forum", though it is clear that the Truckee-Carson Coordination Office (TCCO) plays an instrumental role, coordinating various agencies within the Department of Interior. Bill Bettenberg, the Chief Negotiator for the Department of Interior, seems to be a focal point for carrying forward various sets of negotiations, i.e. the TROA, the WQA. It seems as though each process (and sub-process) identifies their own mechanism. It appears quite decentralized.

For interstate allocations, U.S. District Courts have jurisdiction to hear and decide claims of aggrieved parties. The Alpine Court has jurisdiction over and administration of interstate transfer issues.

"The U.S. District Courts for the Eastern District of California and the District of Nevada have jurisdiction to hear and decide any claims by any aggrieved party against the State of California, State of Nevada, or any other party where such claims allege failure to comply with the allocations or any other provision of this section."

The WQA established an alternative dispute mechanism to deal with contentious issues relating to water quality standard issues.[75]

Indicators of Instability can be found in some of the litigious actions taken by the TCID and others after the 'time out' period expired. Despite their losses in court, there are indications that the TCID continues to be involved in litigation with the Tribe. Please see section IV.B. for more details.

V. RELATIONSHIP BETWEEN PARTIES

A. Reduction in Conflict & Hostility

Is situation escalating or de-escalating?

At the stage of this writing, it is difficult to authoritatively say whether there has been a reduction or escalation in conflict and hostility. For the most part it appears that the relationships between the parties are professional, with a culture of negotiated settlement taking hold. However, given that some parties are dissatisfied with the end result – and given the professional restraint that colors their statements – this author feels that it is too early to do anything but "wait and see."

De-escalating views: Senator Reid states, "Earlier in my career I ended 100 years of water battles between California and Nevada by creating the Truckee-Carson-Pyramid Lake Negotiated Water Settlement…This settlement created a model that has been used for the settlement of dozens of other water disputes throughout the century."[76]

"This effort [Reid's settlement negotiations] was successful in bringing a number of parties together to develop mutual agreements on a series of Truckee River storage, management, and conservation issues; on interstate allocations for the Truckee and Carson rivers and Lake Tahoe; and on restoration of Lahontan Valley wetlands."

"A broad coalition of regional interests formed in support of these agreements, which provided the basis for the legislative proposals ultimately enacted as Public Law 101-618... the Department of the Interior endorsed the agreements and helped shape the final legislation."[77]

"The 1990 passage of the federal Truckee-Carson-Pyramid Lake Water Rights Settlement Act marks a recent milestone for the Truckee River. Among other things, the act achieved an interstate allocation of the water resources, a goal both Nevada and California have pursued for many years..."[78]

Escalating views:

> ...But as current events are demonstrating, that legacy, rather than an end to western Nevada water wars, is turning out to be a great deal more conflict – and very bitter conflict at that.[79]

This view proposes that, in many ways, rather than legislation that actually settled western Nevada water wars, Public Law 101-618 appears instead to have been largely a new and more sophisticated (or, vis-à-vis the general public, deceptive) offensive in those same old wars. In this view specifically, it was an effort to divide up Newlands Project irrigation water for the benefit of an organized coalition of opposing interests. Rather than initiating an era of water peace in the Sierras, this view considers the law as simply an occasion for the deployment of what they say might be described as an entirely new generation of water warfare weaponry, strategy and tactics.[80]

> Jim Bentley, spokesman for FACTS, criticized Reid for claiming his bill would end 80 years of water wars. "It does settle the allocation of water between California and Nevada," Bentley said, "and it does offer to dismiss a rather vague and untested claim by the Pyramid Lake Tribe. But other than that, it doesn't settle anything."[81]

B. Improved Relations

With the lack of clarity regarding the escalation or de-escalation of this conflict, it may be premature to determine whether or not there has been an improvement of relations between the parties. While some improvement in interpersonal relationships may have occurred, it would be precipitous to expand this to the larger community, especially given problems of ripeness and re-entry.

C. Cognitive & Affective Shift

A new attitude engendered by the Preliminary Settlement Agreement led to the Pyramid Tribe being asked to appoint a tribal representative to sit as a voting member of the new Water Planning Commission of Washoe County – an approach of participatory problem solving rather than adversarial encounters. "The local community decided on its own that it was better to work with the Tribe from the outset and to try to find ways of addressing water issues that would benefit the system as a whole, or failing that, at least would not harm the downstream interests of the Tribe, than to proceed on its own and risk more adversarial encounters. We are not aware of any similar arrangement between local government entities and an Indian tribe."[82]

The Pyramid Lake Paiute tribal newsletter noted a change in understanding of the relationship between the environment, economic development, and quality of life:

> In working on implementing the Settlement Act, it is difficult to change a general mentality that making a profit by sacrificing various aspects of the environment is good. However on the other hand there has been an increased awareness that protecting the environment is a key objective in economic development. Awareness has also started to become a role in evaluating quality of life through indicators. Quality of life indicators should be based on resource availability. The challenges we face regionally, and nationally, should be accepted together to accomplish a quality environment. An understanding of limitations for resource availability should account towards the level of development the environment can afford.[83]

D. Ability to Resolve Subsequent Disputes

In her feasibility study for the Second-Generation Negotiations, the mediator noted:

> Based on the interviews conducted, it is clear that there is a strong desire to resolve remaining issues surrounding implementation of Public Law 101-618. All parties expressed the preference to do so through negotiation rather than continued litigation, although parties also feel a significant and understandable degree of skepticism about the difficulties involved. Overall, people express a deep and genuine desire to... develop a shared concept for how to live in this region together and as neighbors.

She also cites the exchanges between Lahontan Valley residents and members of the Pyramid Lake Paiute Tribe as an inspirational example.

> Another significant development is the investment by people in the Lahontan Valley to establish the new linkages between themselves needed to talk about their goals for the future and to plan for these negotiations. The broad-based public participation which characterizes this work can only strengthen the community and help them take on their responsibilities for participation in solving regional issues in a new way. The Lahontan Valley Environmental Alliance [LVEA], created by formal agreement of the local governments, the TCID and the two conservation districts is developing the capacity to represent more diverse needs in the valley than have been part of previous negotiations, while maintaining the active involvement of the agricultural interests that remain at the heart of the community.[84]

The *Truckee River Chronology: Part III* also notes:

> This Lahontan Valley Environmental Alliance signified a more community-wide effort to negotiate a settlement to outstanding water issues and to maintain a viable agricultural industry in the region.[85]

Increasing shift towards negotiation versus litigation

"Litigation ends communication, doesn't it. Has your attorney ever told you not to communicate with the other side? He probably told you he did not trust what your adversary would do with the information. What he really meant was that he did not trust you not to blow a perfectly good legal argument. As negotiators we learned to see the issues through each other's eyes..."[86]

Successive rounds of negotiations indicate an intractable problem or comfort level of the parties in dealing with one another; a good relationship able to handle new conflicts?

The successive rounds of negotiations, particularly the Second-Generation Negotiations, would appear to reflect an intractable problem. At the same time, there seems to be a culture of negotiation developing, and parties and people interested to participate.

In a public statement sent out by the TCCO and signed by representatives of the Conservation Caucus, Fallon Shoshone Paiute Tribe, Lahontan Valley Environment Alliance, Newlands Water Protective Association, Pyramid Lake Paiute Tribe, and Sierra Pacific Power Company, and the Carson Water Subconservancy District:

"Although the parties to the negotiation have not reached agreement on many difficult issues including wetlands, recoupment, operating criteria and procedures for the Newlands Project, and municipal and industrial water systems, the parties will continue efforts on these and other issues, building on the knowledge and relationships gained in these negotiations.

> The WQA seems to be held in high esteem, as a negotiated settlement rather than litigation, to solve outstanding issues from Public Law 101-618 regarding water quality. "This settlement is a remarkable accomplishment and another example of the Clinton Administration's commitment to hammering out local, on-the-ground, common sense solutions to long-standing and contentious problems," said Secretary of the Interior Bruce Babbitt.
>
> "This agreement demonstrates that our environmental problems can be solved through the cooperation of citizens and their representative government," said Lois Schiffer, Assistant Attorney General in charge of the Environment and Natural Resources Division.
>
> "With the participation of all, we now have a solution that is win-win-win: for the environment, for Nevada's local communities, and for the Pyramid Lake Paiute Tribe," said EPA Regional Administrator Felicia Marcus.

"The signing of this agreement is another milestone of cooperation among users of the Truckee River," said Ada Deer, Assistant Secretary of the Interior for Indian Affairs. "This is truly a situation where all the parties involved, including the residents of the Truckee Meadows, the Pyramid Lake Paiute Tribe, and fish and wildlife will benefit from this agreement."[87]

"From an Interior viewpoint, recent years have seen a marked improvement in people working together to solve mutual problems and take cognizance of both mutual and divergent interests in the agreements. The recent successes of with the new O & M contract with TCID and the Truckee River Water Quality Settlement are examples; substantial headway is being made with regard to TROA and a second set of negotiations with the Fallon Tribes."[88]

Calling a mediator back in, exchanging representatives

The successive attempts to negotiate with the farmers and bring them into the process are well known. In Department of Interior Betsy Reike's testimony to Congress, she noted that the Bureau of Reclamation was ready to take over the Newlands Project if the farmers could not negotiate a new contract to operate it.[89]

Any evidence that ongoing problems in the relationship are being handled in a constructive manner?

The WQA, the TROA, and the various other contracts that have emerged are all deemed (relatively) successful by various parties (though interpretation seems to differ widely on criteria for success). Still, the ways in which they are being designed and implemented reflect increasing emphasis on negotiation (rather than facilitation, which the parties seem not to have rated as overly helpful), public meetings and scoping sessions, etc.

> The proposed settlement of a lawsuit involving the Reno-Sparks sewer treatment plant is great news. It continues the spirit of compromise fostered by the Negotiated Settlement, replacing water wars with mutually agreeable decisions to redistribute our most precious resource. Again, Everyone benefits...[90]

E. Transformation

> "We have not settled everything on the Truckee yet, but we are working on additional areas. Our efforts are aimed at moving forward but not allowing any linkage that would undo the progress we have already made. The strategy is a continual building on what has been accomplished so far."[91]

Most people seem to speak of the case not in transformational terms, but as setting a precedent for indigenous peoples, the environment, and endangered species. The CCC Report describes how a relatively powerless minority (Indian tribe) was able to influence the water allocation process through an assertion of federal environmental laws and tribal trust responsibilities. Litigation was successfully pursued to change the balance of power among major water users in the basin and to reallocate water in favor of protecting Pyramid Lake and its endangered fish.

VI. SOCIAL CAPITAL[92]

A. Enhanced Citizen Capacity to Draw on Collective Potential Resources

1. Aggregate of Resources

By late 1996 there was some indication that financial resources needed to implement the agreement had begun to flow. The EPA announced a change in its clean water loans to allow the use of loan funds to buy water rights. This change, announced shortly after the agreement was signed, allowed the local government, state agencies, and others to buy irrigation rights from farmers and to return the water to the Truckee River and Pyramid Lake.[93] The same article notes that while the farmers felt "barred" from the process, at least 900 owners of water rights were expected to sell, removing over 5,000 acres of farmland from production.

Another potential resource may stem from the installation of water meters throughout the Reno/Sparks area. While the documentation leaves the issue of who pays for their installation unclear – perhaps pointing to Federal sources – their impact on the ability of the region to control its own water use and costs during drought years may provide a resource for future litigation, or simply for civic pride. If one can measurably demonstrate changes in behavior – and the resulting savings – then one can reasonably use this to power new efforts in unrelated areas of civic life. Resource power, whether it is an increase in flows on the river or civic pride and ability, has improved somewhat for almost all of the stakeholders and associated parties, with the notable exception of the irrigators as represented by the TCID.

2. Potential Assistance Relationships

One clear new relationship that has formed is the Truckee River Partnership, formed to oversee the application of the new water allocation policy. Covered in more detail below in section B, this body represents an official continuance of some of the unofficial relationships that were created through the mediation process. Other relationships are not clear, although there is evidence of ad-hoc arrangements from the region's handling of recent flood conditions as well as the collaboration between multiple groups to re-seed trees on the riverbanks and to regenerate endangered fish species.[94]

Another indicator that new relationships have grown – though perhaps not to their full potential – comes from an overview of attempts to manage the

Truckee River Basin from 1980 through the 1990s. While some of this examination may not cover all of the negotiated settlement, Cobourn notes that despite the large number of projects and the high level of involvement, the nature of that involvement is still not coordinated enough to be considered integrated.[95]

Part of the improvement in relationships may in fact come from the settlement negotiation process itself. However, it may be argued that the changes in goal setting and accommodation, as well as the willingness to work together, may stem from changed attitudes and perceptions within the general public, public servants, and industry toward both the environment and the rights of Native Americans to their heritage and resources. At this point, given the ongoing status of the settlement negotiations and the numerous court cases being filed, it may be premature to state that all of the parties are willing to work together, though many of them have demonstrated an ability to do so on certain projects.

3. Generalized Reciprocity

Despite some indication that the parties have embraced ideas of general reciprocity in the implementation of the TROA, there are mixed signs that goodwill and trust remain. On the positive side, there was a statement by the Tribal Chairman that despite the fact that the State Engineer granted all unappropriated water to the tribe, it "knows it has to continue its work to address the Truckee's many competing demands."[96]

However on the other side of the coin, the Pyramid Lake tribe incurred the wrath of the Nevada State Department of Conservation and Natural Resources. The tribe requested that the Interior Department block the State's takeover of Carson Lake unless the State accedes to the Tribe's demand that 83 percent of water it purchased be used for wetlands conservation.[97]

B. Increased Community Capacity for Environmental/Policy Decision-Making

1. Aggregate of Resources

The most noticeable effect of an increase in the aggregate of resources is the aforementioned Truckee River Partnership. As detailed below, the organization is given the task of implementing and guiding new water policies. However, the extent of participation, openness of the decision-making processes, and the inclusion of all of the stakeholders is unclear at this time. The creation of the partnership can, however, be seen as a concrete positive effect of the negotiation process; especially one that is designed to provide a forum for conflict resolution regarding the distribution of benefits and impacts of the settlement agreement.

2. Increased System Efficiency

As noted above by Cobourn, although there has been an increase in interest and participation in water management on the Truckee River, there is little evidence that systemic efficiency has increased. In fact, an article detailing the willingness of the U.S. Bureau of Reclamation to take over the Lake Tahoe Dam in order to preserve the TROA angered locals who see the unilateral actions of the government as having a dampening effect on confidence and on the agreement itself. Another example was the February 14, 1999, requirement that the Environmental Impact Statement be redone in order to address new impacts of the agreement.[98]

3. Increased Capacity for Cooperation

There has been some indication that groups of stakeholders have come together to form associations or organizations designed to allow broader participation in decision making processes. As noted by the *New York Times* on November 30, 1997, one effect of the implementation of the Truckee River accord has been the creation of the Truckee River Partnership, a group of business, environmental, and government representatives charged with guiding the new water allocation policy. Interestingly enough, although this article indicates that the farmers of the Newlands project have been isolated, the rest of the water-using community – including the formerly castigated Paiute Tribes – has come together in a new alliance of interests. Namely, they share the goal of taking some of the farmers' huge share of the water.

4. Increased System Capacity for Responding to External Challenges

There is some indication that the regional ability to deal with drought and flood conditions has improved through the negotiation process. As with many of the other measurements of social capital, it is difficult to trace the improvement to the existence or the process of the negotiated settlement.

5. Increased Information Flow

Some increased information flow has been noted with respect to the Washoe County Regional Water Planning Commission. The Commission, which continues to have an interest in the ongoing settlement negotiations, posts information including meeting minutes on its public access website. However, among the other parties to the negotiated settlement, only the city of Reno and the Sierra Pacific Power Co. maintain websites, and these are primarily devoted to public relations rather than informational purposes. Although there may be public service announcements and inter-party communications, the author is unable to ascertain at this time their level or whether the information flow has substantially increased following the negotiated settlement.

C. Social System Transformation

Accessing social system transformation depends on what is included in this category. One could say the SPWC is a part of the social system and that it has undergone a form of transformation. One could also say that the gradual change in community views towards water use, the environment, and Native Americans spurred the change in power relationships between the Tribes and the irrigators, leading to the settlement. In addition, if one views the context of the social system to include the preference for litigation over less adversarial forms of dispute resolution, then this author must assert that there has been little in the way of major social system transformation. However, within the parameters of the Guidebook definition, there have been some improvements and changes that could constitute minor to mid-level social transformations.

1. Assistance and Support Provided to General Community

As noted above in section III.A the negotiated settlement has provided an influx of capital to the tribal communities and to support institutional arrangements for continued water management. Additionally, according to promotional literature, the Sierra Pacific Power Co., through a subsidized foundation, provides between $700,000 and $1,000,000 annually for cultural, educational, and community building projects.[99] It is unclear whether this is generally driven by public relations or has been affected by improved relations with the Tribe and the local communities, but the latter is quite likely.

2. More Resilient Social/Political/Economic System

It seems quite clear that the negotiated settlement does not provide for a more resilient social/political/economic system. Many instances of this are provided by the ongoing litigation both to implement the settlement, and to oppose the settlement by included and excluded parties. One article in the *Reno Gazette-Journal* notes that of the many things the negotiated settlement does not do, one important one is that it fails to "[s]ettle legal battles among the tribe, the federal government and the Newlands Reclamation Project."[100] In addition, the same newspaper reported that Lake Tahoe residents were considering a lawsuit to address reservoir management under the negotiated settlement. Their grievances stem from the fact that during the flood year, managers kept Lake Tahoe's water levels high, refusing to allow extra runoffs even though the higher levels were causing erosion and damage to waterfront properties.[101]

3. Increased Civic Discourse

There are mixed results in this category. On the one hand, it is difficult to find information stating that the current negotiations are held in public view, despite calls for them to be opened. It does appear, however, that while access to the actual negotiations may be limited, their results and reports are generally available on websites or during public review sessions.[102] There are also indications that Sierra Pacific has attempted to improve its image, although this may be mostly a public relations campaign rather than an effort to establish dialogues on issues of community concern.[103]

Despite these improvements, the continuation of the negotiations to implement the TROA, as well as continued litigation related to water management indicates that while civic discourse has increased overall, intra-party discourse is still running much higher than inter-party discourse, especially when the federal government or the irrigators are involved.

4. Perceived Interdependence

In the area of perceived interdependence there were signs of recognition by both Sierra Pacific and by the leadership of the Pyramid Lake Paiute Tribe. However, this does not seem to have extended too far beyond the realm of recognized 'shared interests,' into the deep-rooted recognition that the actions of one party will have an effect upon all of the others.

5. General Trust

In the area of general trust, this author would have to state that although some level of trust may have been achieved, the process of the settlement negotiations – and the departure of the TCID – did little to improve overall trust in the region. Instances of continued litigation – between the Tribe and the State as noted above, and between the farmers, State legislative representatives, and backers of the settlement – show a general lack of trust and a willingness to continue litigation as the best means for either obtaining objectives or forcing opponents to the table. In the second case, Assemblywoman Marcia de Braga (D-Fallon) submitted an assembly bill that would have scuttled the provisions of the negotiated settlement. This prompted Senator Reid to reopen negotiations and provide funds for the Tribe to purchase another $7 million in water rights from Newlands farmers.[104]

This latest agreement might be taken as an indicator of increased general trust, but this author would deem that trust still to be quite tenuous given the manner in which the negotiations were impelled.

CONCLUSION

Overall, there seemed to be several major themes carried in the local papers throughout the negotiating process – both for the negotiated settlement and the mediated talks. One of the most prominent, especially in *Gazette-Journal* editorials, was the admonition to the TCID to continue negotiating and to get the best deal it could, instead of walking out, stalling, or "continu[ing] to believe that…[it] is 1940 and not 1990."[105] This admonition was a clear recognition of a shift away from the farmers in both power and moral authority. By contrast, much of the literature from the TCID, Newlands Water Protection Association, and residents of Fallon charge that the main problem with the negotiated settlement is that it provides yet another example of the federal government 'muscling' its way in and ordering people around, a common sentiment in the U.S. west and mid-west. Therefore, it is reasonable to conclude that, despite the many steps that the negotiated settlement has taken forward, the processes surrounding it have failed to include all of the parties or to create a resolution or future process that is likely to satisfy all of their concerns.

NOTES

[1] Committee on Western Water Management, Water Science and Technology Board, Commission on Engineering and Technical Systems, Board on Agriculture, National Research Council, *Water Transfers in the West* (Washington, D.C.: National Academy Press, 1992), 119-136; A. Dan Tarlock, Symposium: Case Study on Regulatory Integration: Water Policy and the Protection of Endangered Species in the Truckee-Carson River Basin "The Creation of New Ricks Sharing Water Entitlement Regimes: The Case of the Truckee-Carson Settlement," *25 Ecology Law Quarterly* 674 (1999).

[2] California Division of Water Rights Atlas, p. 65

[3] AMP, November 1990, p.17

[4] Editing to a standard case format, and additions of criteria under category six, Social Capital, were done by Landon Hancock.

---

[5] Accounts differ as to whether or not the irrigators "walked out" or whether they decided that, despite their willingness to negotiate, the Reid dialogues did not offer a productive venue for them to move forward. Newspaper accounts indicate the former while interviews strongly indicate the latter; with a Clearwater Consulting Corporation report falling somewhere in the middle.

[6] The Dept. of Interior had to sign any agreement before Congress would approve it.

[7] The effectiveness of the Agreement was seen as contingent upon the US government becoming a party to it. See Article III, Par. 29 (g).

[8] Jeremy Pratt, "Truckee-Carson River Basin Study: Final Report" (Seattle: Clearwater Consulting Corporation), 194.

[9] Ibid. p. 119.

[10] Reid letter, August 11, 1990; Also "Time running out on northern Nevada water pact" 'It's your turn' in *Reno Gazette-Journal*, August 6, 1990.

[11] Jeremy Pratt, "Truckee-Carson River Basin Study: Final Report", p. 119.

[12] As an example, contingencies were placed on several aspects of the legislation. Many of the Agreement's provisions were contingent upon the final resolution of the Pyramid Lake Paiute Tribe's claims; others were contingent upon the settlement of the Tribe's lawsuits against the State of California. Many of these issues were later dealt with.

[13] Bob Pelcyger, the attorney for the Pyramid Lake Paiute Tribe, has noted that the Tribal Council has always rejected the concept of commoditizing the land. Their overwhelming value is to protect the lake - what was given to them and is part of them. They don't feel themselves as owners, but as the lake were given them to take care of. (Healing the Water) The historic and symbolic meaning of the lake to the Tribe can also be illustrated by the fact that they have folklore common to the members. "We the Paiute People have always been taught to share the water. To take what is needed and not waste what has been placed here by Almighty God. We are told not to fight and argue over our water. Water is life, our life is water. To believe this way is to live this way. Our water is placed here for all of us, but we cannot forget that our Mother Earth creates this water and needs this water for her own survival. This way of life is all we have, this belief in water is all we have. We cannot let this slip away by compromising our future. Water is the key to life and all of depending on our water resource should be able to share what is available and decide if building out until there is nothing left is a way to sustain a livable environment."

[14] Mervin Wright Jr. in Pyramid Lake Paiute Tribe's Newsletter, (December 1994): 4-5.

[15] In the CCC Report, a case study with extensive interviews characterizes the farmers' treatment as one of alienation, and that they have generally been scapegoated. The irrigators hold the view that they are only doing what society asked them to do and praised them for doing in the past. The CCC Report implies that the shift in cultural and societal values from "settling the frontier" towards environmental and Native American sensitivity is at the root of the problem whereby the irrigators have not been given equal consideration.

[16] Jeremy Pratt, "Truckee-Carson River Basin Study: Final Report", p. 192.

[17] US Dept of Interior and California Dept of Water Resources, "Draft Environmental Impact Statement/Environmental Impact Report for the Truckee River Operating Agreement," (US Dept of Interior and California Dept of Water Resources, 1998): 5-7. Chapter 5 "Consultation and Coordination" describes the process in detail, beginning with the formal public NEPA/CEQA scoping where a publication of a notice of intent in the Federal Register on July 21, and preparation on July 27, 1991, was accompanied by a press release issued from Reclamation's Mid-Pacific Regional Office. These announced the locations and times for public scoping meetings. Attended by about 130 people over 4 days in 5 locations, com-

ments were recorded and presented to the public an open scoping session held in Washoe County chambers on December 10, 1991.

[18] Ibid., Chapter 5, p. 10.

[19] Jeremy Pratt, "Truckee-Carson River Basin Study: Final Report", 3.

[20] TCID comments on the Draft EIS/EIR of TROA. Prepared by Russell P. Armstrong, April 27, 1998 and sent to David Overvold. Copy obtained from Mr. Armstrong. Cite from page 4.

[21] Russ Armstrong of the TCID, in this comment, goes on to note that there were no minutes of the meetings. (This was independently verified, for when the author spoke with Bill Bettenberg, chief negotiator for the Department of Interior, he mentioned that they were very informal processes, and minutes were not kept, and agendas not formally made.)

[22] Tom Riggins, "Farmers Being Abused," Letters to the editor, *Reno Gazette-Journal*, August 29, 1990.

[23] Editorial "Next Time, How About Making Talks Public?" *Reno Gazette-Journal*, August 9, 1996.

[24] Signatories include the Pyramid Lake Paiutes, Sierra Pacific Power Co., the States of Nevada and California, and the Department of the Interior.

[25] Sierra Pacific Power Company, "1995-2015 Water Resource Plan" (Sierra Pacific Power Company, 1995), 8-87.

[26] Ibid.

[27] Ibid., 9-10.

[28] Ibid., 11.

[29] Healing the Water, Viewer's Guide.

[30] Jeremy Pratt, "Truckee-Carson River Basin Study: Final Report," 82.

[31] See Section IV.A. for more details.

[32] Public Law 101-618 Section 206 (2)(d).

[33] Ibid., Section 708 (a)(2).

[34] Ibid., Section 708 (3).

[35] Ibid., Section 708 (H)(2).

[36] US Dept of Interior and California Dept of Water Resources, "Draft Environmental Impact Statement/Environmental Impact Report for the Truckee River Operating Agreement," Chapter 5, p. 3.

[37] Mervin Write, Jr., Water Resources Director [later Chairman], Pyramid Lake Paiute Tribe Newsletter (December 1994): 2.

[38] Healing the Water, Viewer's guide.

[39] Christensen, *High Country News*, April 3, 1995. For similar views, see Section V.

[40] Paragraph 206 (c)(3).

[41] Sierra Pacific Power Company promotional literature.

[42] Nevada Division of Water Planning, *Truckee River Chronology Part III – Twentieth Century* (Department of Conservation and Natural Resources, May 7, 1998).
Available from http://www.state.nv.us/cnr/ndwp/truckee/truckee3.htm.

[43] Sierra Pacific Power Company, 1998, p. 11.

[44] Nevada Division of Water Planning, *Truckee River Chronology Part III – Twentieth Century,* 1998.

[45] Digest of the Federal Resource Laws of Interest to the US Fish and Wildlife Service.

[46] US Dept of Interior and California Dept of Water Resources, "Draft Environmental Impact Statement/Environmental Impact Report for the Truckee River Operating Agreement," Chapter 5, p. 2.

[47] Ibid.
[48] Pyramid Lake Paiute Tribe's commissioned report, p.21.
[49] Public Law 101-618, Sec.204(c)(3)(h); Truckee River General Agreement Courts; Preliminary Settlement Agreement as modified by the Ratification Agreement. It is additionally stated in Public Law 101-618 that nothing in the (particular) section is to be inconsistent with Nevada and California State Law(s).
[50] Advertisement/Public Service Announcement in the October 16, 1990 *Reno Gazette-Journal.* Paid for by Westpac Utilities (later Sierra Pacific).
[51] "Reno to Carson March to Protest Water Bill," *Gannett News Service*, July 28, 1990.
[52] John S. Miller, "Ranchers, Indians Fight Truckee Bill," *Reno Gazette-Journal*, January 24 1990.
[53] Jeremy Pratt, "Truckee-Carson River Basin Study: Final Report", 191.
[54] Pyramid Lake Paiute Tribe's commissioned report, p. 23.
[55] Ibid. p. 123.
[56] Pyramid Lake Paiute Tribe's Commissioned report.
[57] Ibid. p. 16.
[58] Nevada Division of Water Planning, *Truckee River Chronology Part III – Twentieth Century,* 1998.
[59] Ibid.
[60] "3 Suits Over River Rights Thrown Out," *Sacramento Bee*, March 5, 1997.
[61] Public Law 101-618, Section 704 (d) (1).
[62] This was in fact completed as submitted as a Report to the U.S. Congress in November 1993 – *Water Rights Acquisition Program for Public Law 101-618 and Lahontan Valley Wetlands, Nevada.*
[63] Betsy Rieke, Testimony p. 1.
[64] Jeremy Pratt, "Truckee-Carson River Basin Study: Final Report," 123
[65] Ibid.
[66] *Reno Gazette-Journal*, July 17, 1990.
[67] Ibid.
[68] Ibid.
[69] Nevada Division of Water Planning, *Truckee River Chronology Part III – Twentieth Century,* 1998.
[70] Statement of Betsy Reike to Senate subcommittee, p.19.
[71] Fish and Wildlife Service Response to the Committee's Request for Specific Recommendations Regarding Aspects of the Newlands Project and the Truckee-Carson Pyramid Lake Indian water Rights Settlement Act.
[72] Nevada Division of Water Planning, *Truckee River Chronology Part III – Twentieth Century,* 1998.
[73] Ibid.
[74] Bureau of Indian Affairs attachment to Rieke's testimony.
[75] Department of Justice press release, October 19, 1996.
[76] Senator Reid's Website. http://www.senate.gov/~reid/
[77] Reike testimony, p.3.
[78] California Department of Water Resources, *Truckee River Atlas* (California Department of Water Resources, 1991): iii.
[79] Steve Miller, "Harry Reid Sows Dragon's Teeth, Turning Nevada Farmland Into Desert," *Nevada Journal* 4, no. 12 (1997). Electronic journal copy at http://www.nj.npri.org/nj97/12/cover_story.htm

[80] Ibid.

[81] *Reno Gazette-Journal*, January 24, 1990.

[82] Comments of Pyramid Tribe in Jeremy Pratt, "Truckee-Carson River Basin Study: Final Report", 77.

[83] Pyramid Lake Paiute Tribe Newsletter, December 1994, p.2

[84] Gail Bingham, "Findings and Recommendations on Convening Second Round Truckee/ Carson Settlement Negotiations," (Washington DC: RESOLVE Inc., 1994), 8.

[85] Jeremy Pratt, "Truckee-Carson River Basin Study: Final Report"

[86] Oldham, "Truckee River Communication Process", p.2.

[87] Department of Justice press release, October 10, 1996.

[88] Notes on draft CCC report (Pratt) by Department of Interior, Draft Report, p. 193.

[89] Jeremy Pratt, "Truckee-Carson River Basin Study: Final Report"

[90] *Reno Gazette Journal*, May 19, 1996.

[91] Oldham speech, Truckee River Settlement, p.11.

[92] This section was written by Landon Hancock after the Social Capital section was added to the Guidebook (see Appendix A).

[93] "Novel Use of Clean-Water Loans Brightens Outlook For a River," *New York Times*, A25, October 31, 1996.

[94] For examples see Lou Cannon, "'Low-Tech' Effort Aims to Return Massive Trout to Nevada Waters," *Washington Post*, A3, April 19, 1998; Jon Christensen, "River in Nevada Helps Its Own Restoration," *The New York Times*, C4, September 24, 1996; and Sean Whaley, "Babbit Weaves Fish Story," *Las Vegas Review-Journal*, B7, August 14, 1998.

[95] J. Cobourn, "Integrated Watershed Management on the Truckee River in Nevada," *Journal of the American Water Resources Association* 35, no. 3 (1999): 623-32.

[96] "Engineer Approves Tribe's Applications for Unappropriated Water," *Associated Press State & Local Wire*, December 7, 1998.

[97] "Truckee Water Pact Hits New Snag," *Associated Press State & Local Wire*, February 14, 1999.

[98] Ibid.

[99] From Sierra Pacific's Community Relations section of its Internet home page, www.sierrapacific.com.

[100] From a sidebar article attached to: Faith Bremner, "Truckee River Fight Reopens Unappropriated Water Dispute," *Reno Gazette-Journal*, November 11, 1997.

[101] Jeff DeLong, "Lake-River Citizen Group Hints at Lawsuit to Force Better Management of Lake Level," *Reno Gazette-Journal*, November 11, 1997.

[102] For one example, see the website for the Washoe County Dept. of Water Resources at http://www.co.washoe.nv.us/utilities/meetings.htm

[103] For more information see the company's website at http://www.sierrapacific.com

[104] Brendan Riley, "Lawmakers Give Details of Truckee River Water Deal," *Associated Press State & Regional Wire*, April 16, 1999. In addition, the State of Nevada pledged $4 million and Sierra Pacific $2.5 to help set up the Newlands Project Water Fund. For more information see, "Senate OK's Water Settlement," *Associate Press State & Local Wire*, May 28, 1999; and Sean Whaley, "Senate Addresses Rural Issues," *Las Vegas Review-Journal*, May 29, 1999.

[105] Editorial, "Fallon Farmers Should Negotiate While They Can," *Reno Gazette-Journal*, 1995.

Chapter Six

THE PECOS RIVER CASE

Sharing a Resource Among Old Rivals

Case Researchers: Annette Pfeifer Hanada and Landon Hancock

A little water clears us of this deed. How easy it is then!

Lady MacBeth in Shakespeare's MacBeth

When we make rivers the battleground, rivers lose their essence.

John Thorson, Special Master for the Arizona General Stream Adjudication

Note: This case report illustrates the use of a particular methodological framework. It is not intended to replicate the legal and historical coverage of this case provided in other sources.[1]

Introduction: This interstate water conflict represents a classic dispute over limited water supplies between two jurisdictions. The relevant areas of eastern New Mexico and western Texas are shown in Figure 6.1.

Time period: The dispute dates to 1947 with the signing of the Pecos River Compact. This report will evaluate the litigation period, 1983-1990, with some coverage of the implementation period. This report was compiled in 1998, and significantly revised and expanded in 2000.

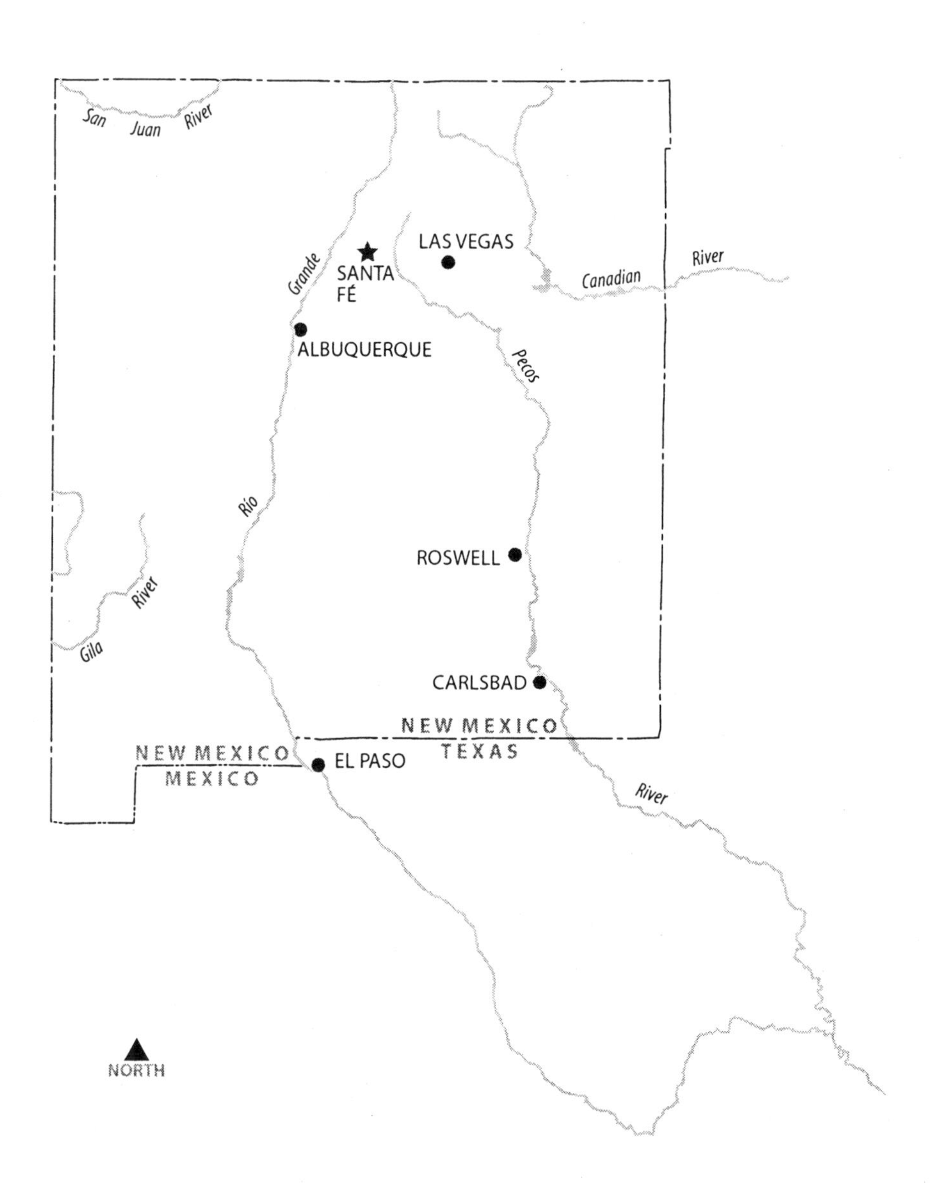

Figure 6.1 Pecos River Case

Basic nature of dispute: The amount of water flow in the Pecos River from New Mexico into Texas according to the provisions of the Pecos River Compact of 1947, and the method of measuring that amount.

Issues: The appropriate method to measure the amount of water to be delivered, payment (in water or money) for water not delivered between 1950-1974, the economic/environmental feasibility for New Mexico to deliver the required water amount, economic hardship, damages to West Texan agriculture due to reduced water deliveries, and sustainability of agriculture in the New Mexico portion of the Pecos River Basin.

Actors and Interests: Since this dispute was carried out mainly in the U.S. Supreme Court, there were a limited number of 'official' parties.

- **State of Texas**, *Interests:* appropriation of water, compensation for deficient water supply, new method for measuring water flow.
- **State of New Mexico**, *Interests:* appropriation of water, payment obligations for inadequate water delivered to Texas.
- **United States** as third party to the compact commission, *Interests:* appropriation of water, equitable resolution of conflict.
- Indirect actors included irrigators & communities on both sides of the conflict. In New Mexico, the Carlsbad Irrigation District was interested in retaining water rights, while irrigators in Texas preferred to be paid for shortfalls in water rather than in cash.

Attempted conflict resolution processes: Litigation (1974-88), out-of court settlement negotiations (1988-89), a form of third party arbitration with a Court-appointed Water Master, ratified by U.S. Supreme Court decisions, (1988-90).

History of the Pecos River Conflict

1909-35: Several court decrees are issued to adjudicate the surface water of various streams in the Pecos River drainage basin.

1948: Texas and New Mexico sign the Pecos River Compact and Congress ratifies it. The compact apportions the water of the river equally between the two states.

1974 Texas files a lawsuit, claiming that New Mexico violated the Pecos River Compact by not delivering as much water as required under the terms of the compact.

1987-8 The U.S. Supreme Court rules against New Mexico, agreeing with the Water Master that New Mexico must pay 'back-damages' of water owed, but disagreeing about the form of payment. The decision stipulates that if an agreement on monetary payments cannot be made in a timely fashion then repayment in water will be ordered.

1990 The U.S. Supreme Court "stipulates" (formalizes a negotiated agreement) that New Mexico shall pay Texas $14 million, in exchange for which Texas will release New Mexico from breaches of the Pecos River Compact between 1952 and 1986.

Background

The Pecos River Compact, an agreement signed between the states of Texas and New Mexico in 1947, governs the allocation of the waters of the Pecos River Basin. The compact requires New Mexico "not to deplete by man's activities the flow of the river at the state line."[2] This should give Texas a quantity of water equivalent to the amount available in 1947 when the Compact was signed and approved by Congress. An "inflow-outflow manual" is used to determine the amount of water Texas should receive over any particular period. The Pecos River Commission, consisting of Texas, New Mexico and the U.S. as a non-voting member, was established to resolve any water allocation disputes by unanimous vote.

In later years it became apparent that the inflow-outflow manual did not accurately reflect the status of the river since the water at the Texas State line was below the predicted amount. After the Commission was unable to agree on a method to determine annual variations in the water flow, Texas filed a lawsuit in the United States Supreme Court. It claimed that New Mexico had breached its obligations under the Compact, and sought a decree ordering New Mexico to deliver water in accordance with the Compact.

A Special Water Master was appointed by the U. S. Supreme Court and concluded that a new inflow-outflow manual was needed to provide an accurate description of the 1947 conditions. He also recommended appointing a third party to participate in the Commission deliberations when the two Commissioners could not agree. The Special Water Master rejected both New Mexico's motion to dismiss the case, and Texas' request for a simplified measurement method to determine the shortfalls in state-line deliveries.

Based on the Compact, the U.S. Supreme Court did not agree to order a third party to participate in the Commission deliberations. The Court decided it could, however, resolve the dispute between the two states itself. A simplified method to determine the water shortfalls was rejected by the Court because it was not similar enough to the inflow-outflow method.

Summary of the litigation

Texas brought the case before the U.S. Supreme Court to gain redress for perceived shortfalls in water delivery on the Pecos River. After appointing a

Special Water Master, the Court affirmed that New Mexico had indeed failed to deliver some 340,000-acre feet between 1952 and 1986. The Court further ordered the appointment of a River Master to oversee water accounting and, if necessary, order New Mexico to compensate Texas for annual shortfalls in water delivery. A 1990 court ruling affirmed a negotiated agreement that allowed New Mexico to make a monetary repayment of $14 million to Texas for the water it failed to deliver between 1952 and 1986.

I. CRITERIA: OUTCOME REACHED

A. Unanimity or Consensus

It is difficult to determine the degree of unanimity within the forum of court adjudication. Several times between the formal filing of the case in 1974 and the final ruling in 1990, one or both sides disputed the findings of the Special Master. However, in each case, the Court decided the veracity of the argument, and often decided against both parties. Although there is little documentation available, the amount of redress paid by New Mexico to Texas ($14 million) was likely reached by unanimous agreement between the two parties. This is surmised from the nature of the Court ruling, which was a 'stipulated' judgment of a joint motion by both parties that was recommended by the Special Master.[3]

B. Verifiable Terms

The resolution of the court case was published in the form of three U.S. Supreme Court cases.[4] The terms of the continuing Pecos River Compact are embodied in the compact itself, but modified by the in-flow/out-flow scheme devised under the guidance of the Special Water Master and ratified by the U.S. Supreme Court.[5]

C. Public Acknowledgement

Other than the ruling of the Court, there appears to be little in the way of publicized media events on the national stage. Additionally, local newspaper coverage of the court decisions was limited to wire coverage and minor editorial notices. *United Press International* had a brief news release in which Texas Attorney General Jim Mattox indicated that "[t]he extra water that will now come Texas' way will be a godsend to West Texas and its people...provid[ing] a substantial long-term boost to the agriculture economy around Pecos..."[6]

The most extensive coverage of the effects of the judgments came in water editorials and stories regarding the environmental impacts and in subsequent cases involving water rights on the Rio Grande. Other coverage, such as radio and video archives, were unavailable and may have been more extensive; however, this seems unlikely given the lack of national coverage by newspaper sources in New Mexico and Texas, along with the location of the decision in Washington D.C.

D. Ratification

Any Pecos River decisions and agreements reached by the parties, or the Special Master, required approval from the U.S. Supreme Court. Because the case involved a compact between two states approved by Congress, the U.S. Supreme Court retains 'original jurisdiction.'[7] In addition, the Court retains jurisdiction to oversee the actions of the appointed River Master to determine the amount of credit or shortfall in Pecos River water that New Mexico owes Texas each year. In doing so, the Court renders quarterly judgements authorizing payment for the River Master's services.[8]

Since this was a legal case settled by the U.S. Supreme Court in original jurisdiction, there was no provision for either ratification or an appeal to other bodies. However, in order for the State Engineer of New Mexico to purchase water rights and ensure adequate water deliveries, the New Mexico House of Representatives was required to pass a bond initiative to fund the effort. This was accomplished by 1995, and has contributed to New Mexico's ability to purchase more than $10 million in water rights.[9]

II. CRITERIA: PROCESS QUALITY

A. Procedurally Just

Although the 1988 Court Order, in finding that New Mexico had violated the Pecos River Compact, clearly made Texas the winner, there was little indication of 'sour grapes' on the part of New Mexico officials. As reported by *United Press International*, Texas officials were pleased with the ruling and felt that it was quite fair. Texas Attorney General Jim Mattox was quoted as saying that, "[t]he extra water that will now come Texas' way will be a godsend to West Texas and its people."[10]

Mattox went on to indicate that he looked forward to the determination from the Court on how much New Mexico would have to pay Texas for the water owed. The Special Master's 1987 report indicated that he believed the amount was 340,000 acre-feet; an amount agreed to by the Court and indemnified against New Mexico in the ruling.[11]

Texas officials also felt that the $14 million settlement for past shortages was an additional victory for them, despite their original estimates of the water value at between $300 million and $1.1 billion.[12]

Texas Attorney General Jim Mattox said it was the first time in United States legal history that one state had been ordered to pay another for illegally taking water. "It is gratifying to bring in $14 million for the state of Texas,' said Mattox. "However, I am even more pleased that New Mexico will have to live by its obligations to keep the water flowing to West Texas farmers and ranchers in the future. Water is their lifeblood."[13]

Curiously enough, New Mexico officials also characterized the $14 million agreement as a victory, compared to the possibility of repaying Texas in water. "This is a total victory for New Mexico," said New Mexico Attorney General Hal Stratton. "This will preclude the necessity of condemning water rights on the Pecos... and shutting down agriculture, at least in the Roswell area."[14]

Neither party expressed dissatisfaction with the U.S. Supreme Court's procedural decision to appoint either the Special Master or the River Master. This is likely the result of the inflexibility of U.S. Supreme Court – and other court – processes and the additional caveat that U.S. Supreme Court decisions cannot be appealed.

B. Procedurally Accessible and Inclusive

There is no information regarding the inclusiveness and accessibility of the process used by the Special Master to determine river flows and compensation. Although there were several Special Masters, who served in succession, no indication was given that any of them were dismissed or that either party to the conflict was dissatisfied with their procedures, even if they disputed the conclusions.

C. Reasonable Process Costs

There is very limited information in the news coverage concerning the costs that were incurred. However, as the loser of the court case, New Mexico has had to bear the burden of most, if not all, of the costs. In the 1990 U.S. Supreme Court ruling, $200,000 of the $14 million settlement was set aside for payment of Texas' legal costs, with the caveat that this payment not establish New Mexico's liability for such costs.[15] In addition to the $14 million settlement, it is mentioned that New Mexico paid Hank Bonhoff, a private attorney, $92,000 for his work on the settlement negotiations.[16]

III. CRITERIA: OUTCOME QUALITY

A. Cost-Effective Implementation

Despite the fact that New Mexico's 'one-time' payment was made in cash, the judgement requires that New Mexico meet its yearly obligations to Texas in the form of Pecos River water. This has created a constant financial burden on the state's Interstate Stream Commission, as it has been forced to buy or lease water rights in order to maintain adequate flow to Texas. This is due to New Mexico's legal system, which only allows the state to acquire the newest water rights first in order to meet its obligations to Texas. Unfortunately, those water rights are owned by individuals who have wells instead of access to surface water.[17] If the River Master judges that New Mexico has not been meeting its obligation, the acquisition of those rights would do little to correct the situation within the three-month time period mandated by the Court. Instead, New Mexico has instigated a program to purchase and lease surface water rights and ensure that enough water reaches Texas yearly. Between the 1991 ruling and February 2000, New Mexico spent $20 million to purchase approximately 8,600 acre-feet a year.[18]

As noted in an editorial in the *Albuquerque Journal*, the state of New Mexico regularly pays farmers $50 per acre-foot for water which they bought for $12 per acre-foot.[19] In addition, New Mexico's acquisition of IMC's (a potash mining company) water rights cost $3 million.[20] It seems clear that while New Mexico has avoided the estimated $200 million cost of shutting down farms and cities to meet its water obligations, the continued cost of purchasing and leasing rights may eventually approach that amount if those costs continue at the levels incurred in the 1990s.

B. Perceived Economic Efficiency

The negotiated settlement of $14 million was perceived by most parties as an economically efficient method for New Mexico to repay Texas.[21] Despite this, some in Texas felt that the 'efficiency' of the one-time payment did little to offset the loss of 340,100 acre-feet of water. For more details, see Criteria IV.A.

Although covered in more detail above in III.A., it seems clear that the combination of New Mexico water law with the judgment's strict requirements for annual shortfall repayment has created a situation in which New Mexico must pay significant sums to ensure adequate water flows to Texas. It could be reasonably argued that these costs are far from efficient, especially if they result in costs that exceed the estimated $200 million of shutting down farms and limiting water to cities in the Pecos River Basin.

C. Financial Feasibility/Sustainability

The reported damage to Texas agriculture as a result of not receiving its share of Pecos River water is significant according to Texas' estimates. However, the judgment along with the Court's ruling that Texas should receive 45 percent of Pecos water have made New Mexico's obligations to Texas possibly quite onerous. Before the 1990 stipulated agreement, which granted a one-time payment to Texas, it was observed that the possibility of paying Texas the 340,100 acre-feet of water would curtail New Mexico's industrial development and possibly shut down some agricultural enterprises. For New Mexico to fulfill its obligation and repay the damage was perceived as possibly "wip[ing] out agriculture in southeastern New Mexico for 10 years."[22]

The stipulated agreement of 1990, however, allowed New Mexico to forsake repayment in water in exchange for a one-time $14 million payment to Texas. This type of 'money for water' payment was seen by the U.S. Supreme Court as an extraordinary, one-time solution to an exceedingly difficult remedial problem. The Court further noted that future performance of the Compact was to be made in kind, forestalling New Mexico's ability to pay Texas money in lieu of water.[23]

The costs to the state of New Mexico to purchase and lease water rights exceeded $50 million between 1991 and 1999, and additional expenditures are expected.[24] However, an editorial in the *Albuquerque Journal* contends that the $50 per acre-foot the state pays to farmers for their water rights is inflated, considering that the farmers pay only $12 per acre-foot, and the taxpayers end up paying the difference.[25]

In addition to the costs of purchasing and leasing water rights, the U.S. Supreme Court ruled that the states would have to share the cost of the continued presence of a River Master to oversee the agreement and determine the amount of water owed to Texas each year, as well as the surplus or deficit owed by New Mexico.[26] Overall, as of late 1999, the Special (or Water) Master and the River Master have billed approximately $400,000.[27]

D. Cultural Sustainability/Community Self-Determination

The ruling will affect both states in different ways. Farmers in Texas will be able to irrigate their crops, while in some areas of New Mexico the economy could suffer. Despite the fact that the settlement agreement precluded condemning water rights on the Pecos and shutting down of agriculture, the extensive leasing and purchasing of farmers' water rights may have some of the same effect, because the water leases and purchases cause reductions in irrigated acreage.

There is little doubt that the water restrictions required by implementation of the court ruling will curtail the rapid expansion of cities and agriculture in

New Mexico's portion of the Pecos River Basin. However, the fact that the state of New Mexico must purchase water rights rather than appropriate them means that, for the most part, the process of water rights transfer and reduced irrigated acreage will be voluntary.

E. Environmental Sustainability

Due to the geographic nature of the Pecos River, a consistent flow cannot be maintained. The deficiency in water flow can be attributed both to natural causes and 'man's activities.' The 'human-derived' activities that impede the normal flow of the Pecos stem from the damming of the river to ensure adequate year-round water for New Mexico farmers and for the state's obligation to Texas. Environmental concerns arise due to this irregular water flow.

"The Pecos River suffers frequent and destructive floods that ruin the channels and fill the reservoirs with silt" say court documents filed by the state of New Mexico. In addition, the region and the river suffer from periodic bouts of drought. Finally, one supposed problem with water supply stems from the introduction of salt cedars into the region sometime during the beginning of the 20th century. Several newspaper stories and a *National Geographic* article contend that an acre of salt cedars are capable of drinking down a million gallons of water a year.[28] However, examinations of water flows in the region contend that water flow and consumption issues are too complex to accurately calculate the effect of the trees. Indeed, according to Fredrick Bruce, a major problem in the Compact was a miscalculation in the amount of salvaged water that planners estimated would be recovered by salt cedar eradication.[29]

The implementation of the Court decision has compounded other environmental problems, leading to the threat of at least one other court case. In addition to environmental problems caused by salt cedars and efforts to eradicate them, artificial controls of the River's flow have affected the Pecos bluntnose shiner, which has been placed on the endangered species list.[30] In 1998, a dispute arose between the federal government, the state of New Mexico, and local farmers over federal plans to increase water flows to sustain the bluntnose shiner. The fish requires between 31 and 35 cubic feet per second to live. However, releasing the extra water required to keep the fish alive was estimated to cost farmers up to 20 million gallons per day. That is water that would be lost both to New Mexico farmers and to water deliveries owed to Texas.[31]

In 1998, a Santa Fe-based environmental group, Forest Guardians, filed a lawsuit against the federal government, charging it with not obeying the Endangered Species Act by ensuring that enough water flows year-round to guarantee plant and animal viability.[32] This suit, along with others filed by the group against the Colorado River Compact, and the Rio Grande and Costilla

Creek agreements, charged the defendants with partitioning river flows with no regard to endangered species. In the Pecos River case, the result was a court order that forced the Fish and Wildlife Service to locate the shiner's habitat in the river. In addition, despite local and state complaints that the federal government was taking over the river, the Bureau of Reclamation and FWS stated their determination to ensure the shiner's habitat remains whole.[33]

These decisions have made it more difficult for New Mexico to provide timely delivery of water to Texas. As a result, New Mexico has been forced to spend more money to purchase water rights from irrigators.

F. Clarity of Outcome

Because of the irregular flow of the Pecos River, the original compact did not specify a particular amount of water delivery to Texas. The measurement to determine the amount has also been controversial. What seemed to be a sensible approach to the changing river flow became the basis of controversy. The current court-ruled agreement could not be said to have much more clarity than the original. The major difference is that the annual amount of water owed by New Mexico to Texas (and any surpluses or shortfalls) is determined by the continued presence of the court-appointed River Master. A second major difference was the acceptance by the court of a new flow measurement designed to gauge more accurately the amount of water flowing through the Pecos and, consequently, the amount of water owed to Texas each year.[34] Ostensibly, New Mexico owes Texas 45% of each year's river flow. However, that flow has to be determined each year by the River Master based on the "inflow-outflow" model, which measures basin inflows and outflows at a number of gauging stations.[35]

G. Feasibility/Realism

The feasibility of the court ruling is difficult to determine. One of the major reasons is the fact that there was no provision for appeal. Regardless of whether or not they felt that the result was feasible, each side (and their constituents) had little choice but to 'pony up' and implement the decision.

With regard to the negotiated settlement of the back claims of water (the 1990 stipulated agreement), however, there is some indication that both state governments were pleased with the outcome and believed that it was both feasible and realistic. Despite this, some parties on each side, namely farmers in New Mexico and political candidates in Texas, felt that the settlement was unfair and would result in unacceptable costs. This issue is covered more in the next section.

H. Public Acceptability

As is evident in the next section, the level of public acceptability varied among the constituencies. Farmers and irrigators in New Mexico were hardly pleased, as were municipalities on that side of the Pecos River basin. However, there are some indications that farmers on the Texas side were also less than pleased with the stipulated agreement, preferring to receive water instead. See section IV.A. for more details.

IV. CRITERIA: RELATIONSHIP OF PARTIES TO OUTCOME

A. Satisfaction/Fairness

There was mixed satisfaction with the 1988 judgement, even within Texas, due to perceptions that the damage payments were too small. As mentioned above, Texas Attorney General Jim Mattox was quite satisfied with the judgement and the following agreement, pledging to use most of the $14 million payment to assist farmers in West Texas. Despite this, a number of individuals in west Texas were unhappy with the $14 million agreement. The most vociferous of these was Republican gubernatorial candidate Clayton Williams, who also owns an irrigated alfalfa farm in the Pecos River basin. His assertion that Mattox "sold West Texas farmers down the river" with the settlement was based on his perception that the farmers wanted water, not money.[36] Additionally, Williams felt that "$14 million [was] just a drop in the bucket compared to what West Texas farmers have lost in water and damages since 1950."[37] Another critic was Billy Moody, Texas' Pecos River Compact Commissioner, who criticized the settlement in a letter to then-Governor, Bill Clements. In his letter, he stated that Mattox owed an explanation to the farmers of West Texas for his decision to lower the settlement amount to less than $50 million.[38]

On one hand, New Mexico was pleased with the chance to pay damages monetarily rather than in kind, but officials remained concerned about the economy in the affected areas. In an editorial, the *Albuquerque Journal*'s Bill Hume contended that as a result of the court rulings, New Mexico's share of Pecos River water was actually shrinking. This perception was based upon the court's requirement that any shortfalls in water deliveries to Texas be made up within one year. Therefore, New Mexico was forced to either purchase surface water rights from original users (such as the Carlsbad Irrigation District), or to force surface users to give up their rights by removing rights from late-comers (who mostly pump groundwater) and original users. Either way, "New Mexico's available water supply is actually shrinking."[39] Additionally, irrigators in Lea County[40] sued the Interstate Stream Commission in an ineffective bid to stop the Commission's plan to purchase and retire water rights. The Lea County farmers indicated that their major problem with the Commission's plans stemmed from wasteful use of water by Texas irrigators, and the fact that they "were never ... invited to the table" to discuss water issues.[41]

B. Compliance with Outcome Over Time

In order to comply with the Court's 1988 decision, the New Mexico Interstate Stream Commission has bought or leased water rights which cost the state over $50 million between 1991 and 1999. To maintain available funding for water purchases and leases, the New Mexico House of Representatives approved a bill in 1995 to allow the ISS to sell bonds for the purpose of purchasing water rights.[42]

Despite the willingness of New Mexico authorities to comply with the judgment, problems have arisen with its implementation and cost. In April of 1993, the state engineer requested an audit of the Pecos River Water Rights Purchase Program for the period from 1990 through 1995. The program was initially established to acquire rights to Pecos water that could increase the amount of water in the river flow to Texas. As of 1995, the Commission had spent about $15 million in acquiring or leasing water rights. The audit revealed that New Mexico had paid $1.5 million more than it should have paid for a block of water to the Pecos River. Further irregularities exposed that the Commission overpaid for water rights from the Hondo Company and several individuals. A consulting contract with Dr. John Hernandez for $52,400 was paid without the formal approval of the Interstate Stream Commission.[43]

The stipulated settlement of New Mexico's owed water (the one-time $14 million payment) was required by March 1, 1990.[44] Although there is little documentation that the payment was made, no indication or report of a lack of payment was found. Therefore, given the willingness of New Mexico to make the monetary payment in lieu of payment in water, it is nearly certain that New Mexico complied, despite the limited time frame.

C. Flexibility

The 1988 Court ruling requires that any shortfall in water deliveries to Texas be compensated within a single calendar year. Therefore, New Mexico has made strenuous efforts to ensure that it maintains a surplus in its water delivery account. The ruling's inflexibility is one of the primary reasons behind New Mexico's purchase of extensive water rights and leases. This is due to the combination of the one-year shortfall repayment with New Mexico's riparian water rights laws. The latter means that New Mexico can only appropriate water rights on the basis of last granted. These rights tend to be based on well water, rather than surface water, and the repatriation of these rights to the state would have almost no impact on river flows, and would certainly not return enough water to the river to erase any shortfalls.[45]

D. Stability/Durability

The stability and durability of the court ruling is dependent upon a number of factors, primarily the erratic weather conditions in New Mexico that often result in years of floods followed by years of drought. Until 1994, New Mexico had managed to keep a surplus in its account of water owed to Texas. However, in 1995 and 1996, New Mexico under-delivered water to Texas, because of drought in the region.[46]

Despite variations in water deliveries, New Mexico has managed to maintain a surplus in its Pecos River water account with Texas. As of February 2000, it had a 21,000 acre-foot surplus. However, according to Norman Gaume, director of the New Mexico Interstate Stream Commission, much of that surplus "has been obtained by leases rather than purchases and permanent retirements."[47]

Environmental concerns have also made it difficult for New Mexico to meet the requirements set by the 1988 ruling. As described above, threats of lawsuits by environmental groups have forced the federal government to increase off-season releases of Pecos River water to support endangered species. These releases were opposed by New Mexico officials as well as by farmers and irrigators, who felt that they would be on the losing end of water distribution. Given that the court ruling was fairly rigid, the farmer's contention that they would have to take last place behind Texas and the Pecos bluntnose shiner was only slightly ameliorated by the federal promise to return water to the river in exchange for increased winter flows.[48]

V. CRITERIA: RELATIONSHIP BETWEEN PARTIES (RELATIONSHIP QUALITY)

A. Reduction in Conflict and Hostility

It is difficult to estimate the level of conflict and hostility between officials of both state governments. There is little evidence of any hostility, either before or after the 1988 judgement or 1991 stipulated agreement. However, continued actions by New Mexico irrigators as well as editorials in the *Albuquer-*

que Journal indicate that the 'losers' in this conflict still feel some resentment towards Texas and the Federal government.

B. Improved Relations

There is little indication that the relationship between Texas and New Mexico officials has improved as a result of the litigation and court rulings surrounding the Pecos River Compact. Numerous incidents of suits, or threatened suits, continue to dog relations between Texas, New Mexico, and other users of water in the region. These center primarily on the Rio Grande and its other tributaries rather than the Pecos, but the continued litigation points toward a lack of improvement in the official relationship.

As for the other parties to the conflict, the farmers and irrigators in the Pecos River basin are unlikely to feel any improvement in relations with New Mexico officials or with Texas as a whole. Their continued contention that they "are being sold down the river" to environmental and Texas concerns is evidenced by the Carlsbad Irrigation District's unwillingness to release water during winter (resulting in a federal takeover of a dam controlled by the district[49]) and by a lawsuit filed by a Lea County irrigator against New Mexico's plan to buy water rights on the Pecos.[50]

C. Cognitive and Affective Shift

There is no indication of a cognitive shift on the part of either party as a result of the settlement of this dispute. Continued litigation on other fronts, as well as resentment in New Mexico attest to a lack of cognitive shift among the various parties. Indeed, the willingness of environmental groups like Forest Guardians to engage in litigation on behalf of their concerns indicates that the only cognitive shift resulting from this case is the efficacy of using the courts to force compliance.

D. Ability to Resolve Subsequent Disputes

Although the judgment modified the inflow-outflow manual to allow for future shortfalls in the river level, it did not allow for methods of payment other than water. As noted in section IV.C., the judgement itself was not flexible, essentially leaving New Mexico with the burden of compliance and encouraging Texas to engage in litigation in the pursuit of other water claims against New Mexico and others.[51]

E. Transformation

There is little evidence to suggest that any transformation processes have occurred as a result of the litigation surrounding the Pecos River Compact. As noted above, Texas continues to pursue litigation as the primary means of assuring its water supply. In addition, various parties associated with the accord or concerned with water issues along the Pecos River continue to pursue litigation as their primary avenue for change.

VI. CRITERIA: SOCIAL CAPITAL

A. Enhanced Citizen Capacity to Draw on Collective Potential Resources

Although one could point toward Forest Guardian's use of the Endangered Species Act as a tool to ensure the survival of the Pecos bluntnose shiner, there is no reason to assume that this citizen capacity stemmed in any way from the litigation surrounding the Pecos River Compact. Indeed, it is unlikely that either the Compact litigation, or the cases that have sprung from it, have enhanced the capacity of the average citizen in New Mexico or Texas to ensure their rights or to participate in water allocation processes.

B. Increased Community Capacity for Environmental/Policy Decision-Making

Like the prior category, there is little evidence to suggest that either of the state's representatives have made any efforts to increase the community's ability to participate in the environmental policy decision-making process with regard to the allocation of Pecos River water or other similar disputes. Complaints by New Mexico irrigators that they were shut out of the decision-making process regarding the bluntnose shiner indicate that governmental bodies continue to be perceived as unwilling to listen to the irrigators' position.[52]

C. Social System Transformation

Unfortunately, one of the primary results of the Pecos River Compact litigation has been to encourage more of the same type of interaction. Changes in social values to emphasize the protection of endangered species and the environment over the needs of irrigation stem from sources other than the Compact litigation itself. Litigation remains the primary method for obtaining an improved status for the environment, within the New Mexico context. This enhances our conclusion that the Pecos River Compact litigation has had little, if any, impact in improving the region's social systems, or in providing for any increase in social capital overall.

NOTES

[1] Douglas L. Grant, "Interstate Water Allocation Compacts: When the Virtue of Permanence Becomes the Vice of Inflexibility," *74 University of Colorado Law Review* 105, (Winter 2003); Emlen G. Hall, *High and Dry: the Texas-New Mexico Struggle for the Pecos River* (Albuquerque, NM: University of New Mexico Press, 2002).

[2] *The Pecos River Compact: Entered Into By the States of New Mexico and Texas*, (Santa Fe: Quality Press, 1948).

[3] *Texas v. New Mexico*, 494 U.S. 111 (1990).

[4] *Texas v. New Mexico*, 494 U.S. 111 (1990), *modifying* 485 U.S. 388 (1988), *modifying* 482 U.S. 124 (1987).

[5] *The Pecos River Compact: Entered Into By the States of New Mexico and Texas,* (Santa Fe: Quality Press, 1948). Ratified by *Texas v. New Mexico*, 467 U.S. 1238 (1984).

[6] "Regional News," *United Press International*, March 29, BC Cycle 1988.

[7] See 462 U.S. 544, (1983). References U.S. Const. Art. III § 2, cl.1 and U.S.C. § 1251(a)(1), granting the US Supreme Court original jurisdiction to resolve controversies between two states.

[8] See 502 U.S. 903, (1991) for the order appointing the River Master and authorizing his or her review and payment for services.

[9] Digest, "House OKs bond sale to buy water rights," *Albuquerque Journal*, March 8, 1995.

[10] "Regional News," *United Press International*, March 29, BC Cycle 1988.

[11] 485 U.S. 388 (1988).

[12] "Regional News," *United Press International*, July 14, BC Cycle 1989.

[13] Helen Gaussoin, "Regional News," *United Press International*, August 9, BC Cycle 1989.

[14] Ibid.

[15] 494 U.S. 111 (1990).

[16] See UPI "Regional News," July 1989 and Gaussoin, August 1989.

[17] Editorial, "Taxpayers Bail Out Pecos Water Users," *Albuquerque Journal*, A20, December 11, 1998.

[18] "State and Regional," *United Press International*, February 25, AM Cycle 2000.

[19] Editorial, December 11, 1998.

[20] "State Buys Back Water Rights From Potash Company," *Associated Press State & Local Wire,* February 25, AM Cycle 2000.

[21] See section III.C. for details.

[22] "Regional News," *United Press International*, February 26, BC Cycle 1990.

[23] Special Master Charles I. Meyers Report, November 30, 1987; 107 5. Ct. at 2285-86.

[24] Corrections, "For The Record," *Albuquerque Journal*, March 10, 1999.

[25] Editorial, "Taxpayers Bail Out Pecos Water Users," *Albuquerque Journal*, A20, December 9, 1998.

[26] 502 U.S. 903 (1991).

[27] Specific amounts are $193,564.67 for the Special Master and 163,407.45 for the River Master, between 2/24/86 and 10/12/99. Amounts vary year-to-year and are billed irregularly two to three times a year.

[28] Cathy Newman, "The Pecos: River of Hard-Won Dreams," *National Geographic*, September 1993, 49.

[29] Fredrick R Bruce, "Salvaged Water: The Failed Critical Assumption Underlying the Pecos River Compact," *Natural Resources Journal* 33, no. 1 (Winter, 1993): 217-228.

[30] Newman, p. 51.

[31] Ian Hoffman, "State: Feds Taking Over River," *Albuquerque Journal*, 1, October 31, 1998. While the article is not clear, it is likely that the water is lost to both Texas and New Mexico because it flows during the off-season for agricultural needs.

[32] "Suit Threatened Over Water Management in Pecos River," *Frontline (Forest Guardians Newsletter)*, # 38, November 12, 1998; and Tom Wolf, "The Law of the River May Finally Meet Its Match," *Los Vegas Review-Journal*, 15B, July 24, 1998.

[33] See Ian Hoffman, "State: Feds Taking Over River," *Albuquerque Journal*, 1, October 31, 1998; and Ben Neary, "Water Release Allowed to Save Threatened Fish," *Santa Fe New Mexican*, April 24, 1999.

[34] See 485 U.S. 388 (1988); and Neil S. Grigg, "Pecos River Compact: Recent Developments," [Internet], *Texas Water Resources Institute Newsletter*, 1987 [cited March 5, 2000]. Available from http://twri.tamu.edu/twripubs/NewWaves/v2n1/abstract-4.html.

[35] Neil S. Grigg, "Pecos River Compact: Recent Developments," [Internet], *Texas Water Resources Institute Newsletter*, 1987.

[36] "Regional News," *United Press International*, August 21, BC Cycle 1989.

[37] Ibid.

[38] Ibid, and see "Regional News," *United Press International*, July 14, BC Cycle 1989.

[39] Bill Hume, "N.M. Share of Pecos River Shrinking," *Albuquerque Journal*, B2, February 23, 1997.

[40] Lea County centers around Lovington, NM, approximately 80 miles N.W. of Carlsbad, NM.

[41] "Judge Throws Out Lea County Water Lawsuit," *Associated Press State & Local Wire*, November 9, PM Cycle 1998.

[42] Digest, "House OKs BondS ale to Buy Water Rights," *Albuquerque Journal*, March 8, 1995.

[43] Karen Peterson,. "Audit Shows State Spent Too Much on Water Rights," *Santa Fe New Mexican*, B4, April 13 1996,.

[44] 494 U.S. 111 (1990). Judgement entered on February 26th, giving New Mexico four days to comply.

[45] Bill Hume, "N.M. Share of Pecos River Shrinking," *Albuquerque Journal*, B2, February 23, 1997.

[46] Ibid.

[47] "State Buys Back Water Rights From Potash Company," *Associated Press State & Local Wire*, February 25, AM Cycle 2000.

[48] "Pecos River Irrigation Water Released to Support Threatened Minnow," *Associated Press State & Local Wire*, November 19, BC Cycle 1998.

[49] "Pecos River Irrigation Water Released to Support Threatened Minnow," *Associated Press State & Local Wire*, November 19, BC Cycle 1998. For continued action on this issue see AP, "Farmers Take On State in Pecos River Water Dispute," *Albuquerque Tribune*, A5, December 10, 1998.

[50] "Judge Throws Out Lea County Water Lawsuit," *Associated Press State & Local Wire*, November 9, PM Cycle 1998.

[51] See 510 U.S. 126 (1993) modifying 501 U.S. 221 (1991) *Texas & Oklahoma v. New Mexico* on the Canadian River Compact and Texas' ongoing battle with Mexico over the latter's

shortages of Rio Grande water (See John Burnett, *US-Mexico Water* [Real Audio], National Public Radio, Segment from *All Things Considered*, June 8, 2000 [cited June 10 2000]. Available from http:/search.npr.org/cf/cmn/cmnps05fm.cfm?SegID=75196.

[52] "Judge Throws Out Lea County Water Lawsuit," *Associated Press State & Local Wire*, November 9, PM Cycle 1998.

Chapter Seven

THE SNOWMASS CREEK CASE

High Country Tradeoffs

Case Researcher: Kristine Crandall

> The West is defined...by inadequate rainfall...We can't create water, or increase the supply. We can only hold back and redistribute what there is.
>
> *Wallace Stegner, 1987*

Note: This case report illustrates the use of a particular methodological framework. It is not intended to replicate the legal and historical coverage of this case provided in other sources.

Introduction: This case study presents a compelling example of how a local instream flow debate, spurred by increased snowmaking demands, led to state-level policy challenges and changes. Snowmass Creek supplies water to the Brush Creek Valley, which includes the Snowmass Ski Area and Town of Snowmass Village, about seven miles southwest of Aspen, Colorado. The relevant areas of the central Colorado mountains are shown in Figure 7.1.

Time period: 1992-1996 (with updates through revisions in 2000)

Basic nature of dispute: Water rights, instream flow protection.

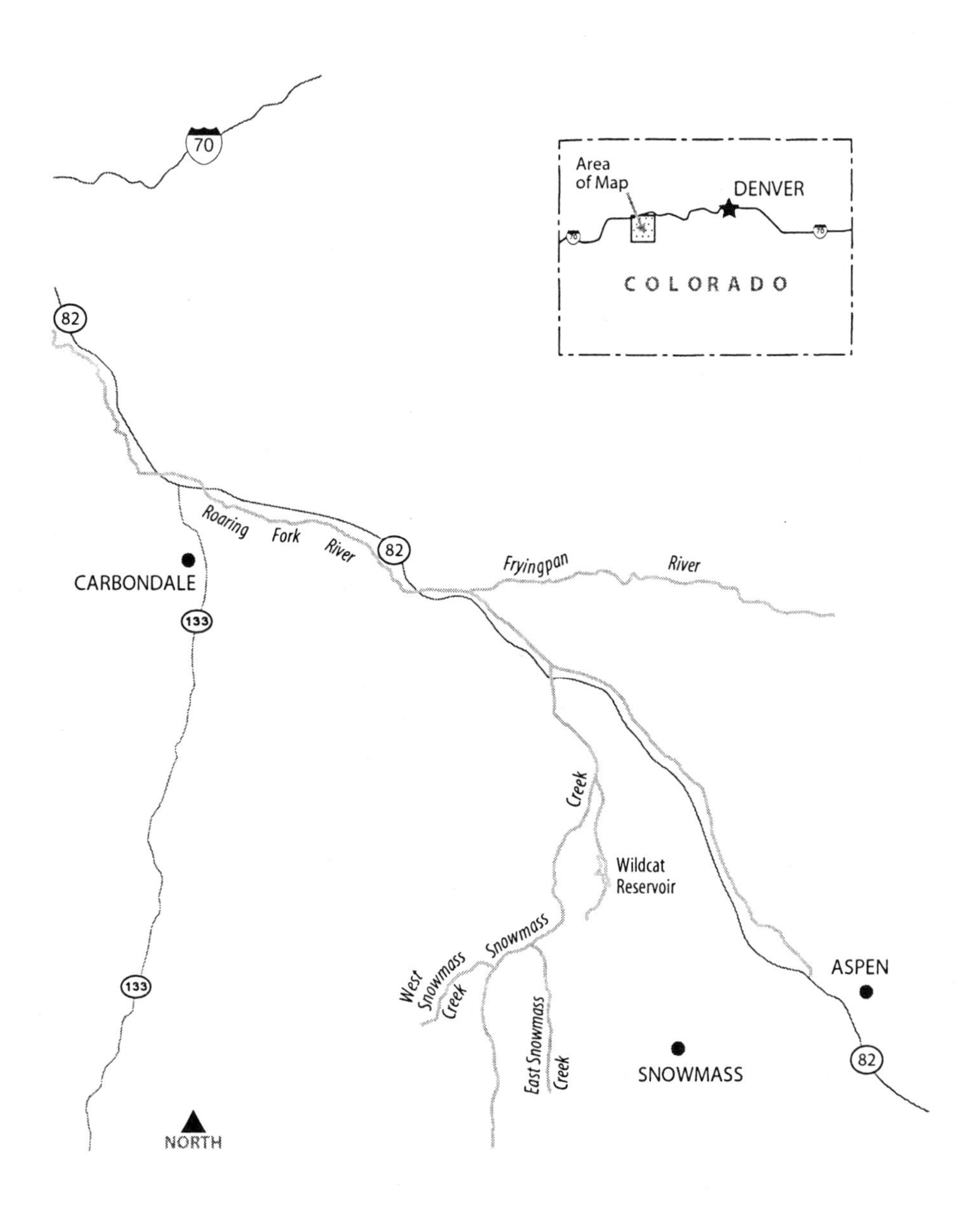

Figure 7.1 Snowmass Creek Case

Issues: The Aspen Skiing Company and Colorado Division of Wildlife attempted to change the standard for minimum instream flow from the level adjudicated in 1976 by the Colorado Water Conservation Board.

Actors and interests:

- **Aspen Skiing Company (SkiCo)**, *Interests:* owns and operates the four ski areas at Aspen/Snowmass.
- **Snowmass/Capitol Creek Caucus (SCCC)**, *Interests:* a group of residents of the Snowmass/Capitol Creek valley advocating for protection of the valleys' environment and rural character.
- **Colorado Division of Wildlife (CDOW)**, *Interests:* the state agency that manages wildlife resources.
- **Aspen Wilderness Workshop (AWW)**, *Interests:* a long-established local group that supports conservation of the natural environment within the general Aspen area.
- **Colorado Water Conservation Board (CWCB)**, *Interests:* the state board that aids in the protection and development of waters for the State, a function that includes appropriation, acquisition and protection of instream flow and natural lake level water rights.
- **Snowmass Water and Sanitation District (SWSD)**, *Interests:* a quasi-municipal organization that supplies water and manages wastewater within its district.
- **Town of Snowmass Village**, *Interests:* not directly involved as a party, but generally supportive of increased revenues from ski tourists.

Attempted conflict resolution processes: The Colorado Water Conservation Board (CWCB) is the state board that decrees, holds, and protects instream flow water rights in Colorado. In 1992, the Aspen Skiing Company (SkiCo) and Colorado Division of Wildlife (CDOW) proposed a modification in an existing instream flow water right. They wanted to lower the minimum instream flow standard for the middle section of Snowmass Creek during winter months. The CWCB held four public hearings to address this proposed change. The Aspen Wilderness Workshop (AWW) and Snowmass/Capitol Creek Caucus (SCCC) participated in some of these hearings and registered their opposition to the proposed instream flow modification. The CWCB, acting as a mediator, evaluated the information and recommended a lower instream flow requirement. This case marked the first time the CWCB modified one of its existing instream flow rights.

The AWW began the litigation phase of the conflict resolution process with an appeal of the CWCB's decision. Eventually, the case ended up in the Colorado Supreme Court, which overturned the decision.

This was followed by the passage of Senate Bill 64, which revised the process by which the CWCB can alter minimum instream flow rights. This bill is referred to as the "legislative agreement" in the case study. Pursuant to the new codified procedures of the legislative agreement, a follow-up agreement was negotiated by the parties. This follow-up agreement, which includes the specific instream flow levels that are to be maintained in Snowmass Creek, is referred to as the "negotiated agreement."

Specific Outcome analyzed: The 1996 Legislative Agreement and the subsequent Negotiated Agreement.

History of the Snowmass Creek Conflict

1976: The CWCB appropriates an instream flow right of 12 cubic feet per second (cfs) for Snowmass Creek (officially decreed in 1980).

1978: The SWSD signs agreements with Pitkin County and the SCCC, incorporating the protection of Snowmass Creek minimum streamflows into its operations.

1991: In August, the SkiCo submits a proposal to the USFS for a master development plan and special use permit for Burnt Mountain expansion of Snowmass area.

1991-92: The Aspen Skiing Co. seeks to establish Snowmass Creek as a source of future snowmaking diversions. It commissions the Chadwick aquatic study.

1992: On September 14, the CWCB recommends modified instream flows, including a reduction to 7cfs winter flow for the 'middle' reach of Snowmass Creek. The next day, the AWW files a lawsuit to oppose the new recommendations.

1993: In July, a Colorado District Court upholds the CWCB's decision. The AWW appeals to the State Supreme Court.

1994: March/November: The USFS issues a Record of Decision on the Burnt Mountain expansion, giving the official "go ahead."

1995: In June, the State Supreme Court overturns the CWCB decision and argues that final rulings on instream flow should be made in water court.

1996: In May, the Colorado legislature passes Senate Bill 64. It requires adequate grounds for modifying instream flow rights, a 60-day public notice of requests for modification, and fair and formal procedures for modification hearings.

In July, new instream flows are set by the CWCB after negotiations between the CWCB, the SkiCo, and the SCCC/AWW. In September, the SCCC activates a January 1995 lawsuit (held in abeyance) against the SWSD that claims the latter violated the '78 agreement by expanding its use of creek water beyond the service plan (e.g. for snowmaking) without addressing the possible use of alternative water sources.

1999: In February, the lawsuit filed against the SWSD is dismissed.

2000: In July, a coalition of environmental groups, including the SCCC and the AWW, petition the Army Corps of Engineers to review the conditions of the 404 permit issued in 1978 for the Snowmass Creek Pipeline diversion.

Background

Snowmass Creek is a natural tributary of the Roaring Fork River, located in Pitkin County in west-central Colorado. As shown in Figure 7.1, it originates in the Maroon Bells/Snowmass Wilderness and joins the Roaring Fork River 17 miles downstream, at the town of Old Snowmass. The Snowmass Creek valley is mainly rural residential with a few ranching operations. Part of the Snowmass Ski Area, operated by the Aspen Skiing Company (SkiCo), extends into the basin from the Brush Creek valley to the east. There are a number of irrigation diversions on the creek that are used in the summer for agricultural purposes, but the main structure that diverts water year-round is the diversion dam and pipeline operated by the Snowmass Water and Sanitation District (SWSD). The SWSD is a quasi-municipal corporation that provides water and sewage treatment services to properties within its service area of Snowmass Village, located in the Brush Creek valley.

Snowmass Creek contains populations of brook, brown, and rainbow trout. It does not see much angling activity since it is surrounded by private land. The Roaring Fork River has been recognized for the excellent quality of its fishery. It has been classified a "Gold Medal" water by the Colorado Division of Wildlife (CDOW).

For over 20 years, there have been efforts to preserve the natural flows and ecosystem of Snowmass Creek. This case study focuses specifically on the modification of the instream flow right on Snowmass Creek – a process that started in 1992 and was resolved in 1996. An important agreement (the '78 Agreement) remains at the heart of much of the controversy over the creek's stream flows.

In 1978, Pitkin County and the Snowmass/Capitol Creek Caucus (SCCC), a group of residents of the Snowmass/Capitol Creek Valley, each developed agreements outlining specific measures that the SWSD would take to protect the instream flow levels of Snowmass Creek (the two agreements are referred to collectively as the '78 Agreement). The County's goal was to ensure compliance with its land-use code, which does not allow development to negatively impact streams, while the SCCC sought to protect the creek's natural ecology.

At the time of the '78 Agreement, the SWSD's main water supply came from East Snowmass Creek (via a pipeline to Brush Creek), but it also held senior water rights lower in the basin on Snowmass Creek. The SWSD was proposing to construct a diversion structure and pipeline to carry supplemental municipal/domestic water for its customers in Snowmass Village (the Snowmass Creek Pipeline). The diversion required approval from the Army Corps of Engineers (Corps) in the form of a 404 permit, because of the dredge and fill activities associated with construction of a diversion dam. The Corps was also concerned about the potential adverse environmental effects of the displacement of water from the Snowmass Creek basin. The '78 Agreement recognized the 12 cubic-feet per second (cfs) minimum streamflow adjudicated by the Colorado Water Conservation Board (CWCB) in 1976. It included agreed-upon methods to protect the natural fishery and ecology of Snowmass Creek, which restricted the SWSD's water rights. The '78 Agreement contained several stipulations directing the SWSD to first maximize water sources other than Snowmass Creek. For example, before drawing the creek down below the minimum flow level, the SWSD agreed that it would use all available water supplies from its other facilities and sources. According to the agreement, the implementation of water conservation practices was another environmentally acceptable means of enhancing existing water. It stated that the SWSD would continue to implement a water conservation program and consider other conservation methods within its discretion.

A third clause addressed the location of additional water sources, whereby the SWSD would agree to undertake a study to assess alternative plans that it could implement to avoid reducing the flow of Snowmass Creek through its diversion through the Snowmass Creek Pipeline. The SWSD, according to the agreement, would pursue alternatives, to the extent reasonably practicable, in order to prevent Snowmass Creek from falling to a level (less than the minimum instream flow) below which the survival of the existing fishery is threatened. The '78 Agreement was not a focal point of the dispute covered in this case study, but it has been the source of a longer controversy which is further described in Part V. Perhaps most importantly, the '78 Agreement provides the general conditions under which the Corps granted the 404 permit to allow the SWSD's construction of the Snowmass Creek diversion and pipeline.

The establishment of instream flow rights is a relatively new way of thinking about water use. Water allocation in the West rests on the "Prior Appropriation Doctrine." This provides water rights for "beneficial use," that is, consumptive use that takes water out of streams and rivers. In more recent times – as agriculture, cities, industry, hydropower production, and flood control operations have dammed, tapped into and re-routed the flowing oases of the west – natural ecosystems supported by flowing streams have suffered. To address this problem, many western states, including Colorado, have developed water policies that include the preservation of instream flows. Montana has taken a very strict policy line on this issue by not allowing any new depletion of winter stream flows.

The Aspen/Snowmass area experiences typical climatic cycles that include dry or late winters. Such winters, without the benefit of snowmaking, result in later opening days and poor skiing conditions during Christmas, the busiest time of the year. This leads to reduced income for ski resorts and nearby businesses that rely on skiers. In an effort to remain competitive with other ski resorts, many of which already had snowmaking capabilities, the SkiCo formulated a specific plan in 1991 for snowmaking at the Snowmass Ski Area. It looked at Snowmass Creek as an ideal water source, and the SWSD agreed to convey extra water to the SkiCo through its Snowmass Creek diversion.

The caveat to the SkiCo's strong interest in obtaining water for snowmaking was that it was planning expansion of the Snowmass Ski Area onto Burnt Mountain. In August of 1991, the SkiCo submitted to the United States Forest Service (USFS) its formal proposal for the expansion. The proposal covered snowmaking activities, which would be supported by the SWSD's provision of raw water from Snowmass Creek.

In response to the SkiCo's interest in using Snowmass Creek water for snowmaking, the Colorado Division of Wildlife (CDOW) was prompted to re-evaluate the creek's streamflows and aquatic habitat. The CDOW is the agency that evaluates stream habitat and fishery requirements for CWCB instream flow adjudications. It had established the 12cfs instream flow level in 1976. The SkiCo hired an aquatic consultant to coordinate with the CDOW's efforts. Together, the CDOW and SkiCo consultant determined that winter flows of 7cfs would be sufficient to protect the fishery in the middle reach of the creek. In justifying this change, the CDOW claimed that the 1976 instream flow level was based on a calculation error.

The SkiCo believed this would allow it to withdraw the difference (5cfs) for snowmaking. In early 1992, based on the CDOW's conclusion and the SkiCo's request, the CWCB began a proceeding to modify the Snowmass Creek minimum streamflows, including a proposal to lower the winter flows in the middle reach from 12cfs to 7cfs.

The Colorado Water Conservation Board (CWCB) is the state agency authorized by statute to adjudicate minimum streamflows in order to "preserve the natural environment to a reasonable degree." The CWCB's establishment of the 12cfs minimum streamflow for Snowmass Creek in 1976 was intended to fulfill this objective. The relationship between the CWCB minimum streamflow and the '78 Agreement is complicated. Although the CWCB's Snowmass Creek minimum streamflow right is junior to any pre-1976 rights (e.g. the SWSD's senior rights), the '78 Agreement requires the SWSD to take the referenced measures to preserve the CWCB's minimum instream flows.

After learning of the CWCB's proposed instream flow modifications from a newspaper article a week before the first hearing on the matter, the SCCC and Aspen Wilderness Workshop (AWW), an environmental group based in the Roaring Fork Valley, became involved. The SkiCo, the CDOW, and the CWCB had not initiated any public forum for discussion of the proposal. Through the hearing process, the SCCC and AWW asked how the CWCB could choose not to uphold an established instream flow right. They sought more in-depth environmental studies of the potential impacts of lower streamflows in the winter.

After three more hearings on the issue, a process that lasted over a six-month period, the CWCB formally recommended a lower winter minimum instream flow for the middle reach of Snowmass Creek. During the hearings, the SCCC and AWW had requested that the CWCB postpone any decision until an adequate evaluation of the stream's ecology and the potential impacts of reduced winter flows could be studied (which would take a winter season). The AWW appealed the ruling, with help from several environmental attorneys who donated their time at the request of the Land and Water Fund of the Rockies. The AWW took the lead in the litigation, working together with the SCCC. After the Colorado District Court upheld the CWCB's decision, these two groups took the matter one step further by appealing to the state's Supreme Court. In the meantime, the Burnt Mountain expansion received the USFS's approval in 1994. The expansion included proposed snowmaking operations, but without a resolution to the issue of water availability.

The Colorado Supreme Court ruled in favor of the AWW in 1995. This ruling reversed the CWCB's instream flow modification on Snowmass Creek. It stated that the CWCB cannot unilaterally reduce instream flow rights and that such issues should instead go to water court for determination. The SkiCo then responded by lobbying the Colorado State Legislature to pass legislation that would overturn the Supreme Court's decision, giving sole authority to make instream flow modifications back to the CWCB. In the meantime, during the winter of 1994/95, the SCCC hired an aquatic consultant to study the winter ecology of Snowmass Creek. Specifically, they wanted to study the potential impacts of lower winter streamflows on the survival of incubating trout eggs.

In 1996, legislative action reinforced the Supreme Court's ruling. Senate Bill 64 (SB 64) was signed into law by Governor Roy Romer in May 1996. It required the CWCB to develop "grounds for modification" of an existing flow right, to properly notify the public at least 60 days in advance of a "request for modification," and to establish fair and formal procedures for the hearings held to address such requests for modification. As part of the fair and formal procedures, all evidence supporting a modification must be open to cross-examination and disclosure, and if appropriate concerns are raised about a reduced flow's impacts on the environment, one year must be allowed to study the possible impacts further[1]. Under the new law, the CWCB still presided over proposed changes to existing instream flow rights, but final rulings were to be made in local water court. The new rules were tested immediately, when the Snowmass Creek case was reheard in September, 1996 before the CWCB. In light of the stream ecology findings of SCCC's aquatic consultant, which raised concerns about the ability of the creek's trout population to successfully spawn and overwinter with lower winter flows, a follow-up negotiated agreement was struck between the CWCB, CDOW, SkiCo and SCCC[2]. The SWSD was absent in these negotiations. It is assumed that the SkiCo, one of its largest customers, represented its interests and that it chose not to become directly involved because the SkiCo took the lead.

The negotiated agreement established seasonally variable streamflows that reflect the survival needs of the fishery in different stream reaches. Table 7.1 summarizes the instream flow modifications negotiated for the creek's middle reach. For example, in a fairly normal hydrologic (moisture) year, which in

Percentile Water Year	Recurrence Interval	Instream Flow Recommendations
50th% or Greater	1:2	12 cfs (10/16 – 11/30) 10 cfs (12/1 – 3/31)
25th% to 50th%	1:4 to 1:2	12 cfs (10/16 – 10/31) 10 cfs (11/1 – 12/14) 9 cfs (12/15 – 12/31) 10 cfs (1/1 – 3/31)
10th% to 25th%	1:10 to 1:4	12 cfs (10/16 – 10/31) 10 cfs (11/1 – 11/14) 9 cfs (11/15 – 12/21) 8.5 vfs (12/22 – 12/28) 8 cfs (12/29 – 12/31) 9 cfs (1/1 – 3/31)
Less than 10th%	1:10	9 cfs (10/16 – 10/21) 8 cfs (10/22 – 10/31) 7 cfs (11/1 – 12/31) 8 cfs (1/1 – 3/31)

Table 7.1 **Final Snowmass Creek Instream Flow Modifications: Middle Reach** (Source: CWCB, 1996[3])

this case occurs one out of every two years, the minimum flow recommendation for the middle reach of the creek is 12cfs between October 16 and November 30, and 10cfs between December 1 and March 31. Along this same middle reach, the lowest flow level in the new decision is a 7cfs flow between November 1 and December 31 in a "less than one in ten" moisture year (the driest scenario). In three out of four years, the new flow rights maintain at least a 9cfs minimum flow. As of 1998, the Snowmass Ski Area had snowmaking equipment serving 130 of its 2,580 acres. The availability of water for snowmaking on these acres depends on the moisture year, as determined by the negotiated agreement. The Burnt Mountain expansion is in progress.

The 1996 negotiated agreement that resulted from the process established in SB 64 ended the battle over the junior instream flow rights, but the SWSD is presently (as of 2000) attempting to claim that the '78 Agreement does not restrict its senior water rights on Snowmass Creek. As already noted, the SWSD did not participate in the conflict resolution process over instream flow modifications, and continued to claim its senior water rights in Snowmass Creek. It is not directly known how the SWSD reacted to the negotiated agreement, but since it believes that its senior rights have priority over any junior instream flow rights, it most likely viewed the negotiated agreement as outside the scope of its water diversion activities. The SCCC sued the SWSD in 1996 over non-compliance with the '78 Agreement. In February 1999, the Pitkin County District Court ruled that the SWSD had not violated any conditions of the '78 Agreement (Case No. 95CV16). In July 2000, the Public Counsel of the Rockies filed a petition with the Corps, claiming that the conditions of the 404 permit granted to the SWSD in 1978 for its Snowmass Creek diversion and pipeline were not being satisfied.

I. CRITERIA: OUTCOME REACHED

A. Unanimity or Consensus

The legislative agreement was not unanimous. The SkiCo lobbied the State legislature to introduce a bill overriding the Supreme Court's decision. According to the local press, the Snowmass Village town council gave preliminary approval to contribute $30,000 to this lobbying effort.[4] While this occurred before the legislative agreement was established, it indicates that the SkiCo did not support the reversal of the CWCB's decision, which was upheld in the form of the legislative agreement. In addition, some parties were not present at the table. A party that was missing throughout the conflict resolution process was the Snowmass Water and Sanitation District (SWSD), since the SkiCo represented its interests. One press article states that the SWSD and SkiCo spent $400,000 on attorneys and other fees during the time of this case study, and through to the dismissal of the SCCC's eventual lawsuit against the

SWSD. The SkiCo is reported to have covered half of this cost.[5] During the time period of the conflict resolution process described in this case study, the SWSD went on record emphasizing its senior water rights on Snowmass Creek – rights which it felt were not restricted by the '78 Agreement. Therefore, one reason the SWSD did not involve itself directly in this hearing process was its own feeling that its senior rights superseded any junior instream flow rights, though the latter were intended to protect the natural environment. This is discussed in more detail in Part IV.

B. Verifiable Terms

The legislative agreement was passed by the State legislature and formally signed. The negotiated agreement was formally signed by the participating parties and ratified by the water court.

C. Public Acknowledgement of Outcome

There was public acknowledgement of both the legislative and negotiated agreements. Press articles and publicly accessible legal briefings covered the legislative agreement.[6] The negotiated agreement received less exposure in the press, but was documented through the CWCB's News and Information website.

D. Ratification

The legislative agreement was ratified by the State legislature on May 23, 1996 and signed into law by Governor Romer. The negotiated agreement was signed by the parties, and was filed with the District 5 Water Court in July 1996 (Case # W-2943). No court was required to approve the legislative agreement. However, one of the important requirements of Senate Bill 64 is that any proposed instream flow right modification made by the CWCB must be filed in writing with the water court. It also allows any member of the public who attends the CWCB hearing to petition for judicial review of the CWCB's decision regarding instream flow right modifications with the water court. Before passage of SB 64, the CWCB was the sole reviewer of proposed modifications of water rights. Thus, the agreement introduces a more formal and active role for the judicial branch to review modifications.

II. CRITERIA: PROCESS QUALITY

A. Procedurally Just

The AWA's appeal of the CWCB decision was based on its view that the CWCB hearing process did not allow publicly requested environmental studies to be done. As part of the public hearing process before the CWCB, the SCCC asked for sufficient time to do winter studies on Snowmass Creek to determine the impact the lower minimum instream flow standard would have on the fishery.[7] Additional time was not granted by the CWCB, which made its recommendation on September 14, 1992 – about six months after the first hearing. In 1995, when the Colorado Supreme Court voted to overturn the CWCB's decision, Sue Helm of the SCCC was quoted as saying: "It's about process, but the big thing to us is all the information will be presented."[8]

B. Procedurally Accessible and Inclusive

1. Public Participation

There were four public hearings held before the CWCB in 1992 to address proposed modifications to the 1980 Snowmass Creek instream flow decree. All the main actors, including the SCCC, AWW, CDOW, SkiCo, and CWCB, participated in the hearing process.

2. Public Access to Information on Issues

The SCCC and AWW did not attend the first public hearing because they did not know about it in time. Public notification of the proposed instream flow modification on Snowmass Creek appeared in the newspaper a week before the first hearing. It is not known whether, and when, such notification appeared in other public forums.

3. Access to Technical and Substantive Information on Issues

It is telling that parties on each side of the debate engaged in separate scientific studies. The SkiCo and SCCC hired separate aquatic consultants to determine limiting habitat conditions for the Snowmass Creek fishery. The CDOW collaborated with the SkiCo, while the SCCC's research was performed independently. Results were shared among the parties, however they provided no basis for agreement until the final compromise was reached in the form of the negotiated agreement's streamflow conditions.

4. Public Education

Less education-oriented and more activist in nature, the SCCC produced two brochures explaining its perspective on the '78 Agreement. Both brochures were publicly disseminated and contained language from the agreement with a presentation of the disputed winter streamflows and the SCCC's view of impacts on fish survivability. The stakeholders did not hold any known public forums or education campaigns.

C. Reasonable Process Costs

1. Costs to Stakeholders in the Process

A study was done by the United Neighborhoods Fund (UNF) in 1996 to examine the cost of the Snowmass Creek conflict to the community and to seek possible solutions. It was estimated, according to information in the local media, that on the side of the SCCC and AWW, pro bono attorney fees amounted to $350,000, paid attorney and consultant fees were $420,000, and volunteer time was valued at $360,000 (13,000 hours)[9].

The SkiCo conducted its own aquatic study and had legal counsel throughout the conflict resolution process. As noted earlier, it has been reported that the SkiCo expended $200,000 and the SWSD another $200,000 on legal and other fees throughout the conflict.

2. Costs to Taxpayers

The UNF study estimated that Pitkin County spent $85,000 in attorney fees. Information is not available on court costs for the conflict resolution process. These were incurred at the state (District and Supreme Courts) level, and also include the time and resources spent by the Colorado Attorney General in representing the CWCB. The Town of Snowmass Village earmarked $30,000 taxpayer dollars to lobby against the legislative agreement.

III. CRITERIA CATEGORY: OUTCOME QUALITY

A. Cost-Effective Implementation

This section describes the costs of implementing the agreement beginning with the first CWCB hearing in 1992 and ending with the negotiated agreement made in September 1996 under the provisions of the legislative agreement. The potential future costs are described in Part III.C., the Community Self-Determination/Sovereignty section.

1. Costs to parties at table

It is assumed that the CWCB bears the extra costs that might arise in altering its public process for changes to the proposed instream flow water rights. The Colorado Water Court will also experience additional costs, borne by Colorado taxpayers and through fees paid by parties appearing in water court proceedings.

There is evidence that parties actually have the ability to pay their share of the costs. These costs arise from activities already performed by the parties. The SkiCo (and the SWSD) will experience opportunity costs into the future because of the inability to use as much water for snowmaking as it might have used without the legislative and negotiated agreements.

Economic effects:

The SkiCo may have had other water sources to explore for its snowmaking, but this was not evident at the time of the conflict. The legislative agreement halted the SkiCo's attempt to secure water for its snowmaking operation from Snowmass Creek, thus preventing expansion of the economic activity cited above. However, the negotiated agreement achieved a compromise that made water available for snowmaking. The SkiCo originally was attempting to gain rights to more water than is available.

2. Costs to public

The most identifiable economic effect of the legislative and negotiated agreements relates to the certainty of ski business provided by snowmaking. According to one newspaper article, the SkiCo's proposed snowmaking capabilities would help draw skiers to the resort for 20 additional days in the fall and 10 days in the spring. In addition, according to the SkiCo, the proposed Burnt Mountain expansion could be financially justified only if snowmaking was included with it.[10]

The article cites a study done by a 70-member citizens' committee for Snowmass Village, which concluded that: "Occupancy in the town's 2,700 rooms could increase by 30 percent for that time, drawing nearly $2 million more in potential gross revenues." The study assumed that tourists spend $1.30 for each $1 spent on lodging, therefore $2.5 million in additional gross revenues could be generated in the retail and restaurant sectors. It was estimated that the SkiCo would bring in an additional $950,000 in gross sales.[11]

In another article, Bill Lund, owner of Oxbow Outfitting in Snowmass Village, noted that the success of retailers in Snowmass Village is directly related to the number of skier days posted at the resort. He stated that, "If skier days are up 5 percent, we'll be up 5 percent – we're very closely aligned."[12] It is important to realize that these are gross economic impacts for the local economy. They do not include the costs associated with servicing additional skiers or the cost (approximately $7 million) of the SkiCo's snowmaking system. They also do not include costs to the community for additional ski-area business, such as additional medical and bus services, as well as traffic congestion and pollution. As indicated above, the increases in gross revenues are assumed to accrue primarily within Snowmass Village. However, Aspen would reap some of these economic benefits as well, as some Snowmass Ski Area visitors stay in Aspen, the location of the SkiCo's main headquarters.

3. Costs to other parties not at table

It is difficult to measure the lost potential gross revenues to the ski industry since the agreement is so recent that there is no clear information on the instream flow levels' influence on snowmaking days. According to the negotiated agreement, in an extremely dry year (one that occurs less than 5 percent of the time) there would be no water available to divert for snowmaking purposes.[13] This is the only scenario in which the SkiCo would have no water for snowmaking, and that is when the ski industry would bear the full cost of missed economic opportunities.

The SkiCo constructed snowmaking facilities on existing terrain in 1996, and presently can provide snowmaking on 130 acres of the Snowmass Ski Area. The main goal is to be able to provide an 18-inch artificial snow base across this acreage by Christmas. The availability of water is a major constraint on this operation. The SkiCo's expanded snowmaking capabilities are reported to have saved two out of the three early winter seasons after the negotiated agreement took effect.[14] This represents a significant preservation of ski industry revenues. During the 1998/1999 winter season, abnormally warm temperatures hampered the effectiveness of snowmaking. At the Snowmass Ski Area, 50 percent of the snow generated from the 1.5 million gallons of water before Thanksgiving melted. This raises a question about the certainty of economic revenues that are generally assumed to be attached to snowmaking activities. Even with snowmaking operations in place, only a small fraction of Snowmass opened by Thanksgiving Day during the 1998/1999 season.

It should also be noted that the preservation of instream flows in Snowmass Creek maintains or increases property values for residents in the Snowmass Creek Valley. A degradation of the creek's natural ecology through reduced streamflows would likely have a negative effect on property values within the valley.

See Part IV for a discussion of the costs resulting from the negotiated agreement, the final result of the conflict resolution process.

B. Financial Feasibility/Sustainability

The State assumed the role of monitoring and enforcing the modified instream flow set out in the negotiated agreement. The Colorado Department of Water Resources is charged with these tasks, and the CWCB is obligated to legally uphold the negotiated agreement's terms. Monitoring and compliance are covered in more detail in Section IV.

C. Cultural Sustainability/Community Self-Determination

Comparative information on longer-term socio-economic variables is available for the greater Snowmass Village area. However, changes in these cannot be linked to the legislative agreement or to the subsequent follow-up agreement. The area is a growing winter and summer resort with economic activities that cannot be separated for purposes of looking at the agreement's effects. In addition, only three years have elapsed since the legislative and follow-up agreements, which is not enough time for a meaningful measure of the change in socio-economic variables.

The legislative agreement influences the decision-making authority/-jurisdiction for altering instream flow rights. Before the agreement, it was possible for the CWCB to unilaterally modify an existing instream flow right without judicial review. With the legislative agreement, proposed modifications can be challenged by the public in local water court, and the CWCB hearing process can be lengthened depending on concerns raised by members of the public who attend the hearings. The decision-making authority for proposed instream flow modifications has thus been amended to include local interests in the form of the water court and can be influenced to a greater extent by local public input.

D. Environmental Sustainability

The legislative agreement specifies procedural requirements for the lowering of existing instream flow rights, but does not address natural contingencies. However, the legislative agreement was put to immediate use when the parties reconvened in the follow-up process and developed a specific modification of Snowmass Creek's instream flows to account for drought years as well as for seasonal streamflow patterns. This follow-up negotiated agreement is described in more detail in the Overview and in Part IV.

The follow-up negotiated agreement sets minimum instream flow requirements, as described in Table 7.1, and Part IV. This essentially commits water to instream flow maintenance, with the quantity depending on the type of moisture year. The Overview and Part IV describe the instream flow minimums and water quantity available for snowmaking under the negotiated agreement.

The AWW and SCCC appealed the CWCB's initial recommendation to lower Snowmass Creek's winter instream flow right because they felt there was a lack of environmental data. The SkiCO, CDOW, and SCCC made assessments of the lower winter streamflows on the creek's fishery and addressed environmental impacts in this manner. There was controversy over the various conclusions of these assessments.

The AWW and SCCC can be classified as environmental groups. The SCCC is also a community group that focuses on issues within the Snowmass/Capitol Creek drainages. The SCCC commissioned its own scientific study on the effect of lower streamflows on Snowmass Creek's aquatic fishery habitat. The CDOW consulted with the SkiCo (which had also done its own aquatic study in cooperation with the CDOW) and the CWCB, and provided its analysis of the impact of lower streamflows on the fishery. The CDOW is the agency responsible for providing biological research and recommendations to the CWCB for proposed new instream flow rights or changes to existing instream flow rights.

E. Clarity of Outcome

There has been no evidence found of later misunderstandings about the agreement. The baseline conditions used as reference points in the agreement seem well-defined. The original 1976 instream flow appropriation was set at 12cfs year-round, establishing the baseline. However, subsequent litigation between the SCCC and the SWSD over senior water rights (senior to the instream flow right) has led to disagreement over the baseline.

F. Feasibility/Realism

1. Legal feasibility

The legislative agreement is consistent with existing law. The negotiated agreement was written in compliance with the legislative agreement.

2. Political feasibility

SB 64 is the legislation needed to allow for the creation of an outcome like the negotiated agreement. It is unknown if representatives from the AWW and the SCCC are ensuring that the provisions of SB 64 are being upheld. Also, there are no known indications that SB 64 is difficult to administer.

3. Scientific and Technical Feasibility

One of the reasons that the CDOW allegedly agreed to lower minimum instream flows on Snowmass Creek was that it made a "computational error" when it first made its flow recommendation in 1976. Scientific studies from both sides became a critical force in the procedural aspects of the proposal to lower flows, and in determining the modified flows contained within the negotiated agreement. In 1996, CWCB Director Chuck Lyle stated:

> Given the complexity of both the scientific and legal issues involved in this case, the state was extremely fortunate to be working with highly professional and technically competent advocates on both sides who were willing to negotiate in good faith to arrive at what we believe to be a recommendation that is based on the best available science and achieves that delicate balance between human and environmental needs.

The most recent effort by environmental groups to restrict the SWSD's Snowmass Creek diversions is seeking the negotiated agreement as the outcome, based on its scientific credibility.

G. Public Acceptability

An editorial by Al Knight in the *Denver Post* criticized the lead-up to the legislative agreement (the Colorado Supreme Court's overturning of the CWCB's initial proposed Snowmass Creek instream flow modification). He argued that the ruling, which requires that instream flow modifications go through the water court, will lead to a more confusing, expensive, and time-consuming process.[15] Several letters to the editor rebutted his comments. No information was found that specifically addressed the public's perception of the two agreement outcomes.

IV. CRITERIA CATEGORY: RELATIONSHIP OF PARTIES TO OUTCOME

A. Satisfaction/Fairness

There is little information on the perceptions of the agreement, as it is defined for this case (i.e. Senate Bill 64). One reason is that so much occurred before the passage of Senate Bill 64 that received attention and scrutiny. In response to the 1995 Supreme Court decision, one press article quoted a member of the CWCB as saying, "I'm certain we'll go wherever we need to go, to water court or to the legislature. I think we'll do what we need to do to have our role clarified."[16] Lori Potter, the Sierra Club Legal Defense Fund (SCLDF) attorney representing the AWW, said in the same article of the Supreme Court ruling: "It's a great victory for Snowmass Creek and for citizens that hold their government accountable." In the article, the SkiCo was noted as having no comment, other than to say it would study the issues and its options. From this and other available information, it appears that the AWW and SCCC felt that the legislative agreement contained a "fair" procedure for instream flow modification.

The legislative agreement was implemented about two months after it was passed, in the form of a specific negotiated modification of Snowmass Creek's instream flow standards utilizing the new procedural requirements. The specifics of this negotiated agreement are covered in the Overview section. In a "News & Information" press release from the CWCB, Lori Potter said of the final instream flow agreement: "This creek is really the big winner here. This is a settlement that protects the stream and ends the controversy."[17] CWCB Director Chuck Lile is quoted as praising the process of good faith negotiation between the parties, and the use of the best available science in achieving a positive outcome.

In this same press release, SkiCo Vice President Fred Smith stated: "Our goal throughout this process has been to improve our snowmaking operations while preserving the environment of Snowmass Creek. We are pleased that the parties have been able to agree upon an approach that will accomplish both of these goals. This will benefit both our community and our environment." And Sue Helm of the SCCC noted: "We're thrilled with the outcome. Our interest has been to protect the ecology of the valley, which includes Snowmass Creek. We feel the stepped hydrographs do balance the needs of man with protection of the environment to a reasonable degree. We hope this agreement in conjunction with the recently passed legislation sets a positive precedent for future modification of instream flows."[18]

Based on these comments, all active parties in this case study were pleased with the outcome. They followed the new procedural guidelines stipulated in the legislative agreement to arrive at this outcome, and felt it was the best solution to the conflict.

B. Compliance with Outcome Over Time

Compliance with the legislative agreement is evident in its immediate implementation through the official development and processing of the negotiated agreement. The legislation specifically states that any instream flow modifications must be filed with the appropriate water court, which provides an additional channel for approval of such proposed modifications.

No information has been found to indicate non-compliance with the negotiated agreement. There was a report of a four-day period in January 1999 when Snowmass Creek streamflows fell below their designated levels based on the instream flow modification (negotiated agreement). The SWSD took responsibility for the low flows, stating that it needed extra water for domestic use, as allowed by its senior water rights. SWSD officials were quoted as saying that the low flow levels were not a result of snowmaking activities in excess of what was allowed by the negotiated agreement.[19] Since the negotiated agreement, under any scenario, mandates a stoppage to snowmaking after December 31st, water diversions that draw Snowmass Creek below the minimum instream flows would presumably be for the domestic system, the SkiCo failed to comply with the agreement. The fact that this issue was questioned in January of 1999, and that the USFS staff investigated the incident, indicates that there remains skepticism within the community about the SkiCo's compliance with the negotiated agreement.

C. Flexibility

The legislative agreement is flexible in the sense that it only governs the process, and allows the introduction of specific scientific recommendations for instream flow and habitat protection, as in with the negotiated agreement. Neither agreement has been modified.

D. Stability/Durability

The legislative agreement contains general provisions for making instream flow modifications in Colorado, so it is difficult to assess the agreement's stability.

The negotiated agreement created a compromise that allowed both for snowmaking and environmental protection of Snowmass Creek. Because of the agreement's stairstep hydrograph, which defines the amount of water available for snowmaking diversion based on four different moisture year scenarios, there are very specific criteria for achieving this balance, which provides for stability. As already noted, the SkiCO, the AWW, the SCCC, the CDOW, and the CWCB embraced this agreement at the time that it was crafted. It is assumed to address the needs of both sides, based on the type of moisture year.

The CWCD is responsible for providing adequate monitoring equipment at the Snowmass Creek pipeline diversion. It has installed a state-of-the-art satellite monitoring station, which will become operational for the 2000/2001 winter season. The satellite gauge, which will generate continuous flow data, will be operated and maintained by the Colorado Department of Water Resources Division 5. Since inception of the negotiated agreement, the Department of Water Resources has participated in measuring streamflows. The day after the specific follow-up agreement was developed between the CWCB, SCCC, AWW, SkiCo and CDOW, the SCCC sued the SWSD for breach of the '78 Agreement. This was precipitated by the SWSD's continuous claim to senior water rights on Snowmass Creek, which it did not feel were affected by the '78 Agreement. The issue has received much attention in the press. In one article, the SWSD Board Chairman, Mike McLarry, was quoted as saying he would bring flows in Snowmass Creek down to dry creekbed "if I had to," (i.e. to service the growing Snowmass Village community). In an *Aspen Times Daily* article, McLarry claimed this quote was taken out of context. The article proceeded to say that he would honor the '78 Agreement with regard to streamflows, although if an emergency arose, he insisted he would "suck" the creek dry.[20] According to the article, members of the AWW and the SCCC expressed shock over McLarry's comments.

The SWSD did not openly participate in the instream flow modification debate, so a new party actively entered the debate at this stage. The resulting lawsuit against the SWSD was ultimately dismissed, with a ruling that it had not violated the terms of the '78 Agreement. The SkiCo was not visibly allied with the SWSD in this new phase of the conflict. With regard to the litigation, Pat O'Donnell, President of the SkiCo, stated in a local newspaper article that "We're bystanders....The skiing company doesn't have any leverage with the (water and sanitation) district."[21]

V. CRITERIA CATEGORY: RELATIONSHIP BETWEEN PARTIES (RELATIONSHIP QUALITY)

A. Reduction in Conflict and Hostility

As of 2000, the situation is escalating between the SCCC and the SWSD, the two parties in the lawsuit decided in 1998, who are now engaged in additional legal proceedings. There is no indication of any hostility between the parties that participated in the instream flow modification conflict. The SkiCo has not been visibly associated with the SWSD in the most recent rounds of litigation, but its allocation of water for snowmaking is threatened if the SWSD loses in any of these proceedings challenging its compliance with the '78 Agreement and 404 permit stipulations. In a newspaper article containing press release information from the coalition of environmental groups taking the latest action against the SWSD, the groups make a point to state that there is enough water to allow for snowmaking during the winter's early season, and they are not out to stop snowmaking activities.[22] An *Aspen Times* editorial highlights the SWSD's unrelenting stance: "District officials have let it be known that they will draw down the creek to whatever level they deem necessary to serve their customers, regardless of the terms of the agreement or the needs of the trout."[23] And the SWSD openly celebrated its lawsuit victory in 1999. Mike McLarry, SWSD's board chair, was quoted as saying: "It's vindication on the district's part. We always knew we were right."[24]

B. Improved Relations

The starting point in this conflict revolves around senior water rights and an agreement made at the time of the Snowmass Creek diversion's approval (the '78 Agreement). The quality of the relationship has changed over time, with the parties involved in the '78 Agreement (the SWSD, AWW, SCCC, and Pitkin County) presently debating its content. Even though it was a written agreement, the fact that these parties have engaged in two legal proceedings shows a lack of trust. The tone of the SWSD has been negatively portrayed in the press (described in Part IV).

C. Cognitive and Affective Shift

In the middle of the conflict resolution process, a Snowmass Village council member was quoted in the media saying: "The residents of Snowmass and Capitol Creek do not see the benefit from ski area expansion." He explains that while snowmaking diversions will make Snowmass Creek less attractive, this use of water is the best decision for the community. He concludes: "I don't think we're going to resolve this thing as friends and neighbors. We're enhancing our property values at the expense of theirs."[25] Although Snowmass Village was not a party in the instream flow modification conflict, this sentiment represents the SkiCo's interests as well. The SCCC is portrayed as inflexible, focused strictly on its neighborhood, and unwilling to support the growing economy of Snowmass Village.

In a brochure to solicit funding to help appeal the CWCB's initial instream flow modification, the SCCC describes the SWSD's interest as shortsighted and driven only by money. The brochure states: "SWSD wants more water. They want to sell it to the SkiCo for snowmaking, and for more development in Snowmass Village. The cheapest way to get water is to divert from Snowmass Creek and pump it over the mountain."[26] Despite these views, the participating parties managed to come up with the negotiated agreement. However, in the long term, these views demonstrate larger differences that have driven the conflict once again into litigation.

D. Ability to Resolve Subsequent Disputes

The fact that the follow-up negotiated agreement was developed shortly after the legislative agreement is a good indication that the parties at the table in this case study proceeded in a constructive manner. They were able to achieve a consensus that required some sacrifice for both sides, which also required a good relationship during the negotiation process. There were no professional mediators involved in the process leading up to the two agreements.

Since the SCCC sued the SWSD in 1996, there have been no signs of constructive methods to solve the conflict over the '78 Agreement. The United Neighborhoods Fund (UNF) entered the conflict in 1995, spending its own privately generated funds to try to develop an alternative water supply that would solve the debate. However, neither the SkiCo nor the SWSD were willing to work toward such a solution.

E. Transformation

The SCCC and other members of the coalition of environmental groups that petitioned the Corps regarding the validity of the 404 permit have acknowledged that there is enough water for early season snowmaking.[27] This reflects an important step in the relationship between the environmental interests and the SkiCo. The continued atmosphere of conflict, however indicates that there has been no progress among the stakeholders (including the SWSD) to resolve this broader debate.

VI. CRITERIA CATEGORY: SOCIAL CAPITAL

A. Enhanced Citizen Capacity to Draw on Collective Potential Resources

1. Aggregate of resources

In 1998, the SkiCo created a Department of Environmental Affairs and initiated the Aspen Skiing Company Environmental Foundation. The latter is a non-profit employee organization dedicated to protecting and preserving the regional environment. This program has contributed significantly to the resources available for addressing community environmental issues and projects. It also has increased citizen participation and the formation of partnerships in such efforts. In its first two years, the Foundation leveraged over

$200,000 for environmental protection in the Roaring Fork Valley. Funds are put toward projects that support environmental education, ecosystem protection, natural resource stewardship, and outdoor recreation. Specific to the area of stream flows, the Foundation has supported water conservation, water quality monitoring, and stream restoration initiatives within the Valley.

While the environmental emphasis did not arise directly from the Snowmass Creek debate, it signified an evolution in the general philosophy of the SkiCo that the natural resource base it relies upon must be preserved. The SkiCo has received several prestigious environmental awards since the inception of its environmental programs, which help its reputation as a "green" ski resort.

The Foundation represents an important collection of resources that are available to various community interests. In this way, it brings together businesses, schools, environmental interests, and governments in collaborative approaches to solving environmental challenges.

The SkiCo's "Green Development" program is one initiative that relates directly to the water conservation issues at the forefront of the Snowmass Creek conflict. This program approaches construction, management, and planning with the goal of reducing the impact of buildings and development on the environment. One of the key considerations in green building is the efficient use of water, energy, and other resources. The SkiCo has also considered environmentally sensitive designs within the Snowmass ski area, through the development of a long-term natural resource management plan that includes wildlife habitat and vegetative diversity. In addition, the Foundation has supported specific stream habitat projects including cutthroat trout restoration, water quality monitoring, and WaterWise school programs.

The CWCB and Department of Water Resources have increased their presence in the area, through increased monitoring capabilities and activity in support of both the negotiated agreement and the instream flow donation made by the Conservation Fund (discussed below). This increased commitment of state resources represents better networking and the availability of more information to the community.

2. Potential assistance relationships

In 1998, the Conservation Fund, in collaboration with the SCCC, donated a summer irrigation right on Snowmass Creek (the lower reach) to the CWCB for instream flow. This donation has been formalized and represents a respectful partnership between actors in the Snowmass Creek conflict, namely the CWCB and SCCC, to enhance instream flows on Snowmass Creek and the Roaring Fork River.

3. Generalized reciprocity

Other than the streamflow monitoring done by the State, and the implementation of the negotiated agreement's stipulations, which are applied to the SkiCo's snowmaking operations, there has been no evidence of regular networking, assistance, or mutual recognition among the participants in the conflict.

B. Increased Community Capacity for Environmental/Policy Decision-Making

1. Aggregate of resources

The parties in the Snowmass Creek debate have not arrived at a point where they share resources, jointly hold or attend meetings, divide tasks, or develop a more unified decision-making structure.

2. Increased System Efficiency

The continuation of the conflict has not created an environment conducive to increased system efficiency.

3. Increased Capacity for Cooperation

The increased capacity for cooperation has been somewhat one-sided in the sense that the environmental interests involved in the issue have expanded. They include the Public Counsel of the Rockies, American Rivers, the Sierra Club, the SCCC, the AWW, Windstar Land Conservancy, the Ferdinand Hayden Chapter of Trout Unlimited, Roaring Fork Audubon, and High Country Citizens Alliance. This coalition represents a mixture of local, regional and national interests that complement each other in providing various functions in the latest stage of the debate. Except for this expansion, there has been little development among the actors in handling the conflict constructively, utilizing broader sources of information, or bringing in objective mediators.

4. Increased System Capacity for Responding to External Challenges

The donation by the Conservation Fund, in collaboration with the SCCC, of the summer irrigation right for instream flow to the CWCB shows the capacity to work together at the outset of a challenging situation. The CWCB is not accustomed to handling such donations, yet the involved partners stuck with the challenge and brought about a positive final outcome. On the other hand, the SWSD's role in the original conflict has become more prominent, which has resumed its role as an adversary against the SCCC's claims.

5. Increased Information Flow

Actual flow data has become more available and accessible as a result of enforcement of the negotiated agreement. In general, instream flow issues in the Roaring Fork Valley have taken on greater importance since the development of the agreements covered in this case study. One result of this is the increased awareness among jurisdictions such as Aspen about the potential effects of snowmaking on other creeks. There is an effort to obtain greater gauge coverage throughout the upper valley to track flow data in relation to this issue.

C. Social System Transformation

The Snowmass Creek issue has led both to a change at the state level, with the development of the legislative and negotiated agreements, and over time has created a level of transformation within the local social system. As previously described, the SkiCo's newly established environmental focus and associated contributions to local environmental causes represents an important support network for the community. Since the agreements were determined, the SkiCo has also invested in additional on-mountain water storage to alleviate its dependency on direct diversions from Snowmass Creek for snowmaking. The general community tone is one of increasing trust of the SkiCo's environmental stewardship philosophy, although close attention continues to be paid to its activities, especially in light of a new proposed residential and commercial development at the base of the Snowmass Ski Area in Snowmass Village.

---

[1] SB 64 is what is referred to as the "legislative agreement" for this case study.

[2] This follow-up agreement, specific to Snowmass Creek, is referred to as the "negotiated agreement" for this case study.

[3] Colorado Water Conservation Board, "Memorandum: Summary Report and Final Recommendation on Snowmass Creek Modifications," (Denver, CO, June 25, 1996).

[4] "Give money to RFTA, not to the SkiCo," *Aspen Times*, November 11-12, 1995.

[5] Jim Pokrandt, "Water District Finds Vindication in Check From Caucus," *Snowmass Village Sun*, June 30-July 6, 1999.

[6] Cameron Burns, Colorado General Assembly; Colorado Environmental Compliance Update, *Denver Post*.

[7] Snowmass/Capitol Creek Caucus, "The Fight to Save Snowmass Creek," October, 1993.

[8] Cameron Burns, "Snowmass Gives Earful to Pitco Commissioner," *Aspen Times Daily*, February 1993.

[9] Jim Pokrandt, "New Solution Eyed For Water Use," *Snowmass Village Sun*, January 1-7, 1997.

[10] Scott Condon, "SkiCo: Snowmaking Must Come First," *Aspen Times*, January 1, 1993.

[11] Ibid.

[12] Janet Urquhart, "Merchants Count on Snowmaking Boost," *Aspen Times*, March 23, 1996.

[13] Colorado Water Conservation Board, "Memorandum: Summary Report and Final Recommendation on Snowmass Creek Modification," Denver, Colorado, June 25, 1996.

[14] Jim Pokrandt, "Water District Finds Vindication in Check From Caucus," *Snowmass Village Sun*, June 30-July 6, 1999.

[15] Editorial, Al Knight, "Snowmass Creek Decision Isn't About Good Versus Evil," *Denver Post*, July 9, 1995.

[16] Cameron Burns, "Snowmass Gives Earful to Pitco Commissioner," *Aspen Times Daily*, February 1993.

[17] Colorado Water Conservation Board, "Memorandum: Summary Report and Final Recommendation on Snowmass Creek Modifications," 1996a.

[18] Colorado Water Conservation Board, "Agreement Reached on Snowmass Creek Streamflow Dispute," 1996b, From *News & Information* section on website, June 28, 1996. http://www.dnr.state.co.us/cdnr_news/cwcb/9703221154958.html.

[19] Jim Pokrandt, "Domestic Uses, Not Snowmaking Spurred Snowmass Creek Draw," *Snowmass Village Sun*, February 17-23, 1999.

[20] Cameron Burns, "Snowmass Gives Earful to Pitco Commissioner," *Aspen Times Daily*, February 1993.

[21] Curtis Robinson, "New Eco Group Causes Some Uneasy Feelings," *Roaring Fork Sunday*, 6-7, January 5-11, 1997.

[22] Scott Condon, "Snowmass Creek Under Attack," *Aspen Times Daily*, 1-A/15-A, July 14, 2000.

[23] Editorial, "The Troubled Waters Of a Beautiful Creek," *Aspen Times Weekly*, 22A, January 4-5, 1997.

[24] Jim Pokrandt, "Water District Finds Vindication in Check From Caucus," *Snowmass Village Sun*, June 30-July 6, 1999.

[25] Cameron Burns, "Snowmass Gives Earful to Pitco Commissioner," *Aspen Times Daily*, February 1993.

[26] Snowmass/Capitol Creek Caucus, "The Fight to Save Snowmass Creek," October 1993.

[27] Scott Condon, "Snowmass Creek Under Attack," *Aspen Times Daily*, 1-A/15-A, July 14, 2000.

Part III

COMPARING THE CASES; TESTING THE FRAMEWORK

Chapter Eight

PATTERNS IN ENVIRONMENTAL CONFLICT RESOLUTION

Comparisons Across Cases

> Coral is set budding under seas,
> Though none, O none sees what patterns it is making?
>
> *Philip Larkin (1922–1986), British poet.*
> *"A Stone Church Damaged by a Bomb."*

In this chapter, we discuss our observations and make note of interesting patterns across the eight cases in our original study, four of which are fully documented in Chapters 4-7. While these cases are not an adequate sample for confidently generalizing at the level of statistical significance about large groups of cases, a comparison among the cases reveals useful patterns and questions for future research. This comparison is organized by criterion. Not all of the twenty-eight criteria are discussed, for two reasons. Either the criterion was not yet in the framework at the time of case analysis, or the criterion was defined, but there was not adequate information to be gathered by case researchers (see Chapter 9 on criterion accessibility).

To further our goal of comparative case evaluation, we found it helpful to give cases ratings or 'grades' on each of the criteria in our framework. These do not represent formal ratings, but provide a summary sense of relative merit and a sense of ranking. Arraying these in Table 8.1 also presents a visual 'report card,' both for the cases (column) and in the sense of distribution on a given criterion (row). Grades of A-E are used, with 'E' the lowest 'grade.'

Not all criteria on which data had been collected had adequate data to permit this summary ranking. We discuss criteria quality further in Chapter 9.

In this chapter, we also offer observations across conflict resolution processes (litigation, negotiation, etc.). These are tentative observations, given the small number of examples of each type of conflict resolution process. However, they suggest areas for further research, exploration by conflict resolution professionals, and strategies for more effective social policies to address conflict.

I. OUTCOME REACHED

This category contains criteria that assess the basic achievement of reaching an outcome in the conflict resolution process. The Outcomes analyzed in the eight cases represent several types of conflict resolution outcomes including: a series of related court rulings in the Big Horn case; an administrative rule issued by a federal agency in the Lower Colorado River case; state legislation (spurred by litigation) in the Edwards Aquifer case; a state agency administrative ruling (also spurred by litigation) in the Mono Lake case; a U.S. Supreme Court ruling and stipulated agreement in the Pecos River case; a negotiated agreement modified and ratified by Congress in the Pyramid Lake and Salt River cases; and state legislation initiated (and followed) by negotiations in the Snowmass Creek case. Most cases involved several different conflict resolution processes that resulted in sequential Outcomes of different types (e.g. litigation and court ruling, followed by negotiations, negotiated agreement and legislation). Nevertheless, the cases can be classified for the purposes of comparison according to the type of Outcome. The Outcomes included two court rulings (Pecos River, Big Horn), two administrative actions (Mono Lake, Lower Colorado River), a pair of negotiated agreements accompanied by federal legislation (Pyramid Lake, Salt River), one negotiated agreement accompanied by state legislation (Snowmass Creek), and one Outcome reached through state legislation (Edwards Aquifer).

It was not easy to decide which Outcome should be the focus of the analysis, given the long history of each case. In general, analyses focused on the most recent formal Outcome as of 1998, when the case analysis process began. Each case involved prior formal Outcomes that were not the focus of analysis. However, these prior court rulings, negotiations, and legislative and administrative acts are included in the case histories. In some cases, formal Outcomes were reached subsequent to the analysis. These may not be reported in this book, due to the necessity of selecting a window of time in which to analyze each case. All of the cases remain active in the sense that at a minimum, issues remain to be worked out regarding implementing the Outcome that we analyze, or resolving an issue that had not been adequately addressed previously. In some cases, litigation, negotiation, or legislation may be ongoing.

A. Unanimity or Consensus

Initially (and throughout our case analysis), the *Unanimity or Consensus* criterion focused on negotiated agreements and sought to assess the strength of approval for the agreement, the degree of dissension at the table, and the absence of key parties. It has been subsequently modified to encompass other forms of Outcomes, such as legislative acts and court rulings.

In three of the four cases that involved legislation (Pyramid Lake, Snowmass Creek, Edwards Aquifer), unanimity or consensus clearly was not achieved since some parties opposed the legislation. There was no active opposition to the federal legislation noted in the Salt River case. While not recorded in our original case analyses, the voting records of the enacting legislative body could indicate whether the legislature was universally supportive or deeply divided over the legislation, and could also reveal regional attitudes toward the legislation.

With respect to court rulings, in at least one of the Big Horn case rulings the five judges were divided over several key issues and wrote separate opinions outlining a majority and a dissenting view.

The ruling of the California State Water Board was unanimous in the Mono Lake case. In the two cases focused on negotiation, there were parties that chose not to attend or to end their participation.

In comparing the cases on the criterion of Unanimity/Consensus, we rank the Salt River negotiated agreement and congressional legislation highest (A), along with the administrative ruling in the Mono Lake case. Not only was the Mono Lake ruling a unanimous decision of the State Board, but it also seemed to reflect consensus-oriented dialogue among the stakeholders. The Lower Colorado River case is ranked B (medium-high), followed by three cases ranked C (medium). Three of the cases (Pyramid Lake, Snowmass Creek, and Edwards Aquifer) made substantial progress in reaching a consensus among disparate stakeholders, even though there was still dissent among them. The Pecos River case (ranked D) also achieved some degree of consensus after active litigation, while the Big Horn case (E) epitomizes the absence of consensus.

B. Verifiable Terms and
C. Public Acknowledgement of Outcome

These criteria assess the level of ambiguity concerning the completion of the process. An acknowledgement that an outcome has been achieved is a basic prerequisite for the closure of a conflict resolution process. The *Verifiable*

Terms criterion seeks to verify that there was consensus on the terms of the Outcome. The subsequent criterion, *Public Acknowledgement of Outcome*, looks for public confirmation that an Outcome was reached.

The Outcomes of all eight cases were formalized as written documents. These documents were made publicly available once the Outcome had been announced (e.g. the legislation had been enacted, the court had ruled). Media sources provided evidence that the Outcome was made public and acknowledged in public forums, such as newspapers and television coverage. There did not appear to be significant differences among the types of Outcomes for these criteria, and they are each ranked medium (C) on this criterion.

D. Ratification

Ratification can take different forms. For instance, the passage of legislation to address an environmental conflict may require court approval. Negotiated agreements generally require ratification by their signatories (e.g., vote of board, referendum or resolution). This action may occur either before or after the Outcome is reached. The *ratification* criterion considers whether the relevant governing bodies have formally approved the Outcomes. This does not include other types of follow-up actions, such as legislative appropriations that provide the money needed to implement an agreement. These follow-up implementation actions are covered under IV. B., Compliance Over Time.

The two cases which featured a court ruling in their Outcomes (Pecos River, Big Horn), did not require a ratification of the court's decision by other bodies of government. The follow-up activities required to comply with these ruling(s) are reported under Compliance (IV.B) and Stability/Durability (IV.D).

Enacted legislation has, by definition, been ratified by a legislative body. Several cases which involved prior court rulings (e.g. Pyramid Lake, Edwards Aquifer) required a legislative Outcome that was consistent with the requirements of the courts. However, the courts do not generally scrutinize the legislation, unless requested to do so by a party that has standing in the legal cases. In the Edwards Aquifer case, the state legislation was reviewed by the courts after parties filed several different lawsuits in objection to the legislation.

Legislation from one level of government may spur related legislation by other governments. This is not considered formal ratification, but does represent a follow-up action by the affected governments, as discussed in IV. B.

Congress ratified the negotiated agreements in the Pyramid Lake and Salt River in a modified form and these are ranked A. The Snowmass case involved a local agreement that needed to be filed with the Colorado Water Court. That filing was not completed until several years after the case was analyzed, which suggests that either lingering issues may have remained or the parties did not appoint someone to undertake this filing in a timely manner.

The absence of ratification by relevant governments may not necessarily indicate a problem. It could simply mean that formal procedures are slow and unintended lapses may occur.

Ratification is not required for administrative rulings, but, as with court rulings, follow-up actions from other governments may be required. As with legislative outcomes, dissatisfied parties may request the courts to evaluate administrative rulings. Such follow-up litigation is reported in IV.D (Outcome Stability/Durability). No cross-case comparisons are provided for ratification because this criterion only applies directly to negotiated agreements.

II. PROCESS QUALITY

This category contains criteria that focus on the quality of the process that was used to achieve an Outcome.

A. Procedurally Just

This criterion seeks to ascertain various *perceptions* concerning the justice of the process. Was the process fair, balanced and complete? Did it thoroughly address the issues and the parties, or did time constraints or power imbalances compromise it? This includes the parties' perceptions of fairness and their satisfaction with the procedure, which research suggests are combined by raters.[1] These measures of procedural fairness are to be distinguished from "Outcome fairness" discussed under the next category, Outcome Quality.

While various types of formal procedures were utilized in each case, few processes escaped criticism. Of the eight cases examined, only one (Mono Lake) seemed to satisfy the parties that the procedure was just in all ways. Legislative Outcomes were perceived as rushed in both cases where they were used to resolve the conflict rather than to ratify previously negotiated agreements (Edwards Aquifer and Snowmass Creek). In contrast, it was specifically noted that administrative processes were not rushed (Mono Lake, Lower Colorado River). Interestingly, there was no evidence of complaints that these administrative processes were "too slow," which would be another potential procedural criticism. Perhaps the formality of their procedures lends credence to the belief that "this is what needs to take place." One litigation case (Pecos River) was seen as long and protracted.

The processes in the Mono Lake and Salt River cases were seen as inclusive and inviting of all views. Cases in which people felt excluded, and which were criticized for the lack of public involvement or debate were mainly negotiations (Pyramid Lake, implementation of Lower Colorado River, implementation of Snowmass Creek), and one case of litigation (Big Horn). This contrast is interesting for a number of reasons. Negotiations often occur behind closed doors to allow parties to consider new ideas without media attention and public commitment. This may give the perception that some parties and/or views have been excluded or neglected. On the other hand, administrative processes are structured to require public comment and to allow for the expression of all views, therefore lending the perception that they have indeed included all views. One negotiated case (Salt River) was a notable exception to this as it was seen as attending to all interests. It would be interesting to determine why the negotiations in this case were perceived as more inclusive than the negotiations in other cases.

Perceptions of bias (Edwards Aquifer, Lower Colorado River implementation, Pyramid Lake, Salt River, Snowmass Creek) seemed to occur more often in cases with negotiation or legislation, while perceptions of fairness were noted in two cases (Pecos River, Mono Lake) that used more formal procedures (litigation, administrative rulemaking). These perceptions of fairness were often affected by the perception of inclusion (Lower Colorado River implementation, Pyramid Lake, Salt River). The more the parties felt they were included in the process, the more likely they were to determine the process to be fair.

Finally, the ability of parties to influence the decision unequally, in short, perceptions of "unequal weight," were also noted (Edwards Aquifer, Lower Colorado River implementation, Pyramid Lake, Salt River). Some might argue that certain parties deserve more weight in a decision, such as the administering agency in a rulemaking procedure. However, it is interesting to note that these cases were not the ones noted as having parties with unequal weight. Rather, the processes of negotiation and legislation, where the structure may be perceived as inviting parties as equals, were seen as unfair. In negotiation there may be an "assumption" or expectation of equal opportunity that is not found in litigation or rulemaking. This assumption can prove to be a source of disillusionment when it does not in fact materialize.

Thus, in comparing across cases, only Mono Lake rates an 'A' for its satisfaction of procedural justice concerns. Lower Colorado River, Salt River, and Pecos River rated a 'B' in that they addressed most procedural justice concerns, though perceptions of unequal weight of some parties' views were noted in the first two, and the latter was perceived as unnecessarily long. Both Pyramid Lake and Big Horn had very mixed reviews on procedure, and thus rated a 'C'. Researchers for the Snowmass Creek and Edwards Aquifer cases documented perceptions of multiple procedural injustices, including haste, bias, and inadequate attention to underlying concerns, and rated a 'D'.

B. Procedurally Accessible and Inclusive

This criterion seeks to ascertain the actual availability of three components that contribute to perceptions of procedural justice. First, did opportunities for public participation exist? Second, did the public have access to information on upcoming opportunities to participate? Third, did the public have access to substantive and technical information on issues? The first asks for details (dates, timing, location, attendance, and effectiveness) of any public hearings, town meetings, surveys, hotlines, citizen boards, or other forms of public outreach and polling. The second component looks for notes on attempts to notify the public, as well as the nature of the contact medium. The third component more specifically addresses public access to information on issues.

In several of the cases (Edwards Aquifer, Lower Colorado River, Mono Lake, Snowmass Creek), opportunities existed for public participation in decision-making, or at least for public "input." Certain procedures actually avoid public comment. Litigation procedures seem to rule out public comment, though hearings may be open for observation and input is sometimes accepted through briefs (Big Horn, Pecos River). Negotiations may intentionally be "kept discrete" or closed to public view (Salt River, Lower Colorado River), or they may 'unintentionally' reduce input by failing to adequately notify parties or the public (Snowmass Creek).

The efforts to notify the public varied widely. Here 'the public' is defined as local citizens who may or may not be direct stakeholders. Administrative rulemaking requires public notice and comment, so these cases (Mono Lake, Lower Colorado River) typically provided extensive opportunities for comment, although publicity of these opportunities may have been narrow and limited (e.g., confined to newspaper announcements and Federal Register). Though legislation cases typically had many hearings or citizen committees, the case reports suggest that information on these hearings and meetings was not always widely announced (Edwards Aquifer, Pyramid Lake, Salt River, Snowmass Creek).

In general, across cases of all types it seems that public notice and access to information was limited. The apparent lack of public notice may stem from two sources. First, it may be genuine because parties may expend minimal effort in order to conserve resources, or simply may not consider the many ways that public views could be solicited and incorporated. Or it may be spurious, in that some forms used to notify the public (radio broadcasts, paid advertisements) may not be archived and thus inaccessible at a later date by researchers.

The Mono Lake and Lower Colorado River cases are rated 'B', with multiple opportunities for public comment and access to information, though possibly not as widely publicized as would be ideal. Edwards Aquifer and Snowmass Creek also presented opportunities for comment and/or information; however, evidence suggests that in these cases public notice or access to information was not adequate, and they have a 'C' rating. Pyramid Lake suffered from inadequate notice, to the point of accusations of 'clandestine' processes, and so receives a 'D' rating. Salt River also had little of its processes open or accessible. We tentatively assign it a 'D', while recognizing that this may have been considered appropriate for the process used. Finally, both court rulings (Pecos River and Big Horn) were not required to use accessible and inclusive processes, suggesting no ratings for these cases. However, the Big Horn case required the publishing of information on "Walton" rights in the newspaper, thus seeming to merit at least a 'C' rating for providing the public with access to related information.

C. Reasonable Process Costs

This criterion examines the costs associated with the *process being analyzed*, and organizes the cost information into three categories according to who bears the costs. Process costs are considered "reasonable" when they are (or are perceived as) proportionate to the magnitude of the conflict and the assets at stake. For a complex multi-party dispute over a large river basin involving millions of dollars in water rights, land, and economic activities, it would not be unreasonable to spend many hours and hundreds of thousands of dollars on a process designed to resolve the conflict. However, these same expenditures would be "unreasonable" for a two- or three-party dispute involving neighboring farmers.

Only costs incurred as part of the process are reviewed here. Costs and cost-sharing agreements related to the Outcome and its implementation are discussed under *Cost-Effective Implementation.*

Costs are broadly defined to include monetary expenditures, staff time, and other resources dedicated to the process. The process to be analyzed is the one that led to the Outcome that is the subject of the research – e.g., the process of negotiating an agreement, of drafting and enacting legislation, of litigating or of promulgating an administrative rule.

Across the cases, comments on the costs of litigation were more common than comments on the costs of the other processes. Actual cost data for some parties, or perceptions about the cost, were noted for litigation associated with the Big Horn, Edwards Aquifer, Pecos River and Pyramid Lake cases.

Costs to state governments were noted more commonly than costs to any other type of party, perhaps because states must answer more thoroughly to their constituents for money they spend in attempts to resolve conflicts than is expected of federal agencies.

While the data obtained about actual process costs was quite sketchy, the litigation cases seemed to be perceived as excessively costly in comparison to perceptions expressed about administrative, legislative and negotiated processes. Due to lack of comparable data across cases, comparisons are not provided across cases for Process Costs.

III. OUTCOME QUALITY

This category reviews criteria that examine the quality of the Outcome provided by the conflict resolution process.

A. Cost Effective Implementation

This criterion focuses on the costs of implementing the terms of the Outcome. It collects information that may be used comparatively to assess whether an agreement/ruling took a cost-effective approach to resolving the technical problems of the conflict and to implementing the terms of the agreement. Implementation is considered cost effective if the actions were undertaken in a manner that considered and minimized the costs of accomplishing what was required. For instance, water market transactions are generally a more cost-effective approach to obtaining the water needed to implement an outcome, compared to constructing new water storage facilities.

The two Outcomes that involved administrative rules (Lower Colorado River and Mono Lake) both had information specifically noted about the costs to taxpayers and ratepayers. Perhaps administrative processes consider costs to taxpayers and ratepayers more carefully than other Outcomes, such as litigation. Or perhaps these effects are simply documented better in these cases. The Lower Colorado River case specifically noted that the cost effective management of water was a goal of the administrative rules. In both of these cases, voluntary water purchases and exchanges were considered an important part of implementation.

The two cases that involved negotiated agreements ratified by Congress (Pyramid Lake and Salt River) contained more detail about the costs to stakeholders than other cases. Perhaps stakeholder costs receive more consideration in negotiations that are ratified by Congress than in other types of processes. Impacts on water rates in some of the affected water districts were also noted for both of these cases.

The cost information for stakeholders, the public and others could not be considered complete in any of the eight cases. The reported cost information included comments on perceived costliness by parties and observers, expectations of future costs for implementation and (rarely) actual data on costs incurred. Both of these cases rely upon negotiated water exchanges and acquisitions to provide water for tribal and environmental needs as a part of implementing the agreements.

The courts did not explicitly consider costs in the two litigation cases. However, the affected parties took steps to cost-effectively comply with the ruling. In Pecos River, the state of New Mexico has complied by purchasing and leasing water from irrigators. In the Big Horn case, the irrigators negotiated a short-term lease of water from the tribes in order to avert disruption of irrigation water supplies.

A cross-case comparison is hindered by the lack of comparable data, so ratings should be viewed as tentative. Ratings are based on noted perceptions of costliness and on the use of water transfers and exchanges, rather than new construction, to procure water. Mono Lake, Edwards Aquifer, Salt River, Pyramid Lake, Lower Colorado River, and Snowmass Creek are all ranked B (medium-high). The two litigation cases are ranked C (medium), due to the absence of cost considerations by the courts.

B. Perceived Economic Efficiency

This criterion assesses perceptions of stakeholders and observers on the balance of costs and benefits of the Outcome, i.e. "was it worthwhile?" At the time of the original case analyses, we had not yet recognized the need to separate out Perceived Economic Efficiency from other economic outcome criteria though this is now done in the Guidebook (Appendix A). Thus data was not specifically collected on this criterion, and no case comparisons are made here.

C. Financial Feasibility/Sustainability

This criterion assesses how the agreement addresses issues of securing funding for implementation and ensuring that economic incentives encourage compliance and support implementation. While the first criterion in this category, *Cost-Effective Implementation*, addresses costliness of implementation, this criterion focuses on how the money will be obtained, such as actual or planned financial arrangements. Financial arrangements specify who pays for what, the monthly or annual obligations, and the time period over which payments are made. Researchers were asked to distinguish financial considerations related to the Outcome from those that would have occurred regardless of the Outcome, and to record those that are attributable to the Outcome.[2]

Researchers were asked to summarize information on the allocation of costs among parties (staff time, money, water, other costs); parties' ability to pay; the spreading costs over time; any large deferred costs; water pricing to promote conservation; other economic incentives to support implementation; loans and cost sharing arrangements; and unfunded mandates, such as requiring monitoring without providing funding.

Cost allocations among parties were noted in the case studies for cases involving legislation, administrative rules and negotiated agreements. The two litigation cases (Big Horn and Pecos River) did not specify cost allocations. Several of the non-litigation cases (Edwards Aquifer, Mono Lake, Pyramid Lake, Salt River) also explicitly provided for loans to specific parties to assist them in paying for costs they had incurred (or expected to incur) and provided specific mechanisms to raise funds to cover costs. These mechanisms include increased water rates, water use fees, the issuing of bonds by public entities and contributions from the federal budget.

Water leases and water acquisitions were authorized in several cases to make more efficient use of regional water sources (Edwards Aquifer, Mono Lake, Pyramid Lake, Pecos River and Salt River). The authorizations for water transactions were provided through legislation, administrative rule making and provisions in negotiated agreements.

Several cases specifically prohibited or limited unfunded mandates. The Edwards Aquifer legislation specifies that there shall be no taxpayer subsidy to implement the legislation. The Lower Colorado River administrative rules specify that unfunded mandates imposed on non-federal and private parties shall not exceed $100 million per year.

Issues about the ability to pay were noted in several cases. In the Edwards Aquifer legislation, agricultural water users are guaranteed lower water use fees than cities, because agriculture has less ability to pay for water. The California Water Board noted (in the Mono Lake case) that the costs to the City of Los Angeles to replace water and hydropower no longer available due to the Board's decision, were reasonable given the assistance provided to the City from federal and state sources and considering the water costs elsewhere in Southern California. The Snowmass Creek case study noted that public entities could cover their increased costs through fee collection mechanisms.

The Edwards Aquifer and Pyramid Lake case studies mentioned water pricing as a tool to encourage more efficient water use.

Overall, based on information provided in the case studies, it appears that litigation does not address financial and economic issues in as much depth as legislation, administrative rules and negotiated agreements. Further research that compares the thoroughness of different Outcomes and the ramifications of various levels of attention to financial and economic matters would be useful.

Comparing across cases, the Mono Lake, Pyramid Lake, Salt River, and Snowmass Creek cases rank 'B' (medium-high) on financial sustainability, given their attention to cost-sharing and financial mechanisms. Edwards Aquifer and Lower Colorado River are given 'A' (high) rankings due to their emphasis on no taxpayer subsidy (Edwards Aquifer) and limits on unfunded mandates (Lower Colorado River). The Pecos River case is ranked 'B' to acknowledge the state of New Mexico's use of bonds to raise funds to comply with the court ruling. The Big Horn case is ranked 'E' because it shows evidence of little consideration of financial matters.

D. Cultural Sustainability/Community Self-Determination

This criterion asks for a record of the communities affected by the agreement/ruling and an assessment of the types of potential effects. These include demographic and economic effects, such as changes in patterns of jobs, income, and taxes. They also include changes in patterns of ownership, decision-making authority or jurisdiction and in the social and cultural "lifeways" of the impacted communities or in the relative balance of these lifeways (the "cultural mix").

Cultural impacts are most prominently considered in the cases in which one of the stakeholders is a Native American tribe. Tribal concerns over cultural resources were noted in the Mono Lake and Pyramid Lake cases. Decision-makers attempted to address these concerns in the ensuing agreements and administrative rules. Federal laws intended to protect cultural resources played a role in both of these cases.

Another type of tribal concern was noted in the Pyramid Lake, Big Horn and Salt River cases. This concern was over tribal government jurisdiction and sovereignty to govern the resources of the tribal reservation. The negotiated agreements in the Pyramid Lake and Salt River cases attempt to address tribes' concerns about protecting their sovereignty. The series of court rulings in the Big Horn case resulted in inconsistent decisions regarding the jurisdiction of the tribal government to manage irrigation diversions by non-Indians whose lands are located within the reservation boundaries.

Every case raised concerns about impacts on ways of life. In addition to concerns about tribal culture, many cases raised issues about reduced water for irrigation and impacts on agricultural communities and life-ways. The Big Horn, Edwards Aquifer, Mono Lake, and Pyramid Lake cases all note concerns expressed by agricultural interests that their way of life would decline as water was transferred away from agriculture to satisfy environmental needs. In Outcomes that involved legislation and negotiated agreements, water was not involuntarily taken from one group for use by another. Instead, voluntary transactions were the mechanism used. In the Pecos River case, the state of New Mexico specifically chose to buffer New Mexico irrigators from the court decision by purchasing water from them with state funds, in order to comply with the court ruling. Alternatively, the state could have chosen to require farmers to cut back water use in order to provide court-mandated water deliveries to Texas. In these legislated and negotiated cases, the voluntary nature of the water transfer arrangements did not ameliorate the concerns of irrigators.

Several cases involved tradeoffs between water for urban areas and water for agricultural use. The Edwards Aquifer, Lower Colorado River, Mono Lake, Pyramid Lake and Salt River cases all involve competition between agricultural and urban users for limited regional water supplies. In these cases, water would move from irrigation use to urban use only when a farmer agreed to sell or lease their water. Again, the voluntary nature of these transitions did not erase concerns about potential impacts on farming and farm communities.

The Snowmass Creek case presents a different type of tradeoff between lifeways. It involves tradeoffs between water for snowmaking to sustain the ski industry, a key employer in the area, and water to maintain ecologically healthy streams and fisheries.

Overall, this criterion does not highlight distinctions among the different types of Outcomes so much as it emphasizes that tradeoffs among communities and ways of life lie at the heart of water conflicts.

In comparing across cases, higher rankings are given to those Outcomes that addressed cultural concerns and concerns about community and way-of-life viability. It must be noted that a strong ranking here may run counter to other criteria. For instance, the protection of agricultural water users generally entails subsidies or restrictions on buying water from farmers, which reduces cost-effectiveness. The Big Horn, Pyramid Lake, Mono Lake, and Salt River cases all explicitly address tribal cultural and sovereignty concerns. They are ranked 'B' (medium-high), except Pyramid Lake, which is ranked 'D' due to the negotiated agreement's failure to accommodate agricultural communities' concerns. In Edwards Aquifer and Pecos River, agricultural water access is diminishing through voluntary transactions. These cases are rated 'C,' along with Lower Colorado River and Snowmass Creek – where the Outcomes affect local economies and ways of life in a manner acceptable to the stakeholders.

E. Environmental Sustainability

Sustainability has been defined in many different ways. In our work, we use "environmental sustainability" to refer to practices that manage current resources in such a way that future generations will have comparable resources available to them. This criterion assesses the degree to which the Outcome considers drought, environmental factors and other natural contingencies, either through direct language in the agreement/ruling or through participation of environmental advocates (agencies and organizations) in crafting the Outcome and its implementation. The criterion also asks what natural resources are committed for implementation, over what time frame, and with what environmental impacts. Finally, this criterion looks for provisions in the Outcome that protect or restore endangered species, water quality and other aspects of environmental quality, and for participation by environmental advocates.

Environmental protection or restoration was one of the key issues motivating the conflict in five of the eight case studies: Big Horn, Edwards Aquifer, Mono Lake, Snowmass Creek and Pyramid Lake. Environmental considerations also are important in the Salt River and Lower Colorado River cases because they affect the options available to the parties to resolve their interjurisdictional water disputes. Environmental concerns did not appear to be prominent in the Pecos River case, possibly because it provides higher flows than had been customary in the river at the New Mexico-Texas border. However, there is an endangered fish species in this section of the Pecos River and environmental advocates are urging changes in river management to protect the fish, in addition to the changes that comply with the Pecos River case court ruling.

Considering the eight cases, the prominence of litigation based on environmental laws is quite evident. At various stages of the dispute, all the cases except Pecos River and Salt River were influenced by litigation over endangered species, water quality or some other environmental consideration. Such litigation appears to be an important force, providing momentum for the parties to achieve resolution either through the courts or through legislation, negotiation and administrative rules.

All of the cases, without exception, involved regional water supplies that were inadequate to satisfy all demands. The two litigation Outcomes, Big Horn and Pecos River, divided water supplies between the competing jurisdictions and those allocations need to be honored during drought years.

Drought was a concern in many of the other cases as well. In the Edwards Aquifer case, minimum spring flows need to be provided even during drought years and water users will need to adjust their own uses to accomplish this. Cities are acquiring water from irrigators to ensure they can satisfy urban water needs during drought. The Mono Lake case ties water use restrictions to lake-levels to protect the lake even during dry years. The city of Los Angeles is acquiring other supplies to ensure it can meet its needs during drought. The Snowmass Creek agreement specifies actions to be taken during normal, wet and dry years to protect stream flows.

In the Salt River case, environmental sustainability was promoted by switching water users from local groundwater (which is being depleted) to imported renewable surface water supplies. However, the imported water requires a large amount of electric power to transport it to central Arizona, raising other environmental concerns, such as the impact of power generation on air quality. During drought, groundwater reserves act as a back-up water supply.

All the cases commit water resources to the implementation of the agreement and consider the environmental effects of changes in water allocation and management. The Salt River case commits electric power to move surface water supplies into the center of the state so groundwater pumping can be reduced.

Environmental advocates and tribes, representing specific environmental concerns, participated in varying degrees to influence the Outcomes of the cases. They were prominent in either initiating litigation or participating in negotiations in the Mono Lake, Pyramid Lake, Big Horn, Edwards Aquifer, Salt River, and Snowmass Creek cases.

All of the cases in which environmental concerns and/or drought were central issues gave considerable attention to these concerns. The complexity of this criterion suggests that cases be compared against one another on several subcomponents of the *Environmental Sustainability* criterion. With respect to provisions in the Outcome that protect endangered species, water quality and other aspects of the environment, Pyramid Lake, Mono Lake, and Snowmass Creek rate 'B' (medium-high). Big Horn ranks lowest, due to the lack of protection for stream flows in the final court ruling. The other four cases are difficult to distinguish from one another based on this criterion and are ranked at 'C' (medium). The Mono Lake, Snowmass Creek, and Edwards Aquifer cases give the most explicit attention to drought considerations and are rated 'A'. The Pyramid Lake, Pecos River, Lower Colorado River, and Salt River cases also provide for dry-year adjustment and supply assurances. They are rated 'B', followed by Big Horn as 'D' (medium-low).

F. Clarity of Outcome

This criterion assesses whether the agreement was clearly worded and specified performance standards. It looks for the presence of any misunderstandings and differences in interpretation and examines Outcome language for ambiguity. It also checks the Outcome and implementation for well-defined baselines and performance standards (e.g., water use, stream levels, and conservation efforts).

Nearly all the cases contain subsequent controversies and negotiations that suggest the Outcome may not have been entirely clear. However, it is not always possible to discern whether subsequent disagreements were rooted in an ambiguity in the Outcome or in a party's dissatisfaction with the terms of the Outcome. Both factors may contribute to the need for further negotiations or litigation. This appears to be true in the Edwards Aquifer case study. Agricultural interests were clearly unhappy with the 1993 legislation and there were disagreements and several lawsuits over its implementation. The Mono Lake case required subsequent rounds of negotiations to work out how to implement the administrative ruling. This suggests that the ruling did not initially include clear implementation details. Irrigators in the Pyramid Lake case specifically criticized the agreement for creating uncertainty for them regarding their access to water.

Baseline conditions and performance standards were explicitly discussed in many of the cases. The legislation in the Edwards Aquifer case set clear goals for groundwater management and standards for assigning pumping rights. The Mono Lake ruling set explicit standards for maintaining lake-levels, as the negotiated agreement in the Snowmass Creek case did for maintaining stream flows. The courts set standards for river flows in the Pyramid Lake case, which had to be honored in the negotiated agreement, which, in turn, also set standards for providing flows to maintain and restore wetlands. The Pecos River ruling set performance standards for river flows from New Mexico into Texas.

With respect to well-defined baselines and specific performance standards, Pyramid Lake, Mono Lake, Edwards Aquifer, and Snowmass Creek rate medium-high (B) for more explicit attention to the issues. Salt River and Pecos River rate medium (C), and Big Horn and Lower Colorado River rate medium-low (D), with less focus on performance standards and baselines.

G. Feasibility/Realism

The *Feasibility/Realism* criterion addresses whether the Outcome is realistic in its assumptions and can be implemented in a practical sense, given legal, political, and technical considerations. Does it consider the legal and political context? Are the scientific and technical assumptions valid? Researchers were not asked to make legal, political, or scientific assessments themselves, but rather to note discussions of such types of feasibility in media and other sources. They looked for public perceptions regarding the political acceptability of the agreement, passage of necessary legislation (political feasibility), the consistency of the Outcome with existing law (legal feasibility), or whether the monitoring/ implementation team was "representative of key interests" (politically realistic). Researchers also looked for discussions about unrealistic commitments and assumptions related to: implementation, financial aid, resource supplies, technology, science, the behavior of other parties, and whether the agreement could be justified to parties' constituencies.

Political feasibility applies most clearly to negotiated agreements since courts and agencies do not require political support to issue a ruling and the passage of legislation is itself evidence of political feasibility for legislative Outcomes. However, political support was shown in the Mono Lake ruling through state legislation that helped fund implementation. In the Edwards Aquifer case, the Texas legislature needed to pass amendments to the 1993 legislation, and by doing so, indicated continuing political feasibility of the Outcome.

The Edwards Aquifer, Salt River, Pyramid Lake, Snowmass Creek, and Mono Lake cases rate highest (B) in political feasibility, due to state or federal legislative support. Lower Colorado River and Pecos River rank medium (C), with implementation supported by legislative actions, while Big Horn is rated low (E).

Legal feasibility was strong in those cases with prior litigation and legal requirements that the Outcome needed to satisfy. The litigation cases (Big Horn and Pecos) are ranked 'A'. The Edwards Aquifer, Lower Colorado River, Mono Lake, Pyramid Lake and Salt River cases all had to consider prior court rulings in crafting the legislative, negotiated and administrative Outcomes. In the Edwards Aquifer and Pyramid Lake cases, dissatisfied parties challenged the legal validity of the Outcome. These two cases are ranked 'C' and the others are ranked 'B'.

IV. RELATIONSHIP OF PARTIES TO OUTCOME

Criteria in this category focus on how parties react and relate to the Outcome of the conflict resolution process, as well as how the Outcome provides structure to the future relationship between the parties.

A. Satisfaction/Fairness – As Assessed by Parties

As noted earlier, research[3] has found that Outcome satisfaction and Outcome fairness are highly related in peoples' minds, so they have been combined in our framework and analysis. *Satisfaction/Fairness* attempts to assess parties' perceptions of satisfaction with and fairness of the Outcome immediately upon its completion, either expressed overtly or through behavior like a refusal to sign or endorse.

Researchers were warned to check across various stakeholders, as this requires the use of multiple sources. It is important to distinguish between expressions of satisfaction with an Outcome (noted here) and expressions of satisfaction with its subsequent implementation (the latter is discussed under criterion IV.D, *Stability/Durability*).

In reviewing parties' satisfaction with conflict resolution Outcomes, the glass can be either half empty or half full. There were few cases in which every party was satisfied. These included the Salt River and Snowmass Creek cases, although in the former case individual farmers were dissatisfied, and in the latter case the satisfaction was "guarded" until modifications were negotiated. By comparison, only one of the cases in our sample revealed no level of satisfaction (Big Horn). Satisfaction was almost always partial or guarded, or, as in Edwards Aquifer, there was satisfaction that all were "suffering" equally.

Sometimes satisfaction by most parties was bought at the expense of one party's dissatisfaction, typically that of the agricultural interests (Pyramid Lake, Edwards Aquifer, a few in Salt River). In the litigation cases reviewed, either one party was satisfied and the other was not (Pecos River), or none of the parties were satisfied (Big Horn). In other processes, satisfaction was mixed.

Since most parties in both Salt River and Snowmass Creek were satisfied, these merit a 'B'. In several of the cases rated 'C' (Lower Colorado River, Mono Lake, Pecos River, and Pyramid Lake), lingering dissatisfactions still did not prevent achieving an Outcome accepted by most. By contrast, all parties were dissatisfied with the Big Horn ruling; we rate this an 'E'. In the Edwards Aquifer case, there was some satisfaction that suffering was equally borne, though protests continue to hamper implementation, meriting this case a 'D'.

B. Compliance with Outcome Over Time

Agreements, rulings, and other Outcomes compel parties to engage in certain behaviors and refrain from others. *Compliance with Outcome over Time* assesses whether parties did act in accordance with the requirements of the Outcome. This includes dissenting parties who may have not signed an agreement (or who opposed legislation or a ruling) but may still be bound by it. Indicators of compliance include any subsequent litigation, initiated or threatened, in order to bring a party into compliance; renewed mediation or negotiations due to perceived noncompliance; records of compliance kept by any monitoring entity; and the inclusion of any provisions in the Outcome for verifying compliance (procedures, mechanisms, or entities).

The researcher was not asked to independently assess compliance, but rather to note the presence of the compliance indicators listed above.

We did not include indicators that asked for ways parties were encouraged to do, or not do, certain things, therefore we only have a few comments on general compliance behavior (more information on incentives can be found under IV.D, *Stability/Durability*). In the Edwards Aquifer case, the EAA found it difficult to comply with groundwater pumping limits, so they tried an alternative strategy of paying farmers to refrain from irrigation. In the Pecos River case, New Mexico leased or bought additional water rights to increase flows into Texas and achieve compliance. Overspending and other irregularities in the New Mexico water rights purchase program were uncovered through an audit requested by the New Mexico State Engineer. The Pyramid Lake case briefly mentions that the Settlement Act included actions and incentives.

Subsequent litigation to bring parties into compliance was threatened in the Big Horn, Edwards Aquifer, and Lower Colorado River cases. (Subsequent litigation was initiated for other motivations in Pyramid Lake and Salt River). Subsequent mediation was attempted only in the Pyramid Lake case.

Most cases did not have records of compliance by a monitoring entity. It appeared that it was often either too early for these to have been kept, or the researcher simply could not find any in existence. In the Mono Lake case, lake level records were available from the DWP. Monitoring records were located in the Pyramid Lake case, but solicited records were for the most part not received. One large record arrived too late to be analyzed. The lack of availability of monitoring records may suggest that few records of compliance are actually maintained.

The Outcomes of administrative rulings and negotiations (Lower Colorado River, Mono Lake, Pyramid Lake, Salt River) included provisions for verifying compliance, but these were not discussed in court rulings (Big Horn, Pecos River) or legislation (Edwards Aquifer, Snowmass Creek). However, the Snowmass Creek case did include procedures for modification. Provisions for verifying compliance came in a variety of forms. The Lower Colorado River case required reports to the Secretary of Interior on records of quantities and credits that were stored and redeemed [Secretary also reports on diversions and consumptive uses]. There were also reports to congressional committees (Pyramid Lake), reports to the State Water Resources Control Board (Mono Lake) on plans for stream and waterfowl restoration, reports on water use (Pyramid Lake), reports on stored and developed water (Salt River), and reports on instream flows and lake levels (Mono Lake). Pyramid Lake's Settlement Act established a framework of actions, schedules, incentives, and coordinating bodies.

One can speculate on why provisions for monitoring compliance seem to be commonly omitted from legislation and court rulings. Perhaps legislation is intentionally vague to allow for easier passage, leaving compliance provisions for the administration stage. Court rulings seem to assume compliance, with any noncompliance handled through later court proceedings.

The ruling in the Pecos River case required follow up actions. The New Mexico legislature passed legislation to help implement the requirements imposed on the state by the court ruling.

In the Edwards Aquifer and Pyramid Lake cases, city governments adopted new water management policies to comply with state or federal legislation.

Arizona passed legislation to create a state water bank, consistent with a provision in the federal rules issued in the Lower Colorado River case. In the Mono Lake case, required follow-up activities were undertaken by various governments, such as the City of Los Angeles. Ratification, per se, was not required.

Comparatively evaluating across cases required trading off evidence of noncompliance (e.g., subsequent litigation to bring compliance) with evidence of structures to encourage compliance. Big Horn had such subsequent litigation, and without clear evidence of compliance mechanisms we rate it a 'D'. Cases earning 'C's included Edwards Aquifer, which had subsequent litigation threatened but also had some minimal mechanisms for encouraging compliance, Snowmass Creek, which had no subsequent litigation but also only minimal compliance frameworks specified, and Pyramid Lake, which had much subsequent litigation but also an extensive framework for encouraging and measuring compliance. Other cases with more extensive reporting, sanctions, and/or incentives for compliance, and little or no subsequent litigation, were Lower Colorado River, Pecos River, and Salt River. These were rated 'B.' Finally, Mono Lake had no subsequent litigation and an admirable and extensive framework for measuring and encouraging compliance, meriting an 'A.'

C. Flexibility

While no Outcomes can be written to anticipate all future contingencies, they can be designed to be responsive and flexible. The criterion of *Flexibility* assesses an Outcome's ability to be adapted to changing conditions. Indicators assess details of any subsequent modifications, the specified process for modification (if any) in the original Outcome, and any unachieved but desired modifications, particularly if the barrier to modification was in the Outcome itself.

Surprisingly, only half of the Outcomes had experienced subsequent modifications. This could indicate that insufficient time had passed for the agreement or ruling's flexibility to have been tested. Few discernable regularities were observed in the pattern of modifications, other than the fact that subsequent modifications were made in all cases of negotiation. In the Salt River case, a petition was filed to adopt a rule specifying procedures for settling federal water rights, while a US congressional committee made modifications in the Pyramid Lake case to prevent irrigators from stalling implementation and to reduce funds allotted to the tribe. The proposed Rules in the Little Colorado River case underwent a period of lengthy feedback and further negotiations when revisions were made. In the Edwards Aquifer case, litigation prompted a change of an appointed board to an elected board.

Were procedures for modification anticipated in the terms of the original Outcome? Almost across the board, all Outcomes either specified a process for modification (Edwards Aquifer, Lower Colorado River, Pecos River, Snowmass) or empowered a certain body to make modifications if necessary (SWCRB in Mono Lake, legislative committee in Pyramid Lake, an arbitrator in Salt River). The Lower Colorado River Outcome required modifications to be made within a certain time period. The cases that encountered barriers to modification understandably involved modifications from one party that were not desired by other parties (Edwards Aquifer, Salt River).

Though modification procedures were often specified, implementation procedures were not. In many subsequent tests of the Outcome, the question was usually not one of revising the original Outcome, but rather interpreting the Outcome in its application. Therefore, though 'clarity' of agreement may initially seem to be at odds with 'flexibility,' these two criteria may actually address different aspects of the Outcome. 'Clarity' may address whether or not things are "spelled out" to make implementation more straightforward, while 'flexibility' may primarily provide processes for subsequent adjustments and modifications. It may be possible to have an outcome that is both 'clear' in the sense of tacking down implementation details, while still 'flexible' by including a process for modification if needed.

Because no significant differences exist across types of Outcomes for this criterion, no formal comparison across cases is warranted.

D. Stability/Durability

Stability/Durability addresses the ability of the Outcome to persist over time. It includes two types of indicators: those that look at characteristics in the Outcome itself and in any accompanying framework for implementation that may affect stability over time; and indicators that actually note evidence of stability or instability over time. Indicators in the first category include stability-promoting incentives in the Outcome, such as penalties, deadlines, or benchmarks, identification of a party (or parties) as responsible for implementation, and provision of an ongoing forum for future conflict resolution. Indicators in the second category, which note actual instability, include noncompliance, resumed litigation or introduction of counteracting legislation, expressions of hostility, communication breakdown, and coercive behavior.

The first indicators are what we often think of as relevant to implementation (incentives, benchmarks, and responsible parties). Positive or negative incentives such as penalties, deadlines, or benchmarks were included in all cases of legislation and negotiation. Incentives were not uncovered in either case of court ruling. One administrative ruling case (Mono Lake) included benchmarks and other incentives, while this was not addressed in the other similar case (Lower Colorado River). One can speculate on why incentives are lacking in court rulings, though they are present in other processes. Perhaps processes such as negotiation, legislation, and rulemaking are more conscious that parties participate in such agreements and their implementation through their own volition, and attempt to support this, while court rulings assume parties have incentives to comply due to the implied potential sanctions of the court.

The types of incentives included penalties (for unauthorized pumping in Edwards Aquifer; for wasted water in Salt River); deadlines (for reducing withdrawals in Edwards Aquifer; for requests, temporary storage, and permits in Salt River; time frames in public law in Pyramid Lake); benchmarks (for lake water levels in Mono Lake; water quality standards in Pyramid Lake; monthly reports in Salt River; Interior has goals, agencies have action plans in Pyramid Lake; detailed contingencies in Snowmass Creek); and rewards (additional water can be diverted in Mono Lake). Given the learning literature on the best way for shaping behavior, more emphasis on positive incentives, i.e., rewards, should be explored.

Nearly all the Outcomes identified the parties responsible for implementation. The party responsible for implementation and/or monitoring typically was a public agency: the Department of the Interior (Big Horn, Pyramid Lake), the EAA in Edwards Aquifer, and the DWP and SWRCB in Mono Lake. One notable exception was in Snowmass Creek, where the skiing company was designated as the monitoring party (for stream flows).

Several of the Outcomes establish or imply that forums used to reach the Outcome should also be used for future conflict resolution (EAA, legislature in Edwards Aquifer; original negotiation forum in Lower Colorado River; SWCRB in Mono Lake; TCCO in Pyramid Lake). This is not addressed in the two court rulings, probably because it is assumed that future disputes would return to the court.

Indicators of actual (as opposed to predictive) instability typically noted were: continued litigation, hostility, threats, lack of trust, and communication breakdowns (Snowmass Creek, Pyramid Lake, Edwards Aquifer, Lower Colorado River). Many of the indicators of actual instability are also things we noted in other parts of the analysis: in *Compliance*, and in section V, *Relationship between the Parties* (which discusses ongoing conflict resolution, subsequent litigation, noncompliance, hostility, and coercion).

In comparing across cases, many of the cases exhibited components encouraging stability, such as incentives, future conflict resolution forums, and parties responsible for implementation. Edwards Aquifer, Pyramid Lake and Salt River had all of these, earning a 'B'. Mono Lake earned an 'A' because its incentives also included the underutilized category of rewards. Lower Colorado River and Snowmass Creek had only some of these stability-promoting mechanisms, meriting a 'C'. The two court ruling cases, Big Horn and Pecos River, had little or no stability-promoting incentives, and no ongoing forum for future conflict resolution other than the implied course of returning to court, rating a 'D'.

V. RELATIONSHIP BETWEEN PARTIES

This category includes criteria that evaluate the relationship between the parties in the context of the conflict and of the Outcome that was produced through the associated conflict resolution process.

A. Reduction in Conflict and Hostility

A common measure of improvement in conflictual relationships is a reduction in hostility. This criterion captures a sense of whether the conflict is deescalating or not, either in actions, rhetoric or tone of communication. Various factors from the literature on conflict escalation, such as the presence or absence of various possibilities for non-alignment (which indicates level of polarization), also are included as indicators.[4]

At first it may seem odd to look for a connection or pattern between the conflict resolution process and whether or not the conflict is escalating. It seems that this criterion would be better predicted by the "stage" of the conflict on the timeline of the resolution process, rather than the type of process used. But since all of the cases were reviewed within the first ten years of the Outcome, their 'stages' were all similar and in a sense 'constant.' Interestingly, one can observe at least a loose association between escalating conflicts and either litigation or legislation having been the dominant process analyzed (Big Horn, Edwards Aquifer, and Snowmass Creek). Cases with other processes either showed no clear rise or fall, or were clearly deescalating. Was this a chicken-or-egg kind of association, in that when conflicts were escalating, litigation or legislation was brought in? This is possible. However, for those cases in which the Outcome was a product of legislation or litigation, the period *subsequent* to the Outcome was marked by escalating hostilities.

Hostilities could take the form of rhetoric, threats, or actions. Hostile rhetoric and climate included the use of in-group sanctioning against members of one's own group that might interact (verbally or economically) across 'conflict lines.' This produced a sense of fear among some parties (Edwards Aquifer). Threats were sometimes issued before the Outcome (issued by state of Nevada in Lower Colorado River), or subsequent to it (issued by utility in Snowmass Creek). An interesting question to explore would be, which parties resort to threats? Is it parties with "power" or those that feel "powerless," or is it powerful parties that feel their power slipping away? Some examples of hostile actions included cutting off water (Big Horn), seeking a temporary restraining order (Edwards Aquifer) and, of course, resuming litigation.

In both cases where the conflict was perceived to be deescalating (Lower Colorado River, Salt River), it was often sensed through changes in tone (more conciliatory in the Lower Colorado River case; friendly, mutually approving in the Salt River case) and the presence of positive themes in communication, such as openness or cooperation (Lower Colorado River). Actions included suggesting various creative solutions, initiating joint projects (e.g., in Salt River on repatriating Indian artifacts and remains).

In both cases where the conflict was perceived to be neither escalating nor deescalating, the relationship between the parties was called "professional" (Mono Lake, Pyramid Lake). This designation is reminiscent of the euphemisms used by diplomats and press secretaries to report to the press on peace talks.

Information on polarization was not often included, unless it was to note that it was absent (Salt River). However, the ingroup sanctioning in the Edwards Aquifer case noted above, as well as the fact that its associated ongoing citizens committee was internally divided, suggests polarization and thus a heightened state of escalation.

In cross-case comparisons, cases with de-escalation (Lower Colorado River, Salt River) were rated 'A', while those perceived to be neither escalating nor de-escalating were rated 'B'. Cases where conflict was escalating were rated lower on this criterion, with Snowmass Creek earning a 'C', and Big Horn and Edwards Aquifer earning a 'D' for hostile actions and threats.

B. Improved Relations

Theorists have sought to conceptualize "peace" or "good relations" as something beyond a lack of hostilities.[5] What represents "good relations" in terms of the presence, rather than the absence, of something? This criterion seeks to capture changes in the way parties see and relate to one another that may reflect the essence of successful resolution. To note change, one also must first note the nature of the original relationship as a baseline for comparison. Indicators to explore for change include discussions of the relationship itself, as well as the tone of communication among the parties (hostile, conciliatory), the effort parties expended to protect themselves, and their sense of trust as indicated by the need for lack of enforcement clauses or other formalities.

The reviewed cases seemed to fall into roughly three general categories on these criteria: Negative relations (no improvement, as in the Big Horn, Snowmass Creek, and Edwards Aquifer cases) are rated 'E'. Fragile but increasingly positive relations, which are reinforced through the presence of some authority or framework (Lower Colorado River, Mono Lake), are rated 'C'. Clearly positive relations where little authority or framework is needed (Salt River, Pyramid Lake) are rated 'A'. As in the previous criterion, the pattern of improvement in relations again corresponds roughly with the type of conflict resolution process used. The first category again includes litigation or legislation (though Snowmass Creek was also negotiation), the second category involves cases of administrative action, and the last category contains cases focused on negotiated agreements. This begins to lend credence to the common wisdom (and theory in legal anthropology)[6] that negotiation processes are the best processes for improving relations among conflicted parties. Further research is needed to confirm this trend.

One indicator cited for improved relations was 'cooperative relations' (noted in the Lower Colorado River, Salt River and Mono Lake cases). In two cases, a norm of cooperation resulted in labeling those being non-cooperative as deviant (Lower Colorado River, Pyramid Lake). Negative relations were indicated by the parties' negative portrayal of each other (Snowmass Creek).

Though not asked for in the indicators we included, an often cited indicator of relationship quality (and particularly of trust) was whether the parties managed their relationship themselves or whether they relied on a third party. In the Big Horn case, parties primarily managed their relationship through the court. In the Little Colorado River case, parties had an authoritative third party in the form of the River Master. Similarly, parties in the Mono Lake case, though being supportive of the process and feeling a sense of fragile trust, felt reassured by the third party oversight provided through the legal process. Cases where relationship was now one of cooperation and respect made no mention of a need for a third party. To clarify, it does not seem that the third party is what produces the reduced relationship; rather it is the tense or fragile relationship that requires an outside 'manager' because the skills, mechanisms for cooperation, and trust are not yet internal to the relationship.

This can be seen as linked to but different from whether or not parties felt they had a 'working relationship.' In certain cases, the parties may not have felt trust and yet still noted a working relationship or a theme of cooperation. This was true in the cases of Mono Lake and Little Colorado River, and even in the Edwards Aquifer case. The presence of a 'working relationship' even before trust exists echoes discussions in the conflict resolution literature about how conflicting parties develop 'a working trust.'[7]

In cases where relations had clearly improved, parties were quick to note changes in their sense of the personal competence of the other (Salt River) and their increased respect for the other (Pyramid Lake). Increased trust was shown, for example, in the switch from quarterly to annual budgets and the dispensing of other formalities (waivers of claims) (Salt River). Some would argue that improvements such as increased respect can only come through a conflict resolution process where the parties are able to learn about each other as individuals,[8] and where the process is one of them sharing information with each other and 'practicing' their relationship.[9]

C. Cognitive and Affective Shift

This criterion is designed to provide evidence of the phenomenon that many practitioners (and even parties) note of a shift in parties' framing of the conflict and/or the relationship. Indicators included noting the ways parties referred to one another and the way they describe or explain the other parties' behavior (pre- and post-agreement). Building on literature from family systems theories, it also included a bit of narrative analysis of the way "stories" are told about the conflict – do narratives change (pre- to post-) in their description of causality, interactions, values, etc.

Several of the cases did not provide information for this criterion (Big Horn, Edwards Aquifer, Lower Colorado River, Pecos River, Salt River). Perhaps this is because it was almost the last criterion, and the last page where we had detailed indicators, and researchers were tired. Perhaps it was because researchers could not find evidence of this criterion. (See next chapter on criteria analysis for assessibility information).

Of the three cases providing information on this criterion, one case noted no evidence of shift (Snowmass Creek) and in fact provided counterevidence that parties had *not* shifted since they each still saw their own views as the ones best for the community. No 'penetration of the other's perspective'[10] had occurred and so this case is rated 'E'.

In both cases where shifts were ascribed, Mono Lake and Pyramid Lake, new attitudes were noted. In one case (Mono Lake), this included admitting of past mistakes. Greater attempts were made to involve other parties from the outset, and greater attention was paid to relationship building. Shifts in language used to frame the conflict and tell its story were also observed. New connections are seen between issues, as between environmental protection and economic development (Pyramid Lake). Both cases also had several examples of parties or observers noting a change in public values and priorities (in the Pyramid Lake case this included the perceived devaluing of irrigators' lifestyle). These two cases are rated 'B'.

D. Ability to Resolve Subsequent Disputes

This criterion addresses the degree to which the relationship between the parties is able to handle subsequent related conflict, such as problems with implementation of the Outcome. Indicators include evidence that problems are handled constructively, evidence that an ongoing relationship has emerged in which it is possible to address future concerns, and possibly the emergence of an ongoing forum for conflict management. This considers the parties' subsequent joint "track record" in terms of actions rather than simply perceptions.

In both cases of legislation (Edwards Aquifer and Snowmass Creek), controversy seems to have continued after the Outcome. In Snowmass Creek, litigation has continued against the utility that was missing from negotiations, and proposed resolution alternatives were considered to have been less than fully considered by certain parties. In Edwards Aquifer, implementation of the Act has continued to provoke controversy. These two cases are rated 'D'.

By contrast, in the Lower Colorado River, Mono Lake, Pyramid Lake, and Salt River cases, parties seem to have developed a commitment to continue working together. In the Lower Colorado River, parties showed commitment to continue to explore mutually beneficial solutions. In the Mono Lake case, parties announced their intent to work together on the next phase (requiring restoration plans), although requesting continued court jurisdiction. In the

Pyramid Lake case, parties expressed a desire to resolve remaining issues through negotiations rather than litigation. Direct negotiations have been increasingly emphasized in this case since the late 1980s. In the Salt River case, parties agreed to forego future litigation (among parties to the Gila River adjudication), and clearly defined a process for future conflict management.

As noted explicitly in the Mono Lake case, these working relationships often seem to be an unintended but useful byproduct of extensive time working together to produce the first Outcome. Lower Colorado River, Pyramid Lake and Mono Lake were rated 'B'. Salt River rated an 'A' for its parties' public stances to work together on future disputes.

The four cases for which information was noted on this criterion involved administrative rulemaking, legislation, or negotiation processes. Information on this criterion was not collected on either of our litigation cases during the research phase of this project. If it had, it would have been useful to examine if once again we observed a split on the performance on these relational criteria, where litigation and legislation score poorly, and administrative actions and negotiation perform well.

E. Transformation

Some argue that conflict presents an opportunity for individual and collective moral growth.[11] More specifically, this moral growth is toward a social vision that integrates individual freedom and social conscience, and integrates concerns over justice and rights with concerns about care and relationships.[12] This moral growth can occur if conflict resolution processes help people to change their old ways of operating and to achieve new understanding and new relationships through conflict. Indicators include evidence of empowerment (i.e., the parties' renewed sense of their own capacity to handle challenges), evidence of recognition (i.e., empathy for and acknowledgement of others' circumstances),[13] and evidence of other major shifts in perception (e.g., of relationship context, of paradigm, of social and political context, of tools and solutions.) This includes perceptions of ability to achieve results and resolve further challenges.

No indicators had been created for this criterion when our researchers used the first version of the Guidebook. Nonetheless, two researchers volunteered evidence in their cases of what they considered transformation. In the Mono Lake case, the theme emerged of a new environmental ethic in place, and new ways of providing resources. Sources noted that a "new era of California politics" had been entered. In the Pyramid Lake case, it was identified as a "paradigm shift" in social values and priorities. It would be interesting to follow up to find out why the change was perceived as so dramatic in these two cases, but yet not noted as such in other cases involving positive results, such as the

Salt River case. These two cases are rated 'B', and for the others no cross-comparisons are provided, due to inadequate data.

SUMMARY

In this chapter, we have examined each of the eight case studies with respect to each criterion in our original evaluation framework. We found quite varied degrees of contrast among the cases. For a few criteria, inadequate data was available and so cases were not rated. For other criteria, cases did not display much contrast and were placed in the same rating category to indicate "no difference." Sometimes cases fell readily into two contrast groups – high and low. For a few criteria, finer distinctions were possible among cases and all five rating categories were used.

In examining case ratings across categories of success, the cases involving negotiated agreements tend to be rated more highly for the category Outcome Reached, due to emphasis on consensus and ratification.

For the Process Quality category, the cases in which public hearings were required and publicly accessible processes were used received higher ratings for procedural justice and for being accessible and inclusive.

Outcome Quality contains so many criteria that it is not meaningful to compare cases across this broad category. However, it is interesting to note that the one case resolved solely through litigation (Big Horn) ranks low on every one of the six Outcome Quality criteria, except Legal Feasibility and Cultural Sustainability.

Similarly, under success category IV, "Relationship of Parties to Outcome", the cases that involved negotiated agreements, administrative actions and/or legislation tended to be ranked higher than the cases emphasizing litigation. For category V, "Relationship between Parties", four cases received overall better ratings: Lower Colorado River, Mono Lake, Pyramid Lake and Salt River. The two litigation outcomes (Big Horn and Pecos) and two cases involving negotiations and state legislation (Edwards Aquifer and Snowmass Creek) seem to perform less satisfactorily in terms of relationships among the parties.

Comparisons across the eight cases highlight the strengths and drawbacks of the various strategies for addressing environmental conflicts. While our findings support the general notion that litigation produces less satisfying results, they also provide specific comparisons across multiple measures for this general hypothesis. The eight case studies provide a rich and diverse body of information useful to those concerned with resolving ongoing and future environmental conflicts.

Note that aggregating scores or grades across criteria within cases (for a comparative "case score") would be inappropriate, as this assumes that all criteria should be weighted equally and none of them are redundant or overlap. Once again, the only way to make an overall judgment is to consciously choose and/or weight criteria before applying them to scores. Any redundant criteria would add more weight to that dimension or factor. Weighting or choosing criteria becomes an exercise of values, as discussed in Chapter 2. In the next chapter we discuss how the criteria themselves have strengths and weaknesses.

Table 8.1. Cross-Case Comparisons

Criteria	Cases							
	Big Horn	**Edwards Aquifer**	**Lower Colorado River**	**Mono Lake**	**Pecos River**	**Pyramid Lake**	**Salt River**	**Snowmass Creek**
I. OUTCOME REACHED								
A. Unanimity/Consensus	E	C	B	A	D	C	A	C
B. & C. Verifiable Terms/Public Acknowledgement	C	C	C	C	C	C	C	C
D. Ratification	NA	NA	NA	NA	NA	A	A	NA
II. PROCESS QUALITY								
A. Procedurally Just	C	D	B	A	B	C	B	D
B. Procedurally Accessible and Inclusive	C	C	B	B	NA	D	D	C
C. Reasonable Process Costs	ID	ID	ID	ID	ID	ID	ID	ID
III. OUTCOME QUALITY								
A. Cost-Effective Implementation	C	B	B	B	C	B	B	B
B. Perceived Economic Efficiency	ID	ID	ID	ID	ID	ID	ID	ID
C. Financial Feasibility/Sustainability	E	A	A	B	B	B	B	B
D. Cultural Sustainability/Community	B	C	C	B	C	D	B	C
E. Environmental Sustainability								
1. Environmental Sustainability	E	C	C	B	C	B	C	B
2. Drought Consideration	D	A	B	A	B	B	B	A
F. Clarity of Outcome								
1. Outcome Provisions	C	C	C	C	C	C	C	C
2. Baselines, Standards	D	B	D	B	C	B	C	B

Legend: **A**: High **B**: Medium High **C**: Medium **D**. Medium Low **E**: Low **NA**: Not Applicable **ID**: Inadequate Data/Not Collected during primary effort

Table 8.1. Cross-Case Comparisons, continued

Criteria	Cases							
	Big Horn	**Edwards Aquifer**	**Lower Colorado River**	**Mono Lake**	**Pecos River**	**Pyramid Lake**	**Salt River**	**Snowmass Creek**
G. Feasibility/Realism								
1. Political Feasibility	E	B	C	B	C	B	B	B
2. Legal Feasibility	A	C	B	B	A	C	B	B
H. Public Acceptability	ID	ID	ID	ID	ID	ID	ID	ID
IV. RELATIONSHIP OF PARTIES TO OUTCOME								
A. Satisfaction/Fairness - as Assessed by Parties	E	D	C	C	C	C	B	B
B. Compliance with Outcome over Time	D	C	B	A	B	C	B	C
C. Flexibility	C	C	C	C	C	C	C	C
D. Stability/Durability	D	B	C	A	D	B	B	C
V. RELATIONSHIP BETWEEN PARTIES								
A. Reduction in Conflict and Hostility	D	D	A	B	ID	B	A	C
B. Improved Relations	E	E	C	C	ID	A	A	E
C. Cognitive and Affective Shifts	ID	ID	ID	B	ID	B	ID	E
D. Ability to Resolve Subsequent Disputes	ID	D	B	B	ID	B	A	D
E. Transformation	ID	ID	ID	B	ID	B	ID	ID
VI. SOCIAL CAPITAL								
A. Enhanced Citizen Capacity	ID	ID	ID	ID	ID	ID	ID	ID
B. Increased Community Capacity	ID	ID	ID	ID	ID	ID	ID	ID
C. Social System Transformation	ID	ID	ID	ID	ID	ID	ID	ID

Legend: **A**: High **B**: Medium High **C**: Medium **D**. Medium Low **E**: Low **NA**: Not Applicable **ID**: Inadequate Data/Not Collected during primary effort

---

NOTES

[1] E. Allen Lind, and Tom R. Tyler, *The Social Psychology of Procedural Justice* (New York: Plenum Press, 1988).

[2] Stakeholder's comments on *Perceived Fairness* of cost sharing are noted under IV.A (*Satisfaction/Fairness*). Economic implications for viability of communities are noted under III.D (*Cultural Sustainability/Community Self-Determination*).

[3] E. Allen Lind, and Tom R. Tyler, *The Social Psychology of Procedural Justice*, 1988.

[4] Jeffrey Z. Rubin, Dean G. Pruitt, and Sung Hee Kim, *Social Conflict: Escalation, Stalemate, and Settlement,* 2nd ed. (New York: Colin McGraw Hill, 1994).

[5] Adam Curle, *Making Peace* (London: Tavistock Press, 1971); Johan Galtung, *Peace By Peaceful Means: Peace and Conflict, Development and Civilization* (Thousand Oaks, CA: Sage, 1996).

[6] e.g., see P.H. Gulliver, *Disputes and Negotiations* (New York: Academic Press, 1979).

[7] E. Babbitt, and Tamra P. d'Estrée, "An Israeli-Palestinian Women's Workshop: Application of the Interactive Problem-Solving Approach," in *Managing Global Chaos: Sources Of and Responses To International Conflict,* eds. Chester A. Crocker, Fen O. Hampson, and Pamela R. Aall, 521-529 (Washington, DC: U.S. Institute of Peace, 1996).

[8] Stuart W. Cook, "The 1954 Social Science Statement and School Desegregation: A Reply to Gerard," *American Psychologist* 39 (1984): 819-832.; Walter G. Stephan and J.C. Brigham, "Intergroup Contact: Introduction," *Journal of Social Issues* 41 (1985): 1-8; See also Robert A. Baruch Bush and Joseph P. Folger, *The Promise of Mediation: Responding To Conflict Through Empowerment and Recognition* (San Francisco: Jossey-Bass, 1994).

[9] P.H. Gulliver, *Disputes and Negotiations,* 1979.

[10] Herbert C. Kelman, "Coalitions Across Conflict Lines: The Interplay of Conflicts Within and Between the Israeli and Palestinian Communities," in *Conflict Between People and Groups,* eds. Jeffry A. Simpson and Stephen Worchell, 236-258 (Chicago: Nelson-Hall, 1993).

[11] Robert A. Baruch Bush and Joseph P. Folger, *The Promise of Mediation,* 1994.

[12] Cf. Virginia Held, *Justice and Care: Essential Readings in Feminist Ethics* (Boulder, CO: Westview, 1995).

[13] Robert A. Baruch Bush and Joseph P. Folger, *The Promise of Mediation,* 1994.

Chapter Nine

EVALUATING CRITERIA FOR SUCCESS

Any evaluation must articulate its criteria. Chapter 2 highlights the diverse array of possible criteria for evaluating success in ECR, each of which can themselves be scrutinized with regard to their usefulness. In this chapter, we evaluate each criterion based on our experience of applying them to case studies of western U.S. water conflicts. In order to inform future case analyses, we also summarize our findings on the best time (over the life of a dispute, an Outcome and its implementation) to measure each criterion, and we provide other methodological notes on operationalizing the criteria. This chapter describes what we learned by applying the success criteria to case studies and it provides recommendations for the future use of these criteria with the framework presented in this book.

Readers may notice that additional criteria beyond those discussed here are presented in Chapters 2 and 3, and in the Guidebook in Appendix A. This is due to the need for criteria revisions and clarification as case studies were being conducted. Some criteria were not developed for use by case researchers until after the studies were completed. Consequently, this chapter evaluates only those success criteria used in the original project case analyses.[1]

We evaluate the performance of the success criteria, as operationalized and applied to the case studies, using three methodological criteria: accessibility, reliability and validity. For "accessibility," we ask if information was obtainable to answer the questions posed in the guide for each success criterion and if not, why. We note the sources of information, the triangulation of information using independent sources, and the costs and difficulties of obtaining information. We also consider whether the criterion and information requested in the guide should be recast to better match the available information sources.

For "reliability," we consider whether other researchers would have uncovered the same information and developed the same impressions and whether researchers sought diverse viewpoints while investigating different questions. We also evaluate the degree to which collected information was influenced by spurious factors such as researchers' personal connections and disciplinary expertise, the presence or absence of cooperative contacts, and the degree of record-keeping by public agencies on specific cases.

For "validity," we first consider whether each criterion is a conceptually valid indicator of the success "concept" being addressed. For instance, is *cultural sustainability* a valid criterion under the category of "Outcome Quality," or is *public acknowledgement of outcome* a valid criterion under "Outcome Reached"? Then we ask whether the indicators used to operationalize each criterion are valid measurements of the success concept we sought to measure. If not, we reflect on whether a different kind of question should have been asked for that criterion. We also assess whether the criterion as currently conceived was applicable to the different types of cases and Outcomes and, if not, how it could be refined for broader applicability.[2] Each criterion's ratings on our three methodological criteria are displayed at the end of the chapter in Table 9.1.

Our analysis also suggested that there are better and worse times to assess each criterion. Possible assessment times include: (1) baseline, before the resolution process, (2) during the resolution process, (3) immediately upon achieving the Outcome, (4) short-term after the Outcome is reached, and finally, (5) long-term after the Outcome has been achieved and implemented (see Figure 9.1).

<table>
<tr><td rowspan="2">BASELINE before resolution process</td><td rowspan="2">DURING the resolution process</td><td rowspan="2">When Outcome Achieved</td><td colspan="2">IMPLEMENTATION</td></tr>
<tr><td>SHORT-TERM after Outcome</td><td>LONG-TERM after Outcome</td></tr>
<tr><td>1</td><td>2</td><td>3</td><td>4</td><td>5</td></tr>
</table>

Figure 9.1 Stages for Criteria Assessment

Knowing which criteria to assess at a given stage can make evaluation easier. For each criterion, we considered when each ideally might be assessed. This information is displayed in Table 9.2 at the end of this chapter, but can fruitfully be consulted as each criterion's evaluation is summarized. In sum, this chapter evaluates each criterion. Each criterion's definition is restated, followed by an assessment of each criterion's accessibility, reliability and validity. We then discuss related methodology issues and timing for measurement for each criterion. The chapter concludes with a discussion of our findings and their implications.

I. OUTCOME REACHED

This category contains criteria that assess the basic achievement of reaching a conclusion.

A. Unanimity or Consensus

Unanimity or consensus assesses the strength of approval for the Outcome, the degree of dissension over it and the absence of key parties.

Evaluation

Accessibility — Information confirming that an Outcome had been reached was readily available from media sources, with little cost or difficulty. However, it was more difficult to verify whether all key parties were represented at the table. Instances in which a key party walked out or objected to the Outcome were noted in media sources. But it is innately difficult to document that all interests were represented, unless an unrepresented party made themselves publicly known. Moreover, the negotiating parties may collectively agree to withhold disagreements from public knowledge. Most researchers did not seek more than one source to verify that an Outcome had been reached. In those cases for which information was triangulated, the sources included court reports, newspapers and web sites.

Reliability — Researchers should be able to verify that an Outcome was reached. An Assessment of the representativeness of the parties might vary, depending on the personal knowledge of the researcher. In some but not all cases, views from diverse parties were cited. Spurious factors influencing reliability might include the varying accessibility of different parties' viewpoints, as well as the researcher's knowledge of the case and its parties.

Validity — The original criterion of *unanimity* did not apply in a straightforward manner to the two cases for which the Outcomes were court rulings, or to the Lower Colorado River case in which the Outcome was promulgation of an agency administrative rule. There likely were negotiations through the course of litigation and during political activities undertaken by parties to influence administrative actions. In fact, in some cases an agency action or legislation might merely be the formalization of an agreement. These negotiations might remain undiscovered if the case researcher focuses on a particular piece of legislation or administrative action. It would be instructive to attempt to analyze these prior interactions along with the litigation, legislation, and administrative actions that were the primary focus of these case analyses.

In cases involving state legislation, media sources provided some coverage of parties' attitudes toward the legislation. The unanimity criterion applies directly to the cases in which a negotiated agreement was analyzed. For the Salt River and Pyramid Lake cases, the criterion was analyzed with respect to the agreement among local parties and not to later congressional debate and ratification. *Unanimity/strength of approval* could have also been applied to the congressional process.

Unanimity/Consensus proved to be more difficult to assess than expected and, as operationalized, is an incomplete and potentially inapplicable measure for some dispute resolution processes of the overall success concept, "Outcome Reached." Moreover, it is not clear exactly whose approval should be assessed for different types of Outcomes. For a negotiated agreement, it is fairly clear to assess the approval of the parties. But for a court ruling or administrative action, it is not clear whether unanimity should be assessed among the parties or among the judges or administrators making the ruling.

Further Methodology Notes — It proved surprisingly difficult to document that all parties were at the table (or that all actually agreed) from readily accessible information. To verify this would require careful interviews of all parties and affected interests.

It is necessary to clarify whose strength of approval should be measured for different types of Outcomes. Should case researchers document the votes of legislatures, panels of judges and administrative boards, or does this criterion apply only to negotiated agreements?

Timing for Measurement — This criterion is best assessed at the time that an Outcome is announced when media coverage is most widespread. However, media sources can also be accessed later through library and internet research. The question of whether all key parties were represented should be examined at the time an Outcome is achieved. However, a key party's absence may not become apparent until problems with implementation arise. Consequently, this issue should be re-examined at later stages of the evaluation.

B. & C. Verifiable Terms/Public Acknowledgment of Outcome

Verifiable terms/public acknowledgement of outcome seeks to verify consensus on the terms of the Outcome and that it was publicly confirmed. Researchers were asked to provide information on whether the Outcome was written and formally signed, on evidence of a common understanding of the terms and on media events surrounding the Outcome.

Evaluation

Accessibility — Information confirming a written Outcome was easy and inexpensive to obtain, as was information on the publicity surrounding the Outcome. The primary sources for this were internet and library records of media coverage. The question of whether the parties shared a common understanding of the terms of the agreement was not readily verifiable from available sources. Researchers generally did not seek similar information from several different sources. A few researchers had difficulties tracking down media coverage, due to either lack of internet expertise or to the fact that some Outcomes were reached prior to the 1990s and media sources from that period are less readily available.

Reliability — Researchers all should have been able to find consistent information to document that a written Outcome was developed and to assess the publicity surrounding it. There are no apparent spurious factors, though local access to press sources may be useful. As noted above, it is difficult to determine that there was a common understanding of the terms among the parties.

Validity — As discussed under *unanimity*, the concept of consensus did not apply to cases where the Outcome was a court ruling, administrative ruling or legislation. Asking whether the Outcome was in writing and whether there was publicity around achieving the Outcome is, at best, an imprecise manner to verify that the parties had a common understanding of the agreement. However, a signed written Outcome is a valid first indicator of consensus on terms. This notion of a "common understanding of terms" should be assessed when any problems with implementation are examined. The "common understanding" indicator overlaps with Criterion III.C, *Clarity of Outcome*. Later misunderstandings and disagreements over terms may be a better indicator of "Outcome Reached" than whether there was initial publicity around the Outcome.

Moreover, when the Outcome is a court ruling, administrative action or legislation, it would be useful to examine whether the parties had a common understanding of the terms of the Outcome. This would best be assessed later, when checking for misunderstandings and differences in interpretation as the implementation of an Outcome progresses.

Timing for measurement — The existence of a written Outcome should be assessable any time after the Outcome is announced. However, as noted above, a consensus on the understanding of the Outcome may be more efficiently assessed after implementation has begun.

D. Ratification

Ratification assesses whether Outcomes were formally approved by relevant governments and courts. Researchers were asked to provide information on ratification by all necessary constituencies (with date and ratification process noted), and on court review and approval (with specific courts and dates noted).

Evaluation

Accessibility — Information on congressional ratification of agreements and on state enactment of legislative agreements was readily accessible from media sources. However, ratification by tribes, cities and water districts was not well documented. It is not clear whether this is due to researchers not knowing that negotiated agreements typically need to be ratified by all signatories that are government entities, or due to the absence of publicity or inaccessibility of public records on ratification by other levels of government.

In cases with a court or administrative ruling, the government entities often later enacted policies to comply with the ruling. Information on this was not explicitly requested under I.D. *Ratification* in our original framework. However, when provided it proved useful in assessing the success of the Outcome and the process. Sometimes, judicial approval of an Outcome is required, so the case study research will need to inquire whether this was obtained. For instance, the Edwards Aquifer, Pyramid Lake and Mono Lake cases each involved litigation which helped stimulate a negotiated agreement and/or legislation. Court approval is sometimes required for an Outcome stimulated by litigation, and court scrutiny may be invoked by an unsatisfied party who later reopens litigation.

Reliability — Some researchers were more successful than others at documenting ratification by multiple levels of government. Ratification by tribal governments was the most difficult to document. Because the occurrence of judicial approval and ratification by constituencies is a matter of fact, not generally subject to differing perceptions, there does not seem to be a need to seek diverse views. The researchers' degree of familiarity with legal and political processes may have influenced their ability to document ratification and court approval. Also, those researchers with local access to documents and media may have an advantage in being aware of ratification by a particular party. The researcher in the Snowmass case lives in the area and was able to visit water court offices to check personally on document filings. Yet the researcher in the Salt River case was able to document ratification by municipalities despite being located far from the case study locale.

Validity — Information that confirms ratification by the relevant parties and court approval, when needed, is an important indicator of the success concept, "Outcome Reached" as it indicates a formal approval of the Outcome. Without formal approval, one might question whether a valid Outcome had been achieved at all. *Ratification* does not apply directly to Outcomes that are court rulings and administrative rulings. Judicial approval does not apply directly to legislative Outcomes and administrative rulings but does apply to some, but not all, negotiated agreements. This criterion was difficult to operationalize because researchers did not necessarily know which governments would need to ratify an Outcome and which, if any, courts would need to approve an Outcome. Moreover, it is not clear how they could be certain if these are needed for any particular Outcome. Locating this information becomes part of the research itself.

Further Methodology Notes — Researchers need guidance on how to ascertain whether ratification and judicial approval is needed and by which governments or courts. Researchers should begin with the presumption that any government signatory to a negotiated agreement (nation, state, city, county, tribe, water district, public utility) needs to ratify the agreement. Researchers need to investigate, first of all, whether any formal ratification or court approval of the agreement is required and, if so, by whom. The researcher then needs to ask if these requirements were satisfied, when (dates), and in what form (tribal council vote, city referendum, water district resolution). These questions are all included in the current Guidebook (Appendix A).

An Outcome that is not a court ruling may have still been stimulated by prior litigation. This is the case for all four of the negotiated agreements and legislative actions in our case studies. In these cases, the Outcome must conform to the provisions of prior relevant court rulings and these provisions should be noted as part of the case documentation process. Prior court rulings and conformance of Outcomes with these rulings are relevant to ratification, but may best be assessed under *legal feasibility* (below, section III.G).

Timing for Measurement — Ratification and judicial approval are matters of public record and should be measurable anytime after they occur, though they may be easiest to track at the time they are accomplished. Adoption of policies by governments in response to the Outcome should be documented. This may be done best when later examining implementation.

II. PROCESS QUALITY

This category contains criteria that focus on the nature of the process used.

A. Procedurally Just

Procedurally just seeks to ascertain the perceptions of parties and others on the justice of the process. Specifically, this criterion asks whether the process was fair, balanced, complete and thorough in the sense of both issues and parties, and not compromised by time constraints or power imbalances. This overlaps with assessment of parties' satisfaction with the process (see Chapter 2).

Evaluation

Accessibility — Information on perceptions of procedural justice is available, but it requires a focused effort to collect. Few researchers were adequately thorough. Some experience by the researcher in searching for such information was clearly an advantage, both in obtaining the information, and in the basic recognition that this was not a simple or short question. In light of the past extensive research on perceptions of procedural justice (see the review of this literature, Chapter 2), several dimensions were included in the Guidebook. Researchers were required to use at least some media sources in order to address these thoroughly. Other sources included texts of the agreements themselves, party newsletters and websites, promotional materials, and texts of public speeches. With a willingness to plunge into media sources, this criterion could be readily verified. Many researchers did not address the underlying dimensions of this criterion thoroughly, however, and instead relied on the impression of only one or two sources, or commented on their own impressions based on inferences from subsequent behaviors of parties.

Accessing information for this criterion requires effort and resources. Local media sources are useful, but because they may not be nationally relevant, they are often not archived nationally or widely available. Reading multiple sources to find diverse perspectives is time consuming. Sleuthing skills are an advantage. Finding complete information on this criterion often required persistence and determination on the part of the researcher.

Reliability — The reliability of this criterion is directly related to attention to the underlying dimensions of the construct of procedural justice (e.g., fairness, timing, voice, thoroughness, inclusiveness, etc.). In cases where researchers based their assessment of procedural justice on overall impressions and only one source, rather than investigating the individual dimensions outlined in the Guidebook and multiple sources, the reliability for the assessment is low. Other researchers could glean very different impressions of perceptions of justice by citing similarly limited (but different) sources. When researchers expend the time and effort required to be thorough on this criterion, their assessments are more likely to be replicable.

Validity — The criterion as set out and detailed in the Guidebook has high validity, being based on current theory and research on procedural justice. If the criterion is not addressed as operationalized in the Guidebook, and only one or two dimensions are examined, then the analysis can be questioned as incompletely examining procedural justice.

In the few cases studies for which investigation of the underlying dimensions of procedural justice was thorough, it became apparent that the information most readily available was on the dimension, "all concerns were/were not taken into account" (inclusiveness). As a result, even thorough assessments of procedural justice may still end up weighting this dimension more heavily than others.

Reliability issues also have a direct impact on criterion validity. If only limited aspects of the construct are sampled and examined, differences between researchers will be exacerbated, and the information gathered loses its predictiveness.

Further Methodology Notes — This criterion may be best assessed through interviews with all parties on the complete range of indicators.

Timing for Measurement — This criterion is probably most easily assessed directly following the process, while it is prominent in the media. Verification from sources other than the media requires more time to pass before the information actually appears (in internal newsletters, promotional materials, and websites). Probably the most *useful* time to assess this information would be *during* the process, so adjustments can be made if the process falls short. Media sources might be useful for this, but direct surveying of parties would be productive during an ongoing process, and the validity and reliability (and therefore usefulness) of such assessments would be high.

B. Procedurally Accessible and Inclusive

The *procedurally accessible and inclusive* criterion sought to ascertain the *actual* availability of two components that contribute to perceptions of procedural justice: first, did opportunities for public participation exist, and second, did the public have access to information on issues and upcoming participation opportunities. The first asked for details (dates, timing, location, attendance, and effectiveness) of any public hearings, town meetings, surveys, hotlines, citizen boards, or other forms of public outreach and polling. The second component asked for notes on attempts to notify the public, and the nature of the contact medium.

Evaluation

Accessibility — For the two cases of administrative rules (Mono Lake and Lower Colorado River), information on public participation was easily accessible. Typically, extensive media coverage could be consulted. Researchers also found information on agency documents and files, promotional materials, summaries of public fora on agency websites, and press releases from agencies. Environmental Impact Statements and other documents produced to satisfy NEPA requirements often describe public meetings. Those researchers who had such cases were able to provide details of attendance and public involvement.

For cases of court ruling, public participation is not a consideration.

The processes for negotiated agreements often were private and confidential, so information on the process may be difficult to obtain since there may actually be no minutes or records. Information on one such agreement in the Salt River case was obtained after the fact from one of the relevant agencies, under the Freedom of Information Act.

It was surprisingly difficult to obtain information on public participation for legislative processes. Presumably, hearings open to the public were held, but evidence of this was found in only one of the two legislation cases, and details were not available.

Information on the other indicator for this criterion, public access to information about issues and meetings, was uniformly difficult to obtain, regardless of the type of process. Likely sources such as advertisements, public postings, radio broadcasts, etc., are ephemeral and not archived in public records.

Reliability — In general, information for this criterion was difficult to gather without being local, and present during the process. In the few cases where public participation was mandated, formalized, and record keeping was required, information was readily available and could be reliably assessed. However, even in these cases a researcher would have to know about the existence and location of such records. The necessity of such specialized knowledge makes reliable assessment less likely. In response to this, we have added an appendix to the Guidebook on policies that mandate public participation (Appendix B).

Validity — When available, these measures are good indicators of the level and quality of public involvement. Again, for some processes (court rulings, private negotiations), public involvement may not be a consideration.

The second component (public access to information) is only measured indirectly, through announcement of information availability. This does not directly describe the difficulty or ease a member of the public might have in obtaining information. Ideally, one would also have to factor in the resources available to parties and the public for obtaining information, and the degree to which resources and status aid access to information and participation in public meetings or consensus-building processes.

A larger question can be raised about public participation and access to information and their centrality to the concept of successful resolution. Cases which some might regard as successful with a "quality process" did not always have open and public processes. It would be interesting to consider ways to measure the quality of a confidential and private process.

Timing for Measurement — Public participation was either impossible to determine, or well documented. If well documented, it could be assessed at any subsequent point. If it was not considered central, assessments could only be made locally at the time of the process, such as attendance at court hearings or legislative sessions. Such information is rarely reported in the media, though it was reported for one of our case studies. As with procedural justice, the researcher's local presence could help to assess this criterion.

As noted above, it was virtually impossible to assess public access to information and announcements of meetings after the fact. When an administrative process required notification, a note was made that it had been advertised, although the only subsequent records were those published in the Federal Register. One researcher found a press release, which presumably led to public announcements. No researchers were able to find other sorts of announcements, notifications, signage, or evidence of media broadcasts. Much more thorough assessments could be made of this indicator if it were assessed locally during the time of notice.

C. Reasonable Process Costs

This criterion examines costs associated with the *process being analyzed*, with cost information organized into three categories according to who bears the costs.

Evaluation

Accessibility — Researchers provided two types of cost information in the case studies. Most commonly, perceptions of the costs of participation were reported as quotes from one or more of the parties. Less commonly, there were estimates of actual process costs incurred.

Complete information on the costs of participating in the process was not available for any case or even for any one party in any case. Cost information is not routinely compiled by case for either public agencies or private sector participants. When such information is compiled internally by a party, they may not choose to make that information available or it may not be available in a convenient form (i.e., scattered records in several different departments). There were a few instances in which agency officials or environmental organizations offered to check their records and provide an estimate of their costs.

Anecdotal quotes from one or more of the parties expressing views about the cost of a process were available for most of the cases primarily from media sources.

One of the difficulties of obtaining actual cost estimates lies in locating a cooperative contact that is willing to check records and gather cost estimates for their party.

Reliability — Researcher skills, persistence and specialized knowledge of the case and its parties affect the amount and quality of information. This, of course, is true for many of the criterion, but particularly for cost data, which may be privileged information or may take time and trouble for parties to compile.

In general, a particular party will only be familiar with their own process costs so there is little likelihood of confirming costs for a specific party through several independent sources. An exception might occur when a government agency is responsible for covering some of a particular party's costs. The US Department of Interior and Department of Justice, for instance, cover some of the costs incurred by Native American tribes to adjudicate and negotiate their reserved water rights and some costs for federal agencies involved in conflicts. Cost information also may be available from the federal Office of Management and Budget, the General Accounting Office, or the Congressional Budget Office if those agencies have analyzed the case from a cost perspective. Public documents from these sources, when such documents exist, provide an opportunity to triangulate cost data.

Validity — Reasonable process costs are a valid component of process quality. However, as operationalized, complete process costs could not be compiled for any case and so this criterion remains an incomplete measure.

III. OUTCOME QUALITY

Criteria in this category examine the quality of the Outcome provided by the conflict resolution process.

A. Cost Effective Implementation

Cost effective implementation assesses whether an agreement took a cost-effective approach to resolving the natural resource problems in the conflict. Collected data include costs to parties at the table, to the public (agencies, courts, other costs paid through taxes) and to others, such as utility ratepayers.

Evaluation

Accessibility — Researchers provided two types of cost information in the case studies. Most commonly, perceptions of the costs of implementing the Outcome were reported as quotes from one or more of the parties. Less commonly, there were estimates of actual costs incurred or of projected future costs based on data and records from parties or an independent observer.

Complete information on the costs of implementation was not available for any case, or even for any one party in any case. When cost information is compiled internally by a party, they may not choose to make that information available or it may not be in a convenient form (i.e., scattered records in several different departments). For negotiated agreements in which cost sharing of implementation is specifically negotiated, this information can be obtained from the agreement itself or from analyses of the agreement. There were a few cases in which a prior research effort had compiled information on costs for one or more of the parties. Since cost data is so difficult to obtain, it generally cannot be confirmed by checking with more than one source.

Anecdotal quotes that expressed views about the cost of an Outcome were available for most of the cases from one or more of the parties, primarily from media sources.

As with process costs (II.C), the difficulty of obtaining actual implementation cost estimates is in locating a cooperative contact that is willing to check records and gather cost estimates for their party.

Reliability — Researcher skills, persistence and specialized knowledge of the case and its parties affect the amount and quality of information that can be obtained. Cost data may be privileged information or may take time and trouble for parties to compile. Familiarity with economic concepts helps researchers identify some of the less obvious categories of costs (opportunity costs, contributions of assets other than money, costs deferred to future years). To assist future research efforts, an economics appendix (Appendix C) has been added to supplement the Guidebook found in Appendix A.

In general, a particular party will only be familiar with their own implementation costs, so there is little likelihood of confirming costs for a particular party by checking several independent sources. However, sometimes costs involving expenditure of federal money will be investigated by a public organization such as the US General Accounting Office, the Congressional Budget Office, or the Office of Management and Budget. In addition, Environmental Impact Statements (and Assessments and Reports) typically include a section on economic and financial impacts of the policy being analyzed.

Validity — Initially, it was not made clear whether costs associated with the process of reaching an Outcome or costs associated with implementing the terms of the Outcome were to be investigated. In general, case researchers ended up doing one of these, but not both, for any one case. The latter is now addressed under the success category "Outcome Quality." The former is examined under "Process Quality."

As a concept, cost-effectiveness is a valid indicator of Outcome quality. Outcomes that provide for solving the water conflict in a least-cost manner are viewed as more "successful" than those that involve unnecessary costs and a waste of water, money and time. For instance, an Outcome that provides additional water for a particular need through water conservation and transfers will generally be more cost-effective than an Outcome that provides for construction of a new dam and reservoir or transportation of water across long distances.

Cost-effectiveness, as an evaluation concept, implies that specific goals have been identified and examines the relative costs of different means for accomplishing those goals. Cost-effectiveness does not explicitly examine benefits.[3] As a criterion, it assumes that the goals agreed to by the parties are "worth" accomplishing. Cost-effectiveness is appropriate for analyzing negotiated agreements, court rulings, administrative actions and legislation in which the goals have been clearly identified.

As operationalized, however, this criterion did not adequately address the issue of whether the Outcome promotes relatively low-cost solutions to conflicts. Moreover, some case study researchers included costs that could not be directly attributed to the Outcome, but were general regional costs of water supply and water management. This incorrect attribution of costs needs to be minimized by the establishment of a well-defined baseline to distinguish costs and impacts that are directly related to a particular agreement. We want to be able to say, for instance, that a court ruling caused a ten- percent increase in water rates. However, this type of comparison requires clear identification of the water rates without the ruling. (This careful identification of a baseline for purposes of assessing an agreement is known as the "with and without principle" and is discussed in detail in Appendix C).

In principle, the concept of cost-effectiveness can be applied to all of the types of Outcomes analyzed, although it is not generally an explicit goal of court rulings. The criterion was difficult to operationalize for all types of Outcomes.

Timing for Measurement — The provisions of an Outcome can be analyzed any time after the terms are available in order to examine the approaches to resolving water problems. However, the actual costs of implementing the Outcome will not be known until implementation is complete, and even then may not be accessible, as cost information is likely to be distributed across many stakeholders and over many years. Parties' perceptions regarding their likely implementation costs could be ascertained easiest when the Outcome is reached and there is media coverage of the affected parties' reactions.

C. Financial Feasibility/Sustainability

Financial feasibility/sustainability assesses how the agreement addresses the issues of cost allocation, funding for implementation and ensuring that economic incentives encourage compliance and support implementation. Researchers were asked to summarize information on cost allocation among parties, their ability to pay, spreading costs over time, large deferred costs, water pricing to promote conservation and other economic incentives to support implementation, loans and cost-sharing arrangements, and unfunded mandates.

Evaluation

Accessibility — Information on cost allocation among parties and across time, water pricing and other incentives to support implementation, and on loans and cost sharing was most readily accessible when the Outcome itself contained these provisions. Some or all of the above items were included in the two negotiated agreements, the two legislated Outcomes and one of the two administrative actions. Citations from the Outcomes and commentary by parties and independent observers were used as sources. Information of this sort was not readily available for the court rulings, though quotes in the media from parties and observers provided some information on the financial implications of rulings. Information was not reported from more than one source, with the exception of quotes expressing opinions, which were reported for more than one party for some cases.

A few cases contained information on the ability of parties to pay and on unfunded mandates directly in the terms of the Outcome. In several other cases, information consisted of quotes from parties expressing their opinion on these matters. In the remaining cases, no information was reported and these issues may not have been relevant to the case.

Reliability — Researcher skills, persistence and specialized knowledge of the case and its parties affect the availability of information on actual costs incurred by parties. Familiarity with economics helps researchers identify some of the less obvious categories of costs and financing mechanisms (contributions of assets other than money, costs deferred to future years). Appendix C supplemental to the Guidebook, was added to assist this.

Validity — The specific questions under the criterion, "Financial Feasibility/Sustainability" are useful indicators of Outcome Quality, although several items are merely descriptive (cost allocation among parties and across time) and are not, as phrased, evaluative.

Timing for Measurement — For written Outcomes, which address the issues raised, the information is available anytime after the Outcome is made public. Quotes reflecting the perceptions of the parties are most available around the time the Outcome is reached, but can be retrieved later from media archives. Problems regarding ability to pay and unfunded mandates may not become apparent until difficulties arise with implementation. These issues should be re-evaluated as implementation proceeds. The nature and magnitude of financial responsibilities also should become more clear as implementation proceeds.

D. Cultural Sustainability/Community Self-Determination

Cultural sustainability/community self-determination asked for a record of communities affected by the Outcome and an assessment of the types of potential effects. These effects include demographic and economic effects, such as changes in patterns of jobs, income, taxes, etc., but also changes in patterns of ownership, changes in decision-making authority or jurisdiction, and changes in the social or cultural "lifeways" of the impacted communities or the relative balance of these lifeways (the "cultural mix").

Evaluation

Accessibility — Researchers were able to find material (from the agreement and commentary on the agreement) regarding consideration of cultural resources and protective measures adopted as part of an agreement. For some cases, parties were quoted (from media sources) on the effect of the agreement on culture and/or way of life. In those cases for which cultural issues were a concern, there seemed to be little cost or difficulty in obtaining information.

Material on community impacts seemed fairly easily accessible, though the sources varied widely. Most researchers used at least some media sources, often including national newspapers or wire service reports. Some also consulted the agreements themselves, as well as journals, party newsletters, publications of national and state water resource associations, and promotional videos. One report cited a draft environmental impact report.

No researchers provided information on changes in socioeconomic indicators over time. For most cases, it was too soon to assess such long-term indicators. For one case (Pyramid Lake), the researcher was able to locate relevant information, but deemed it too expansive an analysis for the current report.

Reliability — When the agreement itself addresses cultural issues, information is readily obtainable and not subject to spurious factors often related to access. In general, only those parties concerned about effects on *their* culture or way of life were cited. Other parties may have little or no comment on the subject.

In general, reports on community impacts use multiple sources and are triangulated and verified, so reliability is likely to be high. Three case reports cited only one source (Big Horn, Edward Aquifer, Pecos River), and so reliability for these reports on this criterion may be lower.

Validity — Conceptually, the fact that an Outcome considers impacts on cultural resources and ways of life is a valid component of Outcome Quality. As operationalized, researchers were able to note when cultural issues were an issue in the process, but were not able to systematically evaluate how well

Outcomes actually addressed these issues. Moreover, in some cases there was no evidence that cultural issues were important and, in other cases, the Outcome may not have sought to address cultural issues even though one or more parties were concerned (i.e. Pyramid Lake, agricultural way of life issues).

In general, most indicators for this criterion appear to be valid measurements of impact on cultures and communities. However, several researchers omitted a listing of communities affected.

It may be difficult to strongly connect changes in socioeconomic indicators to the Outcome or the conflict resolution process itself. Many of these indicators will change naturally over time, as the result of a number of other factors. Therefore, causal conclusions must be extremely guarded.

Timing for Measurement — The provisions of the Outcome itself can be analyzed any time. Parties' perceptions are most likely to be detectable at the time the Outcome is reached or later, when a phase of implementation raises cultural issues.

The *potential* for these effects can be assessed when the Outcome is reached, and any time thereafter. As with most implementation criteria, *actual* assessment of most of these effects requires the passage of time. For most cases, it was too soon to assess actual effects.

E. Environmental Sustainability

The *environmental sustainability* criterion assesses the degree to which the Outcome considers drought, environmental factors and other natural contingencies, either through direct language in the Outcome or through participation of environmental advocates (agencies and organizations) in crafting the Outcome and implementation. This criterion also asks what natural resources are committed for implementation, over what time frame, and with what environmental impacts.

Evaluation

Accessibility — Information on consideration of drought, environmental impacts and projected resource uses over time was obtainable from the Outcome itself and from comments by parties and others reported in the media regarding the Outcome, with little cost or difficulty. However, the participation or inclusion of environmental groups or agencies proved more difficult to verify unless they were key parties whose presence was widely noted in media sources.

Early guidebook questions regarding commitment of natural resources under the Outcome were vague and were not addressed by most of the researchers.

Reliability — Other researchers should be able to uncover the same information regarding terms of the Outcome pertaining to drought, environmental impacts and resource use over time. Collection of information on participation by, and inclusion of, environmental interests may be more subject to spurious factors, such as researchers' connections to parties and prior familiarity with the case. In general, the researchers for this criterion did not seek diverse views. This may not be a problem as drought and environmental impacts generally could readily be verified as key issues or, if not mentioned in media and other sources, would not be considered a central issue in the case.

Validity — The questions raised under the criterion "Environmental Sustainability" are conceptually valid components of "Outcome Quality." As operationalized, however, it proved difficult to verify that an Outcome *adequately* addressed drought, other natural contingencies and environmental impacts. Researchers generally were able to comment on whether or not the Outcome had clauses related to these topics, but evaluation of an Outcome's adequacy to respond to an actual drought or to unanticipated environmental impacts could not be accomplished until these occurred and the parties had to respond. Failure to adequately consider environmental impacts may not become apparent until implementation progresses. Moreover, drought is not a dominant concern in every watershed so an Outcome that does not discuss drought is not necessarily deficient.

The questions under this criterion apply least directly to court rulings, and most directly to negotiated agreements, administrative actions and legislation – all of which can be expected to address practical water management issues such as drought. Courts typically rule only on narrow legal issues and generally do not have a mandate to address contingencies and potential impacts.

Timing for Measurement — Wording of the Outcome can be assessed after it is made public. However, adequacy of the Outcome in addressing drought and environmental impacts can only be measured after these issues have arisen during implementation. Participation/inclusion of environmental interests can best be assessed at the time an Outcome is reached, when media and commentary by affected parties are most accessible.

F. Clarity of Outcome

Clarity of Outcome assesses whether the Outcome was clearly worded and performance standards were specified. Researchers were asked to provide information on misunderstandings and differences in interpretation (if any), and to examine Outcome language for ambiguity. They were also asked to check the Outcome and its implementation for well-defined baselines and performance standards (water use, stream levels, conservation efforts).

Evaluation

Accessibility — The development of subsequent misunderstandings were not immediately discernable in our cases as most of the Outcomes were fairly recent and in early stages of implementation. However, parties' comments on ambiguity and anticipated differences in interpretation were reported from media sources. Most researchers did not independently review Outcome language for ambiguity. Information on baselines and performance standards was cited for a few cases, based on the Outcome and comments regarding the Outcome. However, this item was not reported for several cases in which it is a relevant issue.

Reliability — There was a good deal of variability across cases regarding the depth to which *Outcome Clarity* was addressed, only partially due to differences among cases. For some cases, only one parties' viewpoint regarding ambiguity was cited, leaving it unclear whether other parties also perceived this as a problem. The question about baseline/performance standards was not carefully addressed in several cases where it did apply (Pyramid Lake, for instance).

Validity — The questions concerning later misunderstandings and well-defined baselines and performance standards are conceptually valid indicators of *Outcome Clarity*. However, they were difficult to operationalize. For most cases, not enough time had passed to verify whether ambiguities became apparent. Moreover, when parties complained about ambiguity at the time the Outcome was produced, it is not clear whether there really was confusion over meaning or they simply did not like a particular provision of the Outcome (as in the Big Horn case, a tribal attorney criticized the court ruling). Some parties prefer ambiguity in the Outcome, as this would leave them more flexibility.

Further Methodology Notes — It is not a sound strategy to require researchers to review the language of an Outcome for ambiguity. Most case researchers would not have sufficient familiarity with the case and with technical and legal terms to accomplish this. It is preferable, instead, to search for commentary from parties and independent observers on the matter. For lengthy, complex Outcomes, it is more time-efficient to rely on neutral expert's summaries and analyses. The researcher should not be encouraged to review an entire Outcome for ambiguity, but rather to seek commentary on this matter by others.

An examination of the terms of the Outcome, as well as commentary on the Outcome, can assess the question of baseline/performance standards.

Timing of Measurement — Later misunderstandings cannot be investigated until implementation proceeds, though concerns about ambiguity may be expressed early on. Outcome provisions specifying baselines and performance standards can be assessed any time after the Outcome is made public.

G. Feasibility/Realism

The *feasibility/realism* criterion addresses whether the Outcome is realistic in its assumptions and can be implemented given legal, political and technical considerations. Does it consider the legal and political context? Are the scientific and technical assumptions valid?

Accessibility — The passage of needed legislation (political feasibility) was noted with little difficulty. Some researchers found comments regarding legal feasibility and unrealistic commitments and assumptions, mostly in media sources quoting parties. However, such perceptions likely reflect the party's specific perspective and do not shed light on the actual legal feasibility or realism of the agreement. No researcher found "unrealistic symbolic commitments" and most did not find material addressing scientific and technical assumptions or the ability of parties to justify the agreement to their constituents.

Reliability — Given the difficulty of obtaining some of the information requested, many of these items are difficult to verify from diverse perspectives and may be more influenced by spurious factors such as researcher's personal knowledge and connections.

Validity — Legislation being passed, as a measure of *political feasibility*, applies to Outcomes that are court rulings, administrative actions and negotiated agreements. It also could apply to legislative agreements that need amendments in order for implementation to move forward (as in the Edwards Aquifer case).

Legal feasibility cannot be accurately measured by citing parties' opinions on the matter. However, citations from the Outcome itself that address consistency with existing laws are useful. This question does not apply to court rulings as well as it does to other types of Outcomes.

The indicator of representativeness of a monitoring/ implementation team was not easily addressed due to lack of information on such a team (or its non-existence) *and* the fact that representativeness is a matter of perception, to some degree. Important first questions should be whether a team has been assembled and its composition.

Indicators asking about unrealistic commitments in the Outcome aim at an important issue of "realism" and implementability but proved very difficult to assess.

Timing for Measurement — While some indicators of feasibility can be assessed through reviewing the Outcome itself when it emerges, most feasibility issues will best be assessed during and after implementation.

H. Public Acceptability

Public acceptability assesses whether the agreement was acceptable to the public and political leaders.

Accessibility — Researchers generally did not find material on *public* perceptions regarding the agreement and cited parties' reactions, primarily from media, instead.

Reliability — Given the difficulty of obtaining some of the information requested, many of these items are difficult to verify from diverse perspectives.

Validity — The acceptability of the Outcome to a larger public seems a valid indicator of Outcome Quality, but proved to be difficult to operationalize.

Timing for Measurement — For those items that researchers were able to find information on, most are best assessed around the time the agreement is announced.

IV. RELATIONSHIP OF PARTIES TO OUTCOME

Criteria in this category focus on how parties react and relate to the Outcome, as well as how the Outcome itself provides structure for the parties' future relationships.

A. Satisfaction/Fairness—As Assessed by Parties

Satisfaction/fairness as assessed by parties assesses parties' satisfaction with the Outcome and their perceptions of its fairness, either expressed overtly or through behavior, such as a refusal to sign.

Evaluation

Accessibility — Media sources, including wire services, were readily accessible for most cases. Researchers also drew on parties' newsletters, websites, and other promotional materials. Some parties had commissioned reports. Secondary and scholarly sources that had analyzed a few specific case studies also were drawn upon.

Reliability — In general, the reliability level for this criterion should be high. Potential threats to reliability (and therefore validity) arise in cases where research on this criterion is shallow and not complete across parties, or in cases where distinct Outcomes (e.g., a court ruling and a subsequent implementation agreement) are confounded in the reporting of expressions of satisfaction. In the latter case, each distinct Outcome needs to be analyzed separately.

Validity — As operationalized, this seems to be a useful measurement of parties' satisfaction and perceptions of Outcome fairness. Expressions of satisfaction from parties' representatives in the media or other sources are valid measurements of satisfaction to the degree that spokespeople can actually be representative. Sources seldom, if ever, give a sense of the range of views within particular parties. However, the internal diversity of parties may not be relevant to judging resolution success if representatives have the power to bind their parties to the Outcome.

Party satisfaction may not be a valid measure of success for a court ruling, since this is not a goal for the legal process. It may, however, be a valid measure to apply to the subsequent implementation of that court ruling.

Timing for Measurement — Media sources, if available, were accessible at any point in time. Certain other forms of summary reports, such as website summaries or commissioned reports, would only be available after enough time had passed to assemble them. Therefore, this criterion may best be measured after such documents have been created, unless researchers plan to interview parties directly.

B. Compliance with Outcome Over Time

Agreements, rulings and legislation compel parties to engage in certain behaviors. *Compliance with Outcome Over Time* assesses whether parties did indeed act as prescribed by the Outcome. Indicators include any subsequent litigation initiated or threatened in order to bring a party into compliance, the subsequent renewal of mediation or negotiations due to perceived noncompliance, records of compliance kept by any monitoring entity, and the inclusion of any provisions in the Outcome for verifying compliance (procedures, mechanisms, entities).

Evaluation

Accessibility — Information was readily available on the indicators of compliance. Media sources reported noncompliance and subsequent litigation. Subsequent litigation and/or other procedures (such as mediation) also were noted in academic journals and third party reports. Information on provisions and mechanisms for verifying compliance was detailed in acts and rules, though missed by some researchers. Some parties had additional summaries of provisions.

Researchers had difficulty accessing records of regulatory or monitoring organizations set up to monitor compliance, often because these records had not yet been accumulated.

Reliability — In general, a high level of reliability can be expected on the measurement of this criterion, except in cases where researchers were incomplete (as in not listing provisions for measuring compliance). In such cases, presumably other researchers may find additional or different information. Information on subsequent litigation should be highly reliable.

Validity — These indicators seem to be appropriate operationalizations for measuring compliance. However, subsequent litigation may not indicate noncompliance, but rather dissatisfaction with the original settlement. As a measurement of noncompliance, the subsequent litigation considered here should only be that which was specifically initiated to bring parties into compliance with the original or modified terms.

Provisions and/or mechanisms for verifying compliance also are useful indicators for this criterion, but case study researchers often did not report these.

Indicators that require some passage of time, such as the monitoring of commission records, are valid indicators of compliance. However, their usefulness can only be judged after enough time has elapsed for such information to accumulate. In only two cases (Mono Lake and Pyramid Lake) were researchers able to obtain such information. In the latter case, the information was for only one commission (TROC) and was too complex to analyze given the scope of resources available for this project.

Further Methodology Notes — Compliance indicators fall into the chronological categories of immediate, short-term, and long-term. For example, immediate indicators of (potential for) compliance would be provisions in agreements, while short-term indicators would be mechanisms instituted for measuring compliance (such as water meters in the Pyramid Lake and Edwards Aquifer cases), and long-term indicators would include results of monitoring compliance over time.

Timing for Measurement — The inclusion of provisions and/or mechanisms for verifying compliance can be assessed as soon as the Outcome is available. Subsequent litigation may occur shortly after an Outcome is achieved, or later. Records of regulatory or monitoring organizations, or other records kept to verify compliance, can only be assessed after enough time has passed for those organizations and records to become established.

C. Flexibility

While no Outcomes can be written to anticipate all future contingencies, they can be designed to be responsive and flexible. *Flexibility* assesses an Outcome's ability to adapt to changing conditions. Indicators assess details of any subsequent modifications, the process specified in the original Outcome for modification (if any), and any unachieved but desired modifications, particularly if the barrier to modification was in the Outcome itself.

Evaluation

Accessibility — Information on modification procedures was easily accessible in the text of negotiated agreements, rulings, or legislation. Information on subsequent modifications, or attempted modifications, was available in the media, and in parties' papers, newsletters, and websites. It was also found, in some cases, in board minutes or commission reports. Any barriers to modification contained in the language or structure of Outcomes were never specifically noted.

Reliability — The reliability level of this criterion should be high, especially when assessed through examining the modification procedures specified in the Outcomes. In this case, the presence or absence of such procedures should be easily verifiable by others, although our researchers relied only on the primary sources (the agreement or Outcome itself). The quality of these procedures was not assessed directly by researchers, in part because this would result in significantly less reliability (unless additional criteria for "good modification procedures" were established). Researchers did not note comments by others (e.g., parties) on the quality of such procedures.

The reliability of information on the indicator of modifications should be high and tended to be verified in more than one source. Information on modifications that were desired (by at least one party) may be less replicable. In some cases, modifications that were sought but not obtained were discovered through a researcher's particular extensive knowledge of the case and the parties (Edwards Aquifer and Mono Lake), rather than from easily accessible information.

Validity — Flexibility as measured by the inclusion of modification procedures is really a measure of Outcome Quality itself, rather than a measure of the parties' relationships to the Outcome. As such, it could be measured immediately after concluding the Outcome. The other indicators of flexibility that were analyzed in this section remain as long-term indicators of Outcome flexibility.

The validity of the indicators themselves for measuring the criterion of flexibility can also be assessed. The first two indicators, evidence of modification and existence of modification procedures, have high levels of validity. The last indicator, modifications desired but not achieved, does not necessarily reflect (a lack of) flexibility. In nearly any conflict, some parties will desire to modify the outcome if it does not fully meet their needs. The absence of continuing pressures for modification may indicate that the dispute has been satisfactorily laid to rest. In such cases, the absence of modifications does not indicate inflexibility.

Conceptually, *flexibility* can be a euphemism for ambiguity that leads to future conflict if it leaves issues to some future decision-making process that is not transparent and/or not perceived by parties as procedurally just.

Timing for Measurement — The modification of an Outcome itself can be assessed any time after requests to modify an Outcome arise. The presence of modification procedures in the Outcome can be assessed immediately upon achieving the Outcome. The assessment of the original spirit of the flexibility criterion (if the Outcome can be adjusted to changing conditions) can only be done after time has passed and conditions have changed, creating a desire among parties for subsequent (rather than immediate) modification.

D. Stability/Durability

Stability/durability addresses the ability of the Outcome to persist over time. It includes two types of indicators: those that look at characteristics in the Outcome and in any accompanying framework for implementation that may affect stability over time; and indicators that actually note evidence of stability or instability over time. Indicators in the first category include stability-promoting incentives in the Outcome such as penalties, deadlines, or benchmarks, identification of a party (or parties) as responsible for implementation, and provision of an ongoing forum for future conflict resolution. Indicators in the second category (noting actual instability) include non-compliance, resumed litigation or introduction of counteracting legislation, expressions of hostility, communication breakdown, and coercive behavior.

Evaluation

Accessibility — Judging from the varied degrees of success in assessing this criterion, it appears to be difficult to address. Most researchers who addressed this criterion drew upon the Outcome itself to address the predictive indicators (i.e., provisions for future stability), while drawing on media sources for indications of subsequent hostility, non-compliance and breakdown. It was often too soon to address these latter indicators. Many of the predictive indicators consist of implementation guidelines, and perhaps their inaccessibility indicates a frequent lack of implementation guidelines in agreements. One researcher (Pyramid Lake) actually found much information on implementation guidelines in U.S. Senate subcommittee testimony and addenda, as well as in agency press releases and third party sources.

Reliability — Because the range of reporting information for this criterion is so varied in the case study analyses, the reliability of the assessment of this criterion also varies. Where the information gathering was thorough, the reliability can be high. If the information gathering is inadequate, different researchers may find different answers and develop different assessments.

Validity — Many of the questions in this section involve implementation. It is unclear how valid implementation questions of court rulings can be since they are not designed to spell out future implementation, but rather to rule on questions of law.

The short-term or immediate indicators (incentives, implementation, responsibility, availability of a conflict resolution forum) may be less valid as *measurements* of stability since they are really only *predictors* of future stability. As such, they are also relevant to the category "Outcome Quality." The indicators of instability over time may be more valid at addressing the spirit of stability, although these can primarily be indicators of escalation, which, it can be argued, is conceptually separate from instability. For many of our cases, it was too early to measure these.

Measurements of long-term *stability* should be developed, in addition to measurements of instability. Such "positive" concepts (stability, peace) are hard to measure as anything other than the 'absence' of something else (instability, war, conflict, etc.).

Timing for Measurement — Immediate and predictive indicators can be measured at the completion of the Outcome. Long-term indicators of stability require some time before assessment.

V. RELATIONSHIP BETWEEN PARTIES

This category includes criteria that evaluate the relationship between the parties in the context of the conflict and the Outcome that was produced through the conflict resolution process.

A. Reduction in Conflict and Hostility

A common measure of improvement in conflictual relationships is a reduction in hostility. This criterion attempts to capture a sense of whether the conflict is escalating or not, either in actions, rhetoric or tone of communication. Also included as indicators are various factors taken from the conflict literature on escalation,[4] such as the presence of various hostile tactics, pervasiveness of tension between groups (generality of issues), and the possibility for nonalignment (indicates level of polarization).

Evaluation

Accessibility — Information on this criterion seemed accessible in the media, though it was sometimes sparse. Researchers usually cited brief information and did not use multiple sources (with notable exceptions, see Lower Colorado River and Pyramid Lake), perhaps because they were not readily available. It appears that good relations do not "make news" like hostility does. Most researchers relied on media sources. Parties' websites or newsletters, public relations documents, congressional testimony, and secondary source reports on the conflict were also useful. Some also inferred relations from subsequent litigation behavior or used inside knowledge from their familiarity with parties, although these strategies raise reliability issues.

Reliability — When researchers used multiple publicly available sources, reliability was high. However, most researchers did not use multiple sources for this criterion, relying instead on the impressions of one party, or one party's information on other parties. Where the sources were not clear or extensive, the reliability is only moderate. Others may find contrary evidence of escalation or de-escalation. The nature of the information itself is often contradictory, which will also reduce the reliability of the assessment. The most helpful assessments were those in which researchers also presented information on a baseline from the past for comparison (Lower Colorado River).

Validity — The indicators are valid measures of levels and changes in hostility and of changes in conflict when "conflict" means hostility, antagonism, or adversariness. As indicated above, the measures are most valid when explicitly compared before and after the Outcome. In one case (Mono Lake), media headlines were tracked as an indicator. The validity of this may be questionable if the press emphasizes the perspective of only one or two of the parties instead of reporting many different parties' impressions.

Further Methodology Notes — The indicator "public initiatives, legislative actions" was originally included in the Guidebook to capture possible subsequent hostile tactics by parties who were unhappy with the agreement. No evidence of this was found in these cases; however, some researchers did use examples of positive public initiatives (such as joint projects) as evidence of de-escalation and a collaborative tone among the parties.

Timing for Measurement — This criterion is measurable any time after the Outcome is announced. However, it is advisable to measure before or during the conflict resolution process in order to have a baseline. It is often difficult to assess whether or not the conflict is escalating immediately upon the reaching of an Outcome (such an assessment will likely be mixed), so it may be useful to wait for some time to pass.

B. Improved Relations

Improved relations seeks to capture the changes in the way parties see and relate to one another that may reflect the essence of a successful resolution. To note change, one must first note the nature of the original relationship as a baseline for comparison. Indicators of change include discussions of the relationship itself, as well as the tone of communication among the parties (hostile, conciliatory), the effort parties make to protect themselves, and their sense of trust as indicated by the necessity or lack of enforcement clauses or other formalities.

Evaluation

Accessibility — Information for this criterion was readily available in media sources, and in fact there was much in the media to address the various indicators. In addition, some researchers reported on correspondence between the parties, speeches made by party representatives, party Internet sites, party promotional materials, a report by a third party, and requirements specified in the agreements themselves.

Reliability — Reliability on this criterion will be moderate when compared with other criteria. Because of the subjective nature of relationship assessment, the reliability of that assessment will be linked to researcher's thoroughness in triangulating and verifying information. The varied information for this criterion makes it highly dependent on the type and number of sources consulted. When many sources are consulted, reliability may be high (as in the Lower Colorado River, Pyramid Lake, and Salt River), but this was not done for several cases analyzed for this project. Sources sometimes were not cited, implying that information came from researchers' inferences. In addition, Guidebook indicators for this criterion were not as well developed as others and were left more open to interpretation.

Validity — The criterion "Change in Relationship Quality" or "Improved Relations" should be a highly valid measure of the general category of relationship quality. Criteria indicators themselves are valid inherently but not reliably assessable, which may ultimately reduce their validity. These indicators are a solid beginning but are not complete, and validity would increase with more development.[5] For example, trust was seldom found as operationalized (lack of enforcement mechanisms, etc.). However, researchers found other expressions in the media from the parties that indicated something many would call trust. In these cases, trust (or lack thereof) was found in the tone of and discussion of the question, "will parties follow through?"

Further Methodology Notes — While the indicators specified in the Guidebook for this criterion may have been incomplete, researchers surprisingly were able to find direct discussion by parties of "relationship change" itself as a theme.

As noted above, trust was rarely found as operationalized, although it was sometimes inferred. One researcher suggested that trust is implied when parties are continually working together. Other measures might include the mutual perceived credibility of parties, "acting in a trustworthy manner," and perceptions of good faith in seeking solutions.[6] Building on the legal definition of this last point, "acting in good faith," the operationalization could include the sense that what someone says they are doing is not different from what they are doing. A related dimension, noted earlier, is a sense that a party will follow through on what they say is important. An example of this came from the Lower Colorado River case, when California had to demonstrate to the other parties that it could curtail its water consumption. In the Pyramid Lake case, other parties wanted to see that Nevadans actually would accept a state sales tax to pay for water system improvements necessary to implement the negotiated agreement.

Timing for Measurement — If indicators assess information in the agreement itself or some result of the agreement process, they could be measured immediately. Long term indicators are not developed, but would need the passage of time.

C. Cognitive and Affective Shift

The *cognitive shift* criterion sought to capture evidence of the phenomenon that many practitioners (and parties) note of a shift in parties' framing of the conflict and/or relationship. Indicators include the ways parties referred to one another and they way they described or explained the other parties' behavior (pre- and post-Outcome). Building on some of the literature from family systems theories, the indicators also include questions for narrative analysis of the way "stories" are told about the conflict – do narratives change in their description of causality, interactions, values, etc?

Evaluation

Accessibility — At first glance, it appears that assessing cognitive shift might require a detailed reading of reports on parties' statements during and after the Outcome. Many of our researchers gathered no information on this criterion, perhaps because of the depth of research requested. However, some researchers were able to find direct statements by the parties themselves on the presence of this phenomenon. Though operationalization was difficult, it seemed to have the character of "you know it when you see it" so parties, and thus researchers, reported on it. Information was found in internal party documents, newsletters, commissioned reports, and fundraising brochures. It was also reported in the media, where it was noted when the process and the relationship were perceived as truly representing a new way of doing things and as a shift from past procedures and relationships. Interestingly, it typically took the form of parties expressing changes within themselves, rather than noting changes in others.

Reliability — Reliability for this criterion may be fairly high, though it clearly could be improved through verification from multiple sources. It appears to have a "you know it when you see it" quality. One could legitimately argue that even if only one party sees a cognitive shift, this itself represents an important shift. In other words, this is not a criterion that requires verification across parties and sources, as it is not a characteristic of the agreement or the relationship, but of individual parties. It is often a characteristic of self-realization.

Validity — This criterion should be highly valid as a measure of relationship quality. However, the indicators in the Guidebook used for operationalization of this criterion and based on relevant literature may be of questionable validity. They were at least very difficult to use in practice.

Timing for Measurement — This criterion can be measured upon completion of the agreement, and possibly even during the resolution process if it refers to perceptions of a new process and a new relationship. *Cognitive and affective shift* may be important for catalyzing an agreement.

D. Ability to Resolve Subsequent Disputes

Ability to resolve subsequent disputes addresses the degree to which the relationship between the parties contains the capacity for handling future conflict. Indicators include evidence that problems are handled constructively, that an ongoing relationship has developed which allows future concerns to be addressed, and possibly the emergence of an ongoing forum for conflict management.

Evaluation

Accessibility — Though it could be deduced from information in common sources, such as the media and party newsletters, this criterion appeared to require a bit more detective or inferential skill on the part of the researcher. That is, the indicators proposed in the Guidebook (and also added by researchers) required them to find evidence that the parties constructively handled subsequent issues. Researchers often inferred this from the way the parties addressed the achievement of an Outcome, often noted in the media, or from processes written into the Outcomes themselves. Sometimes it was inferred from subsequent (negative) actions. Parties' press releases, public speeches, and third party reports were also used to assess this criterion. Triangulation and verification of this criterion needs to be stressed to researchers, precisely because of the temptation to rely solely on speculation.

Reliability — Because this criterion often required inference from the researcher, its reliability was compromised accordingly because inference processes are less replicable across researchers. For the few cases in which researchers relied less on inference and instead identified the variety of possible mixed indicators for this criterion (Mono Lake, Pyramid Lake), one can feel more secure in claiming high reliability for the assessment produced.

Reliability is less likely because the assessment often was not based on information on parties' "track record" per se, but rather it was inferred from agreement mechanisms without also assessing post-agreement follow-up.

Validity — In principle, this criterion appears to be a critical component of assessing the quality of the relationship between the parties. Unless parties interact constructively, future conflicts will once again derail relations, and even the current Outcome's implementation may be threatened.

For many of the cases, however, it was too early to assess this criterion directly, so inferences were made from indicators (e.g., agreement wording) that may be less valid in representing future ability to resolve disputes.

Timing for Measurement — By definition, "future" disputes have not yet occurred, so they must be inferred and predicted from current information. These inferences could be made from information available immediately upon agreement, or even earlier, during the process of reaching the agreement. If, however, the interest is in assessing parties' "track record" of subsequent dispute resolution, it must be measured after some time has passed.

CONCLUSIONS

As a result of systematically evaluating the criteria through their application to case studies, we have learned several things about the criteria themselves as well as about our framework methodology and procedures for evaluating success in ECR cases. We have gained insight on how to better evaluate the accessibility, reliability, and validity of the success criteria, and on the best timing for measuring the criteria. We also identified specific challenges encountered in operationalizing some of the criteria, and we discussed strategies to address these measurement challenges.

We confirmed one of our initial research hypotheses: that success criteria vary in the ease with which they can be assessed. Success criteria may exist 'in theory' as part of a conceptual definition of success, and yet be difficult to assess in practice. If a criterion cannot be practically and reliably assessed, it may be unrealistic to consider it as part of one's operating definition of success. One example of this difficulty was the criterion "Public Acceptability." Our literature review suggested that one widely held perspective on success in ECR is that not only the stakeholders must consider the Outcome acceptable, but there must be public acceptance as well.[7] However, researchers in our study found little or no information on reactions to the Outcome beyond the involved parties themselves.

Baselines were another area where difficulties were encountered. Implicit in many success criteria is change in the dimension itself from some baseline state (improved relations, reduced hostility). We learned that such criteria require a clear definition of the baseline against which change is to be measured, in order to be effectively applied to evaluating cases.

With respect to compliance, we observed that it is necessary to identify a baseline performance standard in the Outcome so that compliance can be monitored. It became clear in our evaluation of the case analyses that researchers need to be made more aware of the notion of baselines and performance standards, as well as the importance of identifying them correctly.

The use of multiple sources, or triangulation, was an issue of both accessibility and reliability, and it relates to researcher thoroughness. For many of the criteria, it was essential to consult multiple sources to balance the subjective nature of the data. However, either multiple sources were often not available or the effort was not expended to consult multiple sources. A lack of triangulation produces criteria assessments that are questionable in their reliability and validity.

In a similar vein, when diverse sources were not consulted (either due to inaccessibility or low effort), we found an overemphasis, and sometimes a sole reliance on, media sources for certain criteria. Triangulation of types of data, for example, also including parties' internal publications, external reviews, website information, etc., would be desirable and would improve reliability and validity. Access to different types of information varied greatly across cases and criteria. Our analysis revealed the emergence of the internet as a powerful tool for both researchers and parties in furthering their agendas and accessing information. We found local access to information important for researchers, particularly in locating information that was generated before the internet was widely used or available.

Moving specifically to the topic of criteria reliability, we found that researchers with specialized knowledge presented both advantages and disadvantages. Researchers with specific background knowledge, either in a field such as economics or law, or with prior experience in environmental conflict resolution, had extra access to information and/or better strategies for locating and understanding information. However, they also produced information that would be unlikely to be replicated by other researchers with less extensive backgrounds. Thus, the reliability question of producing similar results across assessments becomes a tradeoff with the quality and thoroughness of the research conducted.

Some criteria were less valid as indicators of success for certain conflict resolution processes because these processes did not have these criteria as goals. For example, litigation does not usually attempt to be inclusive of all stakeholders in its process, and private stakeholder negotiations may not seek public participation. As introduced earlier, Table 9.1 summarizes our evaluation of the accessibility, reliability and validity of the criteria.

One of the more important and useful findings of this research was the identification of strategic times for assessment, which vary with criteria. As introduced earlier, Table 9.2 summarizes our findings on the best times during the lifecycle of an ECR process to obtain the information needed to operationalize a criterion and apply it to a specific case. As outlined above, our analysis suggested that many of the criteria should be assessed at several points in time over the course of a conflict: (1) baseline, before the resolution process, (2) during the resolution process, (3) immediately upon achieving the Outcome, (4) short-term after the Outcome is reached, and finally, (5) long-term after the Outcome has been achieved and implemented (see Figure 9.1 above). Most criteria cannot be assessed at every stage. Knowing which criteria to assess at which stage in the ECR process can make evaluation easier. Any success criteria that involve measuring a change will require a baseline against which to gauge any movement.

The exercise of evaluating the accessibility, reliability and validity of each success criterion proved valuable. We hope that the findings in this chapter will be helpful in future efforts to evaluate ECR. Some of our findings may be influenced by the nature of our cases studies – western U.S. water conflicts. Application of the success criteria to other classes of environmental disputes would yield additional insights on the usefulness and applicability of the criteria. Many of our methodological findings from evaluating the criteria have been incorporated into a revised and updated Guidebook (Appendix A), which contains more complete and explicit instructions to researchers.

Table 9.1 Assessing the Criteria

Criteria	Accessibility	Reliability	Validity
I. OUTCOME REACHED			
A. Unanimity or Consensus	Mixed	High	Mix./Mod.
B. Verifiable Consensus on Terms	High	High	Mixed
C. Publicized media event around reaching agreement	Mod.	High	Mod.
D. Ratification	Low	Mod.	High
II. PROCESS QUALITY			
A. Procedurally Just	Low	Mixed	High*
B. Procedurally Accessible and Inclusive	Mixed	Mod.	Mixed
C. Reasonable Process Costs	Low	Moderate	High*
III. OUTCOME QUALITY			
A. Costs of Implementing Outcome	Low	Mod.	Mixed
B. Sustainability			
1. Financial Sustainability	High	High	Mod.**
2. Cultural Sustainability	High	High	Mod./High**
3. Environmental Sustainability	High	Mod.	Mod.**
C. Clarity of Agreement	Mod.	Mixed	Mod.**
D. Feasibility/Realism	Low	Low	Mod./High
E. Public Acceptability	Low	Low	Mod./High

*If researched well

**Need some passage of time

Table 9.1 Assessing the Criteria, continued

Criteria	Accessibility	Reliability	Validity
IV. RELATIONSHIP OF PARTIES TO OUTCOME			
A. Satisfaction/Fairness - As assessed by parties	High	Mixed	Mixed
B. Compliance with Agreement Over Time	High	High	High
C. Flexibility	High	High	High
D. Stability/Durability	Low	Mod./Mixed	Mixed
V. RELATIONSHIP BETWEEN PARTIES			
A. Reduction in Conflict and Hostility	High	Mod./High	High
B. Changes in Relationship Quality/Improved Relations	High	Mod./Mixed	High*
C. Cognitive Shift	Mixed	High	High/Mod.
D. Ability to Resolve Future Disputes	Low	Mod./Low	Mod.**

*If researched well

**Need some passage of time

Table 9.2. Timing over Lifecycle of ECR Process for Assessment of Criteria

Criteria	Baseline Before ECR Process	During ECR Process	When Outcome Achieved	Short Term	Long Term
I. OUTCOME REACHED					
A. Unanimity			X	O	O
B. Verifiable Consensus on Terms			O	O	O
C. Publicized media event around reaching agreement			O	O	O
D. Ratification			X	O	O
II. PROCESS QUALITY					
A. Procedurally Just		X	X	O	O
B. Procedurally Accessible and Inclusive		X	O	O	O
C. Reasonable Process Costs		X		O	O
III. OUTCOME QUALITY					
A. Cost Effectiveness			O	O	X
B. Sustainability					
1. Financial Sustainability			X	O	X
2. Cultural Sustainability/Community Self Determination			X	O	X
3. Environmental Sustainability			O	O	X
C. Clarity of Agreement			O	X	X
D. Feasibility, Realism			O	O	O
E. Public Acceptability			X	O	O

Key: **X**= Best time to Measure **O** = Good time to measure

Table 9.2. Timing over Lifecycle of ECR Process for Assessment of Criteria, continued

Criteria	Baseline Before ECR Process	During ECR Process	When Outcome Achieved	Short Term	Long Term
IV. RELATIONSHIP OF PARTIES TO OUTCOME					
A. Satisfaction/Fairness - As assessed by parties			O	X	O
B. Compliance with Agreement Over Time			X	X	X
C. Flexibility			X	O	X
D. Stability/Durability			X	O	X
V. RELATIONSHIP BETWEEN PARTIES					
A. Reduction in Conflict and Hostility	X			X	X
B. Changes in Relationship Quality (e.g. trust)	X	O	O	X	X
C. Cognitive Shift		O	X	O	O
D. Ability to Resolve Future Disputes		O	O	X	X

Key: **X**= Best time to Measure **O** = Good time to measure

NOTES

[1] For the last criterion in the original project case analyses, "Transformation," information was collected for only two cases. Thus this criterion was not able to be evaluated.

[2] For details on the specific indicators used to operationalize each criterion, please consult the Guidebook in Appendix A.

[3] The relationship between benefits and costs is examined under the criterion *Perceived Economic Efficiency*, which was subsequently added to the Guidebook (see Appendix A). Because this criterion was not part of the comparative case study, it is not analyzed here.

[4] Jeffrey Z. Rubin, Dean G. Pruitt, and Sung Hee Kim, *Social Conflict: Escalation, Stalemate, and Settlement,* 2nd ed. (New York: Colin McGraw Hill, 1994).

[5] See also Tamra Pearson d'Estrée, "Achievement of Relationship Change," in *The Promise and Performance of Environmental Conflict Resolution,* eds. Rosemary O'Leary and Lisa B. Bingham, 111-128 (Washington, DC: Resources for the Future, 2003).

[6] Gail Bingham, *Resolving Environmental Disputes: A Decade of Experience* (Washington, DC: The Conservation Foundation, 1986).

[7] Christopher W. Moore, *The Mediation Process: Practical Strategies For Resolving Conflict* (San Francisco, CA: Jossey-Bass, 1987).

Chapter Ten

REFLECTIONS ON RESOLVING WATER CONFLICTS

The eye sees not itself
But by reflection.

William Shakespeare (Brutus in Julius Caesar)

REVIEW OF THE BOOK

In Chapter 1, we introduced the benefits of developing a systematic framework for evaluating environmental conflict resolution. We created this framework with encouragement from conflict resolution practitioners, elected officials, environmental advocates, public agencies, researchers and university instructors. We hope the book helps to craft a collective understanding of effective strategies to address environmental conflicts. Chapter 2 demonstrates the various fields of inquiry and literature (from economics to family therapy to social psychology) upon which we draw to identify criteria for evaluating environmental conflicts. Many of the twenty-eight criteria are not new, but have been applied in other evaluation contexts. Most of them have not previously been applied to environmental conflicts nor to comparisons across cases. The criteria are integrated into a cohesive framework for evaluating environmental conflict resolution. We go beyond the existing body of work by taking multiple success definitions – criteria – organizing and elaborating them within conceptual categories, operationalizing those criteria and then applying them to actual cases. While other lists of criteria exist as well as and many other case studies of conflicts, a full set of criteria have never before been brought together and applied to multiple cases. All of the criteria are not equally applicable to every type of conflict. Our intent was to create a comprehensive menu from which readers could choose the subsets that best suit their own cases and interests. The comprehensive menu itself presents a discussion list of the multiple and often conflicting goals desired in the resolution of water conflicts.

Chapter 3 summarizes the eight cases and our methodology for evaluating them in a consistent manner. Four of the eight cases are presented in detail in Chapters 4-7. Chapter 8 highlights patterns across the cases and Chapter 9 directly assesses the validity, accessibility and reliability of the success criteria themselves.

Here, in our concluding chapter, we glean last insights from the case evaluations for: a) innovative problem solving in environmental conflicts, b) questions needing further research and c) the interplay between public policies and successful environmental conflict resolution.

INNOVATIVE PROBLEM SOLVING

One of the virtues of accumulating case studies is that these cases can provide examples of challenges well-solved. Our detailed analysis of the cases revealed some interesting practical approaches to achieving various elements of success. We highlight these approaches below, organized by success criterion.

Cost-Effective Implementation

In the Lower Colorado River case, we find elements that both violate and encourage cost-effectiveness. Arizona subsidizes the agricultural use of CAP water as part of its strategy for managing its Colorado River water within the federal and interstate management framework for the river. Such subsidies generally go against the idea of cost-effectiveness by making water cheaper to farmers, thus undermining incentives for conservation and more careful management; subsidized prices also may obstruct voluntary transfers of water to urban or environmental uses.

However, the rules adopted in this case do encourage a more cost-effective allocation of water between the states by providing mechanisms for interstate water exchanges. The rules explicitly state that U.S. taxpayers will not subsidize any water banking and exchanges between the parties. The cost-effective management of water was specifically noted as a goal of the federal administrative rules for the Lower Colorado River.

Financial Feasibility/Sustainability

Among the cases, many examples exist of innovative approaches to this criterion. In the Edwards Aquifer case, the agreement mandates that the agricultural pumping fees be limited to one-fifth of urban fees given agriculture's limited financial resources. This clearly makes the Outcome more affordable to farmers, although it also may prevent them from facing the real economic cost of water and may shift those costs over to urban water users. The Edwards Aquifer Outcome specifically requires that the costs be borne by aquifer users with no taxpayer subsidy, and that compensation must be provided for the "taking" of any private property. In order to make its terms more palatable, the legislation defers the most severe cuts in groundwater pumping to future years. However, water providers are required to establish pricing structures that encourage conservation. Loans are available to promote conservation and the reuse of wastewater, while cities propose to assist farmers with agricultural conservation costs.

The Lower Colorado River case provides examples of water leases between tribes and cities, and it mandates greater water re-use. The Rules specifically require the parties to consider financial impacts by limiting unfunded mandates imposed on all non-federal governments and private parties to less than $100 million per year.

In the Mono Lake case, state and federal agencies provided the funds to assist the LADWP in replacing water supplies it can no longer take from the Mono Lake Basin. Cost estimates were developed for conservation and water reclamation programs that will provide replacement water and replace the hydropower that will no longer be generated along the aqueduct. The Board specifically noted that the costs of the replacement water and power were "reasonable," considering the benefits of environmental restoration and comparable water costs elsewhere in urban Southern California.

In order to satisfy the Pecos River agreement, New Mexico sells bonds to raise funds for acquiring water rights. In the Pyramid Lake case, portions of the agreement specify how much money the state and federal authorities must contribute to protect the wetlands and fisheries, to the tribal economic development fund and towards municipal water conservation measures. The Snowmass Creek case study noted that public entities could cover their costs through fee collection mechanisms.

The Salt River case also devotes specific attention to this criterion. The agreement specifies: a) water users must share the costs of various infrastructure and settlement funds, as well as payment of any compensation (including debt forgiveness) for water and water storage capacity; and b) substitutions of renewable water use (effluent and surface water) for groundwater pumping. The terms of water leases from the tribe to the cities are also specified.

Looking at the body of cases as a whole, several of the non-litigation cases (Edwards Aquifer, Mono Lake, Pyramid Lake, Salt River) explicitly provide loans to specific parties to assist with implementation costs. These cases also establish specific mechanisms to raise funds to cover costs, such as increased water rates, water use fees, the issuing of public bonds and contributions from the federal budget. Water leases and water acquisitions were authorized through legislation, administrative rule making and negotiated agreements in several cases to make more efficient use of regional water sources (Edwards Aquifer, Mono Lake, Pyramid Lake, Pecos River and Salt River).

A couple of cases (Edwards Aquifer and Lower Colorado River) specifically prohibited or limited any unfunded mandates. Ability to pay issues were addressed in several cases (Edwards Aquifer, Mono Lake and Snowmass Creek). For the implementation of the Edwards Aquifer and Pyramid Lake agreements, water pricing reforms were required to encourage more efficient water use.

Cultural Sustainability/Community Self-Determination

Each case placed different degrees of emphasis on this criterion. Tribal concerns about cultural resources are most notable in the Mono Lake, Salt River and Pyramid Lake cases. Decision-makers attempted to address these concerns in the agreements and administrative rules, and federal laws intended to protect cultural resources played a key role. The Mono Lake agreement requires the development of a plan to protect all known (and yet to be discovered) cultural resources and to provide access to traditional native uses of resources around the lake. The negotiated agreements in the Pyramid Lake and Salt River cases attempted to address tribal concerns about their sovereignty.

Many cases involve tradeoffs between urban areas and water for agriculture. The Edwards Aquifer, Lower Colorado River, Mono Lake, Pyramid Lake and Salt River cases all involved a competition between farms and cities for limited regional water supplies. Those cases provide mechanisms to transfer water from irrigation use to urban when farmers voluntarily agree to sell or lease their water. However, the voluntary nature of these transitions has not eased concerns about the impacts they may have on farming and farm communities. Tension between rural and urban ways of life, and fears that rural culture will be diminished, permeate many western U.S. water conflicts.

Environmental Sustainability

This criterion is central to many of the cases. The Edwards Aquifer agreement specifically addresses water needs for drought and endangered species. In response, urban areas committed themselves to water conservation and re-use in order to reduce groundwater pumping. The requirements of the Mono Lake ruling are tied to maintaining environmentally acceptable lake levels, which addresses drought effects on the ecosystem. The Pyramid Lake agreement mandates urban water conservation and specifically considers endangered fish and wetland water needs. Environmental organizations and agencies have been involved throughout the implementation of the Truckee-Carson Basin Settlement. The settlement contains specific provisions dealing with drought and environmental groups and agencies have been buying water rights to provide water for environmental needs.

In the Salt River case, environmental sustainability was promoted by switching water users from local pumped groundwater (which is being depleted) to imported renewable surface water supplies. However, it requires a large amount of electric power to convey the imported water to central Arizona, which raises other environmental concerns such as air quality impacts from power generation. During drought, groundwater reserves act as a back-up water supply. Limits on agricultural groundwater use are included in the Salt River agreement, along with provisions to reduce groundwater overdraft and deal with unplanned shortages of Central Arizona Project (CAP) water. The agreement attempted to remain consistent with the Arizona water management, which limits groundwater overdraft. The follow-up negotiated agreement in the Snowmass case makes specific considerations for dry, normal and wet years.

Overall throughout the cases, several provide for cities to acquire water from irrigators to ensure that they can satisfy urban water needs during drought. The Mono Lake, Snowmass Creek and Pyramid Lake cases all include provisions for protecting water-dependent habitat during dry years. The Salt River case encourages a transition from groundwater overdraft to renewable surface water.

Clarity of Outcome

One key factor for this criterion was the establishment of baseline conditions and performance standards. The cases illustrate various strategies to address this. The legislation in the Edwards Aquifer case sets clear goals for groundwater management as well as standards to follow in assigning pumping rights. In the Edwards Aquifer case, baseline conditions are specific in terms of who is eligible to apply for a groundwater pumping permit and the basis for granting permits. However, the actual process of issuing the permits has not proved to be as clear-cut, since each party interprets the rules in their favor.

The parties in the Mono Lake case each argued for different lake levels to serve as the performance standard and the Environmental Impact Report examined the ecological implications of maintaining different lake levels. The Mono Lake ruling set explicit standards for maintaining lake levels, as the negotiated agreement in the Snowmass Creek case did for stream flows. The courts in the Pyramid Lake case set standards for river flows and these are incorporated into the negotiated agreement, which also set standards for providing flows to maintain and restore wetlands. The Pecos River ruling set performance standards for river flows from New Mexico into Texas. Overall, many different strategies were used to set performance standards for implementing the various Outcomes.

CONTINUING QUESTIONS

Accumulating cases can also highlight both patterns and gaps in our understanding. The case analyses suggest a number of avenues for further inquiry. These are highlighted below, organized by success criterion.

Procedurally Just

At least two future research topics regarding procedural justice emerged. Negotiations often occur behind closed doors to allow the parties to consider new ideas without public commitment, so observers may perceive that some parties and/or views have been excluded, or neglected. In contrast, administrative processes require public comment and allow the expression of multiple views, lending the impression that they are accessible to all stakeholders. Only one negotiation case (Salt River) was seen as accommodating to all interests, so it would be interesting to determine why these negotiations were perceived as more inclusive than others.

The ability of certain parties to unduly influence an Outcome, or perceptions of "unequal weight," was noted in several cases (Edwards Aquifer, Lower Colorado River implementation, Pyramid Lake, Salt River). It is interesting to note that each case involved multi-party negotiations, in which the structure presumably invites all parties as equals. Perhaps there is an "assumption" or expectation of equal opportunity in negotiation, an expectation that is not present in litigation or administrative rulemaking. Therefore, when this equality does not in fact materialize it may prove to be a source of disillusionment.

Procedurally Accessible and Inclusive

Several of the cases (Edwards Aquifer, Lower Colorado River, Mono Lake, Snowmass Creek), included opportunities for public participation, or at least public "input," in the decision-making process. The efforts made to notify the public varied widely with each case. For our purposes, 'the public' is defined as citizens who may or may not be direct stakeholders. Administrative actions require public notice and comment, so Administrative Outcomes (Mono Lake, Lower Colorado River) typically had extensive opportunities for comment, although the publicity may have been narrow and limited in scope. (e.g., confined to newspaper announcements and Federal Register). Though cases with legislative Outcomes typically featured many hearings and possibly even citizen committees, the case reports suggest that information on these hearings and meetings was not always widely dispersed (Edwards Aquifer, Pyramid Lake, Salt River, Snowmass Creek).

In general, across cases of all types, public access to information and public notice seemed limited. This apparent lack of public notice may stem from two sources. Further research would be necessary to clarify the relative contributions of each. First, there may genuinely have been a lack of public notice. The parties may expend minimal effort in order to conserve resources, or simply may not consider the many ways that public views could be solicited and incorporated. Alternatively, the apparent lack of public notice may be spurious, in that some forms used to notify the public (radio broadcasts, paid advertisements) are not archived and thus inaccessible at a later time by case researchers.

A larger research question can be raised about public participation and access and its importance to the concept of successful resolution. Cases that some might regard as successful, with a "quality process," did not always include open public processes. Researchers need to consider methods to measure process quality for confidential and private conflict resolution processes.

Financial Feasibility/Sustainability

Based on information from the case studies, it appears that litigation does not address financial and economic issues in as much depth as legislation, administrative rules and negotiated agreements. Further research would be useful to compare the thoroughness of different Outcomes and the ramifications of differing levels of attention to financial and economic matters.

Compliance with Outcome Over Time

Almost every case was missing records of compliance from a monitoring entity. For most cases, it may have been too soon in the implementation process for these to have been established. Or the researcher simply could not find such records. In the one case for which monitoring records were located (Pyramid Lake), they had to be repeatedly requested and one large record arrived so late it could not be analyzed. Our case research suggests that few records of compliance actually are kept, but this deserves further investigation.

Stability/Durability

Various types of incentives were used in the different cases to promote this criterion, including: penalties (for unauthorized pumping in Edwards Aquifer; for wasted water in Salt River); deadlines (for reducing withdrawals in Edwards Aquifer; for requests, temporary storage, permits in Salt River; time frames in Pyramid Lake legislation); monthly reporting in Salt River; public agency action plans in Pyramid Lake; and rewards (additional water can be diverted under specific circumstances in Mono Lake). Given that the literature on effective strategies for shaping behavior demonstrates that positive reinforcement is more effective than negative reinforcement, more research on positive incentives should be explored, and more examples of positive incentives should be documented.

Reduction in Conflict and Hostility

Hostilities take the form of rhetoric, threats, or actions. A hostile rhetoric or climate included the use of in-group sanctioning against members of the group who might interact (verbally or economically) across 'conflict lines.' This produced a sense of fear among some parties (Edwards Aquifer). Threats were sometimes issued before the Outcome (by the state of Nevada in Lower Colorado River) or subsequent to it (by the utility in Snowmass Creek). Further research is needed to understand which parties resort to threats and why. Is it the parties that traditionally have "power," those who feel "powerless," or powerful parties that believe their power is slipping away? Some examples of hostile actions included cutting water off (Big Horn), seeking a temporary restraining order (Edwards Aquifer), and resuming litigation.

Transformation

In the Mono Lake case, the theme of a new environmental ethic emerged along with new ways of managing resources. Sources noted that a "new era of [California] politics" had begun. In the Pyramid Lake case, the transformation was characterized as a "paradigm shift" in social values and priorities. Further research is needed to find out why such a dramatic change was perceived in these two cases, but not in other cases that had positive results, such as the Salt River case.

In sum, numerous avenues for future research were suggested by our initial findings and analyses. In addition to practical questions of how to best address criteria, more basic research questions emerged from these comparisons, as well as areas for future methodological refinement. In our final section we turn to questions for policy.

PUBLIC POLICY AND ENVIRONMENTAL CONFLICT RESOLUTION

Public policies set the stage on which conflicts emerge and are addressed. Policies can aggravate conflicts through ambiguous or divisive provisions. Polices also have the potential to prevent and manage conflicts through encouraging alternative dispute resolution, through clarity in the processes used to set policies and in the material content of the policies themselves. Policies can be structured to address conflicts early, and to anticipate and prevent some conflicts. Public policies can facilitate effective use of information, provide forums for multi-party negotiation, and provide bargaining power for interests which society wishes to promote and which otherwise might not have an influence on the outcome. Policies also heavily influence the strategies that parties employ to resolve conflicts, by affecting the relative costs and benefits of litigation, or of participation in multi-party negotiations.

Several expert groups have recently considered how water policies might more effectively prevent and manage conflict. One report directed toward states recommends that state policy makers do the following: focus on integrated watershed planning and management rather than on political boundaries unrelated to watershed geography (county lines, etc.); use incentives to encourage agency participation in watershed initiatives; facilitate improved communication and coordination among agencies; provide state monies for watershed efforts; establish standards that initiatives must meet to obtain state recognition and funding; and refrain from transferring accountability and authority that now resides with the state to local initiatives.[1] Along similar lines, the Western Governor's Association developed the "Enlibra" principles to guide governments in addressing environmental conflicts. Among the eight principles are recommendations to use collaborative problem solving processes, to rely on market incentives to achieve environmental goals, to address problems across political boundaries so that whole watersheds are considered, to solicit independent scientific expertise and reduce the problem of dueling experts, to assess costs and benefits of problem solving options and to encourage state and local planning to comply with national environmental standards.[2]

The public interest can be greatly enhanced by policies that facilitate effective conflict resolution. In this section, we summarize observations from our case analyses on the role of public policies in environmental conflict resolution.

Policies That Encourage Alternative Dispute Resolution and Consensus Building

Some policies encourage use of Alternative Dispute Resolution (ADR) such as negotiation and mediation for disputes involving public agencies. In 1998, U.S. President Clinton specifically directed each federal agency to take steps to promote greater use of mediation, arbitration, early neutral evaluation, other ADR techniques and negotiated rule making; and established an inter-agency ADR working group to encourage and facilitate agency use of ADR. Many state governments are working to encourage ADR in conflicts involving state agencies.[3]

Public agencies have funded ADR efforts for specific conflicts and have provided trained agency staff to facilitate ADR efforts. For example, the Department of Interior provided the time and expertise of a top federal official for several years to facilitate settlement of the Bay-Delta dispute in California. The Salt River Project (a large water and power provider in Arizona) dedicated the time of its lead water attorney to facilitate resolution of rural-urban disputes over water in the 1980s. Legislatures and public agencies sometimes offer financial incentives to encourage ADR instead of litigation. For instance, the California legislature provided funds for Mono Lake restoration that were contingent on the stakeholders reaching consensus on plans to restore the lake and its environs.

Public notification and public involvement requirements direct agencies to undertake actions that may help build consensus. For example, the U.S. Forest Service must solicit public input when revising and updating forest management plans. State and federal agencies increasingly work with local resource users and citizen groups to resolve problems. It has been argued that agency participation is essential to the success of local initiatives that seek to build consensus on water management among diverse interests.

State and federal agency participation can bring credibility, continuity, money, technical expertise and management and regulatory experience to the process. Moreover, agencies need to take the lead in implementing management strategies for resources under their jurisdiction and for which they are held accountable. There is now a well-placed emphasis on local stakeholders and local needs in resolving conflicts that involve state and federal lands and water. Recognition of the essential role that public agencies play moves local consensus-building initiatives beyond a simplistic "anti-government" paradigm to a more holistic and implementable public-private partnership model.

Policies That Encourage Litigation and Discourage Multi-Party Negotiations

Federal agency budget incentives discourage participation in consensus building and facilitated negotiations when an agency can spare its own budget and staff time by turning a conflict over to a separate agency for litigation. For instance, federal agencies refer conflicts under their jurisdiction to the U.S. Department of Justice to handle litigation. Litigation may be more attractive to agency officials than addressing the conflict through lengthy participation in multi-party negotiations, given tight budgets and overworked staff.

Policies that require public access to meetings, memos and meeting records involving public agencies (open records laws and the federal Freedom of Information Act) can discourage private parties from participating in negotiations for fear that financial matters they would prefer to keep confidential will become a matter of public record. This particularly is a concern of private firms, such as water and power utilities, which operate in a competitive market and which may not want internal financial matters to become public. The Federal Advisory Committee Act (FACA) is intended to ensure open processes and adequate representation on committees that advise federal agencies. Since its passage in 1972, it has helped promote greater balance of interests on councils and committees working on federal matters.[4] However, FACA's chartering and approval requirements create costs and delays for both agencies and participants. Litigation over FACA provisions has made some agency personnel wary of collaborative processes.[5]

Policies can exacerbate conflict when representation in water decision making is out of line with modern demographics and political realities. For instance, there is a lack of urban and recreational water user representation in many states' water management institutions. It is still quite common for state water commissions to favor representation from traditional resource user groups, primarily agriculture. While agricultural representatives are essential, given that agriculture is a major water user, over-representation leads to increased conflict as public officials make choices inconsistent with new social values that view urban growth, recreation and environmental protection as important water uses.

Policy Characteristics That Help Prevent and Manage Water Conflicts

The previous section gave examples of policies that affect resolution of water conflicts. This section discusses generic characteristics of policies that can help prevent and manage disputes.

Clarity and Consistency. Policies can reduce uncertainty for parties by clearly stating the protection afforded to water right holders, water user responsibilities, and agency criteria and procedures for addressing water problems. For instance, state procedures for transferring water rights generally specify the process that must be used, information to be provided and valid grounds for objecting to a proposed transfer.

Policies need to be meaningfully and consistently enforced. Inconsistency and contradictory signals from different levels within an agency or from differing agencies have been a classic problem among federal agencies. Inconsistency can aggravate water conflicts because parties do not know what rules apply or what to expect.

Interjurisdictional Coordination. For rivers, lakes and aquifers extending beyond the borders of a single state, three means of formally resolving interstate conflicts over water allocation have evolved in the U.S.: (a) negotiations among the states resulting in an agreement (an interstate compact) that is ratified by Congress; (b) legislative allocation through an act of Congress in the absence of a negotiated agreement among the states; and (c) litigation in the federal courts resulting in judicial allocation among the states. Negotiated agreements formalized as interstate compacts are common in the western states, where there are nearly two dozen such compacts.[6] Compacts specify each state's right to use water from the source governed by the compact. For compacts to be effective they must be enforceable, with penalties for violation. This, in turn, depends on the costs of measuring water use, monitoring compliance and taking enforcement action when violations are suspected. Enforcement has proved costly. When an injured state brings action in federal court, interstate compact disputes are heard by a Special Master appointed by the U.S. Supreme Court. Several of the eight cases we evaluated involve interstate compact conflicts (Pecos River, Pyramid Lake, Lower Colorado River).

Compact violations that continue for decades and cumulatively involve illegal uses of large amounts of water are not rare. Compact violations can be economically rational when the expected penalties are small and a lengthy court process defers the moment when a violating state actually will be ordered to "pay up." Moreover, if the marginal value of water is relatively low in the state being shorted, it will not be worth their while to undertake expensive prosecution of the violator.

Like other elements of public water policy, interstate water allocation compacts need to specify penalties for violation and specify monitoring and enforcement mechanisms that do not require lengthy court proceedings to implement. Compact provisions also need to account for uncertainties so that the parties do not have to go to court each time a drought or other emergency occurs.

Equitable cost sharing. Once technical solutions to water conflicts are identified and the parties agree on the problem-solving approach, the next challenge is how to divide up the costs of implementing a solution. If the costs are non-trivial and there are no deep pockets that readily can be tapped, then cost-sharing becomes a highly conflictual issue. Cost-sharing was problematic in many of our case studies (Edwards Aquifer, Mono Lake, Pyramid Lake, Salt River). Parties hold quite different perspectives on what constitutes "fair" allocation of costs. For instance, in conflicts over providing water for environmental needs – environmental interests may argue that those who have used a lot of water in the past (farmers, industry and cities) are the cause of stream dewatering and so ought to bear the costs of a solution. Historical water users, on the other hand, will argue that they were using water in a manner legitimized and encouraged by past water policies and ought not to be penalized retroactively. They will argue for a "beneficiaries pay" principle of cost sharing under which environmental interests and public agencies with fish and wildlife responsibilities bear most of the cost.

These diverse views were cogently articulated in several case studies (Mono Lake, Pyramid Lake, Edwards Aquifer). Environmental advocates argue that the "beneficiary pays" principle ignores the critical issue of "how we got here in the first place" and criticize cost-sharing plans that ignore decades of environmentally damaging water development activities. Public policies need to provide cost sharing principles to guide parties in addressing such dilemmas. In the past, federal money helped ease water conflicts by developing new supplies at little cost to local water users. Now, much of the cost burden falls on state and local governments and water users, making disputes over cost sharing more intractable.

Bargaining power for interests representing social values. There are four principal sources of bargaining power in water conflicts: wealth, political connections, property rights and law. The latter two are directly determined by public policy. Water rights initially were recognized by, and are governed by, state water law. Laws to protect particular interests arise from the courts, legislatures and administrative agencies. The structure of U.S. western water law provides strong bargaining power to historic consumptive users of water, through prior appropriation water rights, water use permits and long term contracts to use water and electric power provided by public water projects. Those interests which have money to buy water also have bargaining power – thus the influence of growing cities, resort developers and industry on water allocation and water policy. The bargaining power that has given tribes and environmental interests an influence on water management comes from yet another source – federal and state laws and related court rulings that require protection of endangered species, water quality and wetlands. Any type of water use that has value to society (from growing cotton to restoring fish populations) needs to possess some form of bargaining power in order to influence water management. Public policies can deliberately create bargaining power for previously under-empowered interests.

Building Better Problem Solving Capacity. Policies that encourage coalition building and extended dialogue (as often occurs when negotiating an agreement or drafting legislation) build better working relationships than litigation, though parties left out of legislative processes or negotiating sessions continue to work to undermine implementation even after an agreement is signed or legislation is passed. Policies that require parties to work collaboratively build better and more enduring working relationships. Social investment in collaborative processes is more likely to build up social capital that can yield tangible long term benefits (more cost effective problem solving, for instance) than adversarial processes.

The Influence of Public Policies on Achieving Success

Public policies influence the ability of conflict resolution processes to satisfy the varying success criteria. Here, we refer to specific information in case analyses to highlight the interaction between policies and specific success criteria.

First, we consider the "Outcome Achieved" category and its specific success criteria. Public policies have a direct influence on the criterion *Unanimity or Consensus.* Laws and administrative policies dictate the procedures by which judges issue an opinion, agencies promulgate rules and legislative bodies pass new laws. Policies also indicate what percentage of a legislative body must approve proposed legislation for it to become law.

With respect to *Verifiable Terms*, policies require that court and administrative rulings, as well as legislation, be in writing. In addition, agencies have internal guidelines about publicizing agreements to which they are a party.

Ratification procedures are direct matters of policy. Policies dictate whether formal ratification and judicial approval is necessary for a particular type of agreement.

Public policies also are important in the Success Category "Process Quality." Different types of conflict resolution processes were perceived quite differently with respect to the criterion *Procedurally Just.* Policies that require input from affected interests and notice to parties, as in water right adjudications, help make processes more accessible and inclusive. Analysis of the eight case studies reveals that different types of problem solving processes are "ranked" differently on procedural justice. Legislative processes were perceived as rushed and incomplete. Litigation and court rulings were seen as fair, but protracted and inefficient and removed from public debate. Administrative processes were perceived as thorough and not rushed, but potentially subject to the influence of more powerful actors.

Policy requirements and regulatory deadlines that impose a time line on formal problem-solving processes may leave inadequate time for negotiations and consensus building. However, these policies also give environmental advocates a standard of timeliness and responsiveness to which they can hold public agencies accountable. Policy deadlines help prevent "stalling" by agencies reluctant to take action against politically powerful water users, such as cities and irrigation districts. Deadlines also can provide strong incentives to achieve a settlement.

Based on the case analyses, the implications of policy-mandated deadlines for water dispute resolution are mixed. On the one hand, it is advantageous for agencies to have flexibility in implementing the Endangered Species Act, the Clean Water Act and other national mandates in order to accommodate local problem solving. On the other hand, environmentalists are concerned that agencies and resource users will take advantage of added flexibility to avoid their responsibilities to protect species and water quality. In addition, flexibility and absence of deadlines may reduce the pressure on the parties to achieve a negotiated solution.

In theory, our legal system and litigation processes are set up to address concerns about procedural justice. In practice, these concerns often go unaddressed for at least some of the parties. Some parties may say such formal processes are "fair," yet feel unsatisfied with the results.

In cases where public participation is mandated by policy rules, the process at least appears inclusive. Perceptions may follow that the process has indeed been inclusive, which is an important dimension of procedural justice. At a minimum, public participation rules may catalyze some sort of process to draw the parties into conversation. However, there is also the danger that mandated processes may be considered "good enough," precluding any further attempts to unearth all views and include all parties who have a potential interest. Agencies may feel that they have done the minimum required of them and therefore "have been inclusive."

By contrast, in cases in which the negotiations are not structured by an existing policy such as NEPA, involving parties and identifying concerns becomes a political process and a matter for public debate. This may result in a more thorough job of identifying concerns, unless certain interests dominate the process, or the pressures of the majority exclude a minority voice. A case may actually be perceived as both more and less thorough (as in Pyramid Lake) – more thorough in identifying a broader range of concerns, while less thorough in actually being able to address this broad range. The political process may be perceived as inadequate, ironically, because it has been broader and more thorough and inclusive. In this sense, the values of inclusiveness and achieving consensus and finality seem to be at odds.

With respect to the criterion *Procedurally Accessible and Inclusive*, not all conflict resolution processes (court rulings, negotiated agreements) require public participation. In cases where it was required, extensive provisions were made to ensure that this process was completed thoroughly and in a way that could be documented.

The *Costs of Dispute Resolution Processes* are strongly affected by public policies. Policies need not seek to minimize dispute costs, as non-trivial costs provide an important incentive for conflict resolution. This is evident in the many cases of litigation being settled before the matter goes to a full trial. However, policies should attempt to provide low cost venues for settlement – such as informal hearings, prior to escalating the forum to full public hearings or to court proceedings. Disputes involving small amounts of water among neighbors should be addressed in a lower cost venue than complex disputes over large water projects or interbasin water transfers. Our cases exhibit a range of venues, from local negotiations in a watershed (Snowmass Creek), to negotiations and legislation within a state (Edwards Aquifer, Mono Lake), to lengthy multi-jurisdictional litigation and negotiations (Pyramid Lake, Pecos River).

Under the success category "Outcome Quality," there are many public policy interactions. The costliness of solutions to water problems is one concern that federal policies attempt to address when federal money is being committed through legislation, and federal participation in negotiated agreements. With respect to *Cost Effective Implementation*, the federal Office of Management and Budget (OMB) examines the federal share of costs for agreements to which the federal government is a party and advises the administration on whether the agreement is economically reasonable from a federal point of view. For instance, the OMB recommended a presidential veto of nearly every proposed negotiated settlement of tribal claims in the 1980s and 1990s. It did so on the grounds that the federal costs were too high, that more economical means could be found to address water needs and that non-federal parties should bear a larger share of the costs. Political pressure from the region that benefits from the settlement has generally prevailed over the OMB's veto recommendations. However, OMB's objections have forced more local cost sharing and reduced federal financial obligations. Federal requirements for a cost-benefit analysis of new water projects and the examination of costs, benefits and regional economic impacts for an Environmental Impact Statement all put pressure on parties to pay attention to costs when crafting an agreement. Court rulings do not have to comply with economic criteria, though parties testifying before the courts argue for rulings that do not create undue economic hardship for the interests they represent. Courts may be urged by testimony from affected parties to consider costs, but are not bound to do so. Some states have legislation requiring assessment of costs, benefits and other impacts of agency actions on private property. Such legislation seeks to control regulatory activities that would diminish private property values or unreasonably constrain owners' use of their lands.

Some public policies are at odds with cost effectiveness. For instance, federal policies that subsidize agriculture through long term contracts to use water and electricity help to maintain rural economies and ways of life. However, such contracts also make it more difficult and costly to move water out of agriculture in response to conflicts over water needed for endangered fish or for urban growth.

Many types of public policies (local, state and federal levels) affect *Financial Feasibility/Sustainability*. Cost allocations, water pricing, conservation and availability of loans are all influenced by federal and state water policies and federal obligations to different stakeholders. The acceptability of unfunded mandates also depends on the policy "climate" at the time the agreement was reached. Some of our case studies specifically addressed the problem of unfunded mandates (Edwards Aquifer, Mono Lake).

For *Cultural Sustainability/Community Self-Determination*, federal laws require federal agencies to consider the impacts on cultural resources, as do the policies of some states and tribes. The federal National Environmental Policy Act (NEPA) requires that impacts of federal actions on the human environment be considered.[7] Economic impacts on potentially affected communities are examined in the economic impact section of Environmental Impact Statements. However, "ways of life" are rarely explicitly protected unless they are tied to other policy matters like tribal sovereignty or racial discrimination. Policies that supported inclusive, open processes tended to produce agreements that addressed communities' sovereignty and self-determination. When conflict resolution processes left parties out, their sovereignty/cultural sustainability was often perceived as compromised, and it became a contentious issue.

In our eight cases, the Outcomes reflect a general trend of less support for the preservation of rural and agricultural interests, and more support for tribal self-determination and cultural preservation. Some would argue that this corrects the imbalance of historical support for the former at the expense of the latter.

Under *Environmental Sustainability*, myriad public policies affect the consideration of environmental impacts. Examples at the federal level include NEPA, the Clean Water Act and the Endangered Species Act. Many states and tribes also have environmental policies to which conflict resolution Outcomes must conform.[8] Public policies, in general, do not require a consideration of drought (above and beyond environmental impacts associated with drought – such as reduced stream flows or declining water quality). However, water users in regions prone to drought themselves are motivated to consider drought scenarios.

Clarity of Outcome is affected by public policies in different ways. Specific policies may require an agreement to carefully address endangered species or water quality concerns, mandating clarity on these matters. However, the policy/political process itself may encourage ambiguity as parties seek to "gloss over" tough issues in an attempt to pass successful legislation. The politics of making legislation often leads to vague language because parties can more easily agree on vagaries than on details. Court rulings also frequently are unclear with regard to implementation – leading to follow-up litigation, legislation or agency actions to fill in the details. Court rulings can settle matters of who has what rights, giving stakeholders a clearer basis for further negotiations. In addition, baseline conditions (such as habitat protection or streamflow levels) often are clearly defined by courts in setting restoration standards, giving water users clear parameters to govern future negotiations. This is evident in the Edwards Aquifer, Mono Lake, Pecos River and Big Horn cases, which involve a mix of court, legislative and administrative actions.

With respect to *Satisfaction/Fairness,* legal procedures and public policies certainly have implications for the likelihood that parties will be satisfied and will perceive the outcome to be fair. At least one party, and possibly all (as in the Big Horn case), is likely to be dissatisfied with a court ruling. Appeals and subsequent rounds of litigation are almost inevitable. Legislation does not require that all parties be satisfied (and often they are not). Administrative rulemaking is a public process and is open to the participation of all parties. Negotiated agreements implicitly require active participation by key parties. So these processes may be more likely to produce agreements that satisfy most or all parties. However, the Pyramid Lake case provides an important test of this generalization.

Regarding *Compliance with Outcome over Time*, some cases revealed public policies that provide clear steps to check compliance. Other cases were silent on these provisions. It is not clear whether the presence of compliance monitoring provisions is due to differences in policies, or to differences in the skill and persistence of the researchers to uncover such provisions. Court rulings do not necessarily specify ways to verify compliance, but rather leave monitoring to the interested parties who have legal recourse if others fail to comply with the ruling. The Pyramid Lake case specifically required compliance measures and also mechanisms for future negotiations.

With respect to *Reduction in Conflict and Hostility*, in looking across the cases, court rulings are associated with escalating conflict, while negotiation and legislation are associated with decreasing conflict. However, if a key party feels left out of the negotiations or legislation, some parties' issues will be solved and others may be worse off. Consequently, any change in hostility may be unclear. When a process does not require the participation of all parties, it does not encourage the building of relationships (which leads to the probability of future litigation and struggles with implementation). Similarly, if the process required participation, yet was not a collaborative process, the hostility will remain.

If a settlement or ruling results in increased tension, many would say it had failed. A court ruling that "settles" certain questions of law may have achieved its stated purpose of preserving individual rights, but may not facilitate social harmony. As such, questions of success depend upon which of the many success criteria one emphasizes.

In some situations (e.g., Lower Colorado River), the conflict resolution process has actually made hostile rhetoric pointless. Those who use such tactics are shunned and excluded from the working relationship.

With respect to *Improved Relations*, if coalitions were needed to achieve the agreement or settlement (e.g., as in legislation rather than in a court ruling), then relationships seemed to persist afterward in some constructive form. If the process for reaching a settlement requires working together, then relationships are built. Policies that encourage constructive engagement can facilitate this. Working together also increases parties' perceptions of each other's competence. Certain kinds of agreements imposed the need for a later coordination and the necessity for an ongoing relationship. However, some cases required collaboration to achieve the initial agreement, yet the relationships continued to be rocky afterwards. In these cases, one party had typically been left out (or chose to stay out) of the negotiations (Pyramid Lake, Snowmass Creek). The party who stayed out of the process had not formed working relationships with the others. Consequently, mutual trust had not been established and there is no (relational) barrier to initiating further hostile action against each other.

Some processes included a third party mediator who appeared to hold the trust that was fragile between the parties themselves. This finding reinforces the idea that third party neutrals can often serve as a "repository of trust" for parties who cannot yet trust each other.[9]

With respect to *Cognitive and Affective Shift*, policies that encourage parties to engage directly with each other, rather than keep their distance (through lawyers and the court), cause assumptions to be tested and increase the possibility that parties will encounter enlightening information. Both of these are essential to creating cognitive shifts.

When a party is left out of the process (Snowmass Creek legislation, Pyramid Lake), it does not "buy in" to the solution or the relationship. Policies cannot mandate cognitive shifts, but they can help create them by compelling parties to work together.

With respect to *Ability to Resolve Subsequent Disputes*, implementation seems to provide the true test for many agreements and new relationships. As in the previous criterion, if the public policies governing the dispute resolution process required building a working relationship, it persists after the agreement is signed (and it facilitates implementation of the agreement).

Summary on the Role of Policies

This section has examined some of the ways in which public policies affect resolution of water conflicts and has identified policy characteristics helpful in preventing and managing conflict. However, a larger question remains. It is not clear what degree of effort should be expended in policy attempts to manage and mitigate conflict. While conflicts can be costly and destructive, they also have a socially productive role. New water uses and new values first are articulated through conflicts between those seeking changes and those who benefit from existing patterns of water use. Incentives to accommodate new needs are created by the costs and uncertainties associated with litigation and other forms of conflict. Without such incentives, current water users, managers and policymakers would have little reason to change. Consequently, public policies need not seek to entirely avert conflicts over water. Rather, the goal should be to manage conflicts so that the productive outputs of conflict are generated cost effectively, working relationships are built or reinforced and the costs of addressing conflicts are distributed in a manner consistent with public values.

Public policies have a far-reaching impact on the manner in which water disputes are addressed and the degree of success in achieving satisfactory dispute resolution. Some policies appear to encourage litigation through ambiguity and failure to establish alternative processes. Other, more recent policies direct public agencies to pursue alternative dispute resolution. Policies articulated through court decrees and legislation very directly affect the relative bargaining power of environmentalists, tribes, cites and irrigators embroiled in water conflicts. Public policies can assist in managing water disputes by being clear and consistently enforced, through providing low-cost interjurisdictional coordination, building better problem solving capacity, articulating public support for constructive processes, monitoring costs and progress on implementing proposed solutions, and offering conflict resolution forums appropriate to the complexity of the conflict and the values at stake.

CONCLUSION

When confronted with yet another long book, one may ask, "Why bother? What is the reward for working through such a lengthy framework?" This framework offers multiple uses and benefits for researchers, practitioners, and dispute resolution program managers, as well as those who fund dispute resolution efforts and environmental stakeholders. The framework and associated analyses provide a broader understanding of each parties' individual goals and can enhance participants' ability to craft more effective agreements and to implement those agreements. On a broader social level, the framework and the analysis and reflection it encourages will facilitate the development of better public policies, while also helping agencies to more effectively manage conflicts. Substantial gains can be made for each of these groups from a systematic examination of any single case, and even more so for the consistent cross-case comparison that our framework encourages. We hope that as effective strategies for resolving conflict become an integral part of our cultures and our public institutions, systematic reflection will become second nature as well. Success in environmental conflict resolution will then become a self-fulfilling prophecy.

NOTES

[1] Frank Gregg, Doug Kenney, Kathryn Mutz, and Teresa Rice, *The State Role in Western Watershed Initiatives* (Boulder: Natural Resources Law Center, University of Colorado, 1998).

[2] http://www.westgov.org/wga/initiatives/enlibra, Western Governor's Association, 1998.

[3] Policy Consensus Initiative, "States Mediating Change: Using Consensus Tools in New Ways" (Portland, OR: Policy Consensus Initiative, 1998).

[4] Thomas C. Beierle and Jerrell C. Cayford, "Democracy in Practice: Public Participation in Environmental Decisions," (Washington, DC: Resources for the Future Press, 2002).

[5] Douglas D. Morris, "Alabama-Tombigbee Rivers Coalition v. Department of Interior: Giving Sabers to a "Toothless Tiger," the Federal Advisory Committee Act," *Environmental Law* 26(1), (Spring 1996): 393-417.

[6] Lynne Lewis Bennett, Charles W. Howe, and James Shope, "The Interstate River Compact as a Water Allocation Mechanism: Efficiency Aspects," *American Journal of Agricultural Economics*, 82(4) (November, 2000): 1006.

[7] See Appendix B for more details.

[8] See Appendix B for more details.

[9] Herbert C. Kelman, "Informal Mediation by the Scholar-Practitioner," in *International Mediation: A Multi-level Approach to Conflict Management*, eds. Jacob Berkowitz and Jeffrey Z. Rubin, 64-96 (London: Macmillan, 1992).

APPENDIX A

GUIDEBOOK

FOR ANALYZING SUCCESS IN ENVIRONMENTAL CONFLICT RESOLUTION CASES

APPENDIX A TABLE OF CONTENTS

OVERVIEW

REASONS FOR GUIDEBOOK

In order to provide a framework for considering the important elements for successful environmental conflict resolution, this Guidebook provides a template for organizing case information according to logical categories. These categories are based on various criteria for assessing the effectiveness of conflict resolution processes and outcomes in the realm of environmental and public policy disputes. In Chapter 2, we summarize criteria for evaluating successful environmental dispute resolution. These criteria were developed from a comprehensive review of existing literature,[1] practitioner interviews, and adaptations from related fields. We propose that these criteria fall into six conceptual categories: (1) Outcome Reached, (2) Process Quality, (3) Outcome Quality, (4) Relationship of Parties to Outcome, (5) Relationship Between Parties, and (6) Social Capital. These criteria are listed in Figure 1.

GUIDEBOOK GOALS

Guidebook goals include the following:

(1) To lead users through the process of searching for and organizing the information needed to assess each criterion for specific case studies
(2) To allow evaluation of outcomes and processes based on the criteria, both on a case by case basis and across multiple cases
(3) To provide a framework for discussing conceptualizations of success for environmental conflict resolution.

We have organized the criteria into a framework for understanding and evaluating alternative concepts of success in conflict resolution. In addition, this guide may assist parties and facilitators in identifying the variables they want to consider in developing and evaluating their own conflict resolution processes.

It is important to know that this Guidebook was used to analyze cases by graduate students using accessible public information, without personal interviews of mediators and stakeholders. This framework was designed to be non-intrusive and low cost in collecting and analyzing data. It need not create burdens for process facilitators and parties.

GUIDEBOOK PRODUCTS

Within the Guidebook we describe each criterion along with specific indicators to be assessed, specific sources for the information, and the best times for assessment.[2] This Guidebook provides multiple useful products.

Standardized case studies

First, it provides for standardizing the format of case study reports. Our list of criteria and categories are all-inclusive, attempting to reflect the diversity of variables linked by various scholars and practitioners to understanding effective resolution. Guidebook users may decide to use only a subset of criteria depending on their focus and their assumptions about success. However, the presence of all likely criteria in one framework allows for ascertaining information on all these variables so that comparisons across cases can be made. Not only can cases be comparatively analyzed, but resolution processes (such as litigation, researching strategy, informal bargaining, and mediation) also can be compared through this framework.[3]

Researching Strategy

Second, the Guidebook provides a strategy for researching cases. What are the dimensions that various writers and practitioners have considered important to assess? Where would one go to find such information? When should it be assessed? The Guidebook is structured in a user-friendly way to provide guidance in what can be a daunting form of research.

Organizational framework

Thirdly, the Guidebook provides a framework for storing and organizing the many pieces of information relevant to documenting and analyzing these complex cases. It serves as "filing system" or a series of "file folders" that, once familiar, can make the information organizing task easier and yet more thorough.

Education & research tool

Finally, the Guidebook serves as a way to teach students about the multiple issues involved in these cases substantively, but also about the methodological issues involved in doing case analysis and/or comparative research.

EFFECTIVE ENVIRONMENTAL CONFLICT RESOLUTION CRITERIA CATEGORIES

I. *Outcome Reached*

- Unanimity or Consensus
- Verifiable Terms
- Public Acknowledgement of Outcome
- Ratification

II. *Process Quality*

- Procedurally Just
- Procedurally Accessible and Inclusive
- Process Costs

III. *Outcome Quality*

- Costs of Implementing Outcome
- Financial Feasibility/Sustainability
- Cultural Sustainability/Community Self-Determination
- Environmental Sustainability
- Clarity of Outcome
- Feasibility/Realism (legal, political, scientific)
- Public Acceptability
- Efficient Problem-Solving

IV. *Relationship of Parties to Outcome*

- Satisfaction/Fairness — As Assessed by parties
- Compliance with Outcome over Time
- Flexibility
- Stability/Durability

V. *Relationship Between Parties*

- Reduction in Conflict and Hostility
- Improved Relations
- Cognitive and Affective Shift
- Ability to Resolve Subsequent Disputes
- Transformation

VI. *Social Capital*

- Enhanced Citizen Capacity to Draw on Collective Potential Resources
- Increased Community Capacity for Environmental/Policy Decision-Making
- Social System Transformation

Figure 1. Criteria Categories

DEVELOPMENT OF THE GUIDEBOOK

Braving the Currents and this companion Guidebook grew out of a multi-stage research project funded by the Udall Center for Studies in Public Policy at The University of Arizona. Our goal was to develop criteria and methods for assessing "success" in environmental conflict resolution.

We employed several strategies to develop comprehensive criteria: (1) identifying and refining the various criteria in existing literature from several fields: conflict resolution, planning, economics, public administration, psychology, etc.; (2) augmenting this list through interviews with researchers and practitioners; and (3) reasoning by analogy from related fields where similar evaluation research had been done. Our project's first year of work produced an extensive review of these criteria, and a comprehensive framework for organizing and analyzing the multiple dimensions of success.[4] (See Figure 1.)

In addition to suggesting conceptual categories for the criteria, our analysis suggested that criteria should be assessed at several points in time over the course of a conflict: (1) baseline, before resolution process, (2) during the resolution process, (3) immediately upon completion/signing, (4) short-term after agreement or settlement, and finally, (5) long-term after agreement or settlement. Most criteria are not assessible at every stage, however knowing which criteria to assess at which stage can make evaluation easier (see figure 2). Any criteria assessing change will require a baseline assessment against which to gauge movement. Most criteria are best assessed at one or two different points in time, and they are more difficult to assess at other points. This information is noted under each criterion.

			IMPLEMENTATION	
BASELINE before resolution process	DURING the resolution process	Immediately upon COMPLETION or signing	SHORT-TERM after agreement/ settlement	LONG-TERM after agreement/ settlement
1	**2**	**3**	**4**	**5**

Figure 2. Stages for Criteria Assessment

Accomplishing the overall objective of the project (to develop and evaluate criteria for evaluating success in environmental conflict resolution) required applying this framework to case studies. It was through analyses of actual cases that we examined the feasibility of taking abstract success concepts (agreement reached, agreement quality, process quality, etc.), of developing specific criterion under each success concept and of crafting questions designed to direct researchers to gather information to assess each criterion.

In order to examine systematically the chosen cases with our criteria conceptual framework, each criterion required operationalization. How does one restrain a case review from becoming merely an impressionistic snapshot, determined by the particular "lens" of the reviewer? The review becomes systematic by setting out common conceptual categories in advance, and by specifying how each concept (or in this case each criterion) will be measured. The guide attempts to identify the information needed to assess the case studies on each of the success criteria.

APPLYING AN ANALYTIC FRAMEWORK

We began the process of considering successful resolution by identifying a broad set of criteria and evaluating how to apply them to real cases. With the process we have developed, we can note how cases may rate according to various criteria. However, we cannot comment on overall judgments of success for these cases because to do so requires the additional step of prioritizing (and weighing) the different components of success. This is ultimately not a research question but rather depends on the purpose and values of the guidebook user.[5] In the broadest sense, the weighing of the different success criteria is also a public policy question.

However, once personal or community values have been articulated, the Guidebook framework could be used to generate judgments of success. Once a user (or researcher) has identified criteria priorities (for example, comparing cases on environmental and cultural sustainability, without regard to cost or justice of process) and made explicit what would be most and least desirable on each indicator of the criteria, s/he could then compare cases on only those criteria and produce a "judgment" of success.

As described in Chapter 2, Innes[6] and others have called for more comparative case study analyses. If a body of cases was analyzed using this framework, and if new cases were recorded and reported according to this framework, cross-comparison of cases would be greatly facilitated. Such standardized reporting would also allow for comparison and hypothesis-generation across criteria and related conceptual dimensions.

USING THE GUIDEBOOK

We realize that on first viewing, the Guidebook may seem overwhelming in its scope and number of categories. We have two responses to this. First, the user is encouraged to see this as a new filing system. A new filing system may initially seem difficult to internalize, but once it becomes familiar, it actually shortens one's task rather than lengthens it. After the user invests time up front to become familiar with the various criteria and indicators, this framework will better serve the user.

Second, not everyone will find all criteria necessary or equally relevant to their purposes. Users may focus on the subset of criteria most suited to their purposes. We encourage the use of the exhaustive list of criteria to further the goal of comparative case analysis.

GUIDEBOOK ORGANIZATION AND TERMINOLOGY

The Guidebook is organized into three levels. At the top level, the criteria are clustered into six umbrella categories (see figure 1) designated in the Guidebook by Roman numerals. At the middle level, the criteria themselves are under these categories, and designated by capital letters. Each criterion is accompanied by a general definition. Then, in order to make actual measurement possible, each criterion also must be operationally defined via specific, measurable indicators. We list several possible indicators for each criterion, as well as notes of clarification, cross-referencing, best sources, and best times for assessment. The user is encouraged to maintain this structure for organizing information, as well as for reporting the information, as this standardized template then can provide for easy comparison across cases.

The Guidebook can be used for several types of cases. One of the original goals of this project was to provide a framework that would not only allow for comparison across cases, but also across conflict resolution methods. Thus we have developed it and tested it on cases of negotiation, administrative rulemaking, legislation, and litigation.

This allows for comparative analysis and discussion of the virtues and challenges of various methods of conflict resolution.

However, because each of these methods of conflict resolution may have different goals and outcomes, it has made the development of a general terminology difficult. For example, certain processes aim to produce "agreement," while other processes do not aim for this goal. Therefore we have had to adopt the general term of "outcome" to cover the output of whatever process is being considered — whether that be an agreement, a court ruling, legislation, etc.

TOOLS

Case Report Template

Each of the criteria presents a list of questions to be answered. This framework is meant to provide a template for the case report and for taking notes. Careful noting of sources increases the value of the case report. Once the Guidebook has been worked through, the information can be reported in a textual case study report that follows the outline provided in this guide. We suggest including headings as listed in this outline with answers in narrative form.[7] This guidebook is meant to serve as a form to organize notes and information and as a guide for research, but it is in most cases not meant to be or even likely to be filled in exhaustively and completely. Not every case will provide information for every category herein, but one should note when information is inaccessible or not relevant to a case.

Supplemental Appendices

Appendices following the Guidebook provide additional useful tools for researching and organizing case information. Appendix B reviews federal acts that provide the framework for many steps in the conflict resolution process that may be mandated or directed by law. Users unfamiliar with administrative law should review this appendix, and also consult it when completing Guidebook sections where such acts may have supplied a framework for procedure. Appendix C provides concrete guidance on applying the economic and financial criteria to case studies. Specifically it addresses the criteria of Reasonable Process Costs, Cost-Effective Implementation, Perceived Economic Efficiency, Financial Feasibility/Sustainability, and Cultural/Community Sustainability.

STRATEGIES

Choose a case

A good place to start is becoming familiar with a particular case. The first section of the guidebook, called the case "cover sheet," provides questions one should ask, and ultimately be able to write on for the chosen time period of one's chosen case. It is very important to applying the Guidebook framework properly for one to choose a *specific outcome* of one's case for further analysis.

The Guidebook *cannot* be successfully applied to a case's whole history of conflict and its resolution. Because of the many specific questions herein, one must choose a particular outcome on which to focus. This can be, for example, a negotiated agreement, a court settlement, the passing of legislation, or even the establishment and functioning of an ongoing commission or board. However, one must choose to focus on only one for the purpose of this analysis. Neglecting to define one's focus would result in each criterion being applied to potentially multiple objects simultaneously, with no true, useful measurement of any one thing. One cannot apply assessment without defining boundaries around what one is assessing.

If desired, one could do a separate analysis on the same case for a different outcome or time period and compare. One could focus on two contemporary processes in the same case, or two related court rulings. But these must be treated as two separate applications of this framework in order to give each process and its associated outcome its own assessment.

Become familiar with the case

Before jumping into the specific questions of the Guidebook, it is best to become at least generally familiar with the case. Consult secondary reports or summaries of the conflict, if possible, keeping in mind the particular perspective or bias of the source. Consult newspaper or other media summaries (see below). Working through the Guidebook's case "cover sheet" is a good way to get a "mapping" of the conflict, its players, issues, development and history. (In the conflict resolution field, we actually call this process "conflict mapping" or "conflict analysis.") In doing such an analysis the reporting should be as objective as possible and an outside reader should not be able to detect if the analyst has a particular viewpoint on the conflict. By doing this preliminary analysis, one can get a sense of the availability of information on the case of interest, and thus its feasibility as a case of study. Some cases may actually not be appropriate for this type of analysis, either because information is extremely difficult to collect, or because the window for focus is too vague to define adequately.

Become familiar with the Guidebook

As described earlier, the Guidebook serves as an analytic tool and as a new filing system. As with any new filing system, one has to *use* it to start to remember its many categories. This may mean extra trips to a particular file initially, but once its categories are familiar, new information can be stored easily, and one will even subconsciously begin to look for the existing categories of information.

Choose criteria that matter for your purposes

The Guidebook provides an exhaustive menu of criteria. This is desirable for comparative case analysis. However, case analysis as well as other evaluation goals may best be served by using a subset of criteria.

Consult multiple sources

Research is a scientific enterprise that requires an open, duplicable process. Otherwise, it easily becomes just opinion and conjecture. Results must be both *valid* (i.e., indicators must closely reflect the true concept one is trying to operationalize and measure), and *reliable* (i.e., not due to chance fluctuations or to the vagaries of a particular source, but able to be dependably assessed by the same person at another time and by other persons as well). For this reason it is important that for most indicators in the Guidebook, one needs to crosscheck information in more than one source. Some sources may corroborate each other. Other sources may be contradictory. If the latter occurs, such divergence of opinion should be reported and is itself of great interest.

A list of likely sources is listed for each criterion. These lists of suggested sources emerged from the substantial piloting that the Guidebook has already received as a tool for research and analysis. The media are an important source, and researchers should consult relevant stories. These are both typically available in microfiche form through most university libraries, and often in public libraries. Another rich source is the internal and external public documents of parties, such as promotional materials, annual reports, internal reports, and especially newsletters. Often these will be supplied through a simple phone call or fax to a party. The federal Freedom of Information Act requires that many documents be made available, however invoking it is usually unnecessary, as most government agencies will happily provide copies with a simple courteous request. Amazing information can be obtained through a phone call to a Senator or Representative's office staff, who make extra effort to see that constituent requests for information (particularly government documents) are satisfied. Finally, the internet has emerged as a powerful tool for research. Most parties have their own websites and use them to express their opinions and policies. Government agencies use them for making documents available. Draft settlements or working papers may even be posted.

Weigh the advantages and disadvantages of using interviewing

In our original piloting of this project, we discouraged the use of interviewing in order to limit costs. We wanted to demonstrate what could be accomplished under time and budget constraints that precluded personal interviews. Interviewing also can be an intrusion into the system being studied that can have its own effects. However, for gathering some information, interviews have no substitute, and researchers may decide that for their purposes it is worth the costs. However, the Guidebook has been designed to be able to be completed without the use of interviewing. Strategies for doing this type of research are not listed.

Be a sleuth

Often this information may not be available in obvious places. It may take a bit of sleuthing and persistence to follow possible leads for certain sources. It may take reflection and creativity to realize possible new sources to consult.

Be concise, but thorough in reporting

Once one discovers the wealth of information available with a mere keystroke, it is tempting to cut and paste vast documentation into one's report. Alternatively, one may also be tempted to answer the large number of questions in as brief a way as possible.

Neither extreme is ultimately useful. Both analysis and comparison are best served by concise summarizing for each indicator, usually from multiple sources, supplemented with representative direct quotes. Please consult Chapters 4-7 for examples of cases with what we consider to be an appropriate level of detail. Appropriate level of detail may vary according to users' needs.

Cite properly and completely

Because of the importance of verifiability in ensuring reliable information, the judicious use of direct quoting is encouraged, and correct and complete citation is critical. Often the source of the information is as important a piece of data as what the source actually said; in other words, analyzing what was said as an indicator of a criterion cannot occur apart from considering who the source of the information was. In piloting this Guidebook, it was found that the most helpful referencing style for this type of analysis was the use of footnotes. Consult the case studies in Chapters 4-7 for models of thorough citation.

ONE FINAL NOTE…

One final note before beginning: We as authors would like to receive your feedback. This Guidebook has been an organic product, continuing to change with successive rounds of feedback. Also, one of our general goals is to increase the number of cases available for comparison, therefore we welcome the opportunity to view your completed cases and to make them available to others.[8] Best wishes for the journey.

CASE REPORT FRAMEWORK

CASE INTRODUCTION

COVER SHEET

(1-3 pages; use bullets for lists)

Case Study:

Time period of case study evaluated:

Basic nature of dispute: (water rights, endangered species, etc.)

Issues in dispute:

Actors, and associated interests of each:

Attempted conflict resolution processes: (e.g., mediation, litigation, policy dialogues) — Include details: If litigation, note court; if administrative rulemaking, note agency promulgating rules; if collaborative processes, note who convened, financed by whom, who facilitated, who staffed; if legislation, note legislative body involved.

Specific agreement, ruling, or other outcome focused upon here:

HISTORY

1. *Case History Overview (In narrative form; 3-4 pages)*

   Provide historical context and chronological unfolding of case developments.

2. *Timeline of Major Events (1 Page)*

   What portion of this timeline will be focused on in this report?

CRITERIA CATEGORY I – OUTCOME REACHED

NATURE OF THE OUTCOME ANALYZED:

For clarity in this section, identify once again the nature of the conflict resolution outcome being analyzed (i.e., negotiated agreement, court ruling, legislation, administrative action) and the date, time (or period) during which outcome was produced (e.g., court ruling by X federal district court on May 10, 1998; negotiated agreements in 1997 and 1998, signed on May 10, 1998).

CHARACTERIZE THE OUTCOME FOCUSED ON IN THIS ANALYSIS:

☐ Court Ruling (provide date and specific entity that took action)

☐ Agency Ruling (provide date and specific entity that took action)

☐ Agency rule promulgation

☐ Legislation or other actions by elected bodies (city councils, elected district boards, etc).

☐ Negotiated Agreement

☐ Other, Specify: ....

A. *UNANIMITY OR CONSENSUS*

This criterion seeks to assess the strength of approval for the outcomes reached, the degree of dissension at the table, and the absence of key parties.

SOURCES TO CHECK: Media, web sites, newsletters, court reports.

BEST TIME TO ASSESS: At time agreement is announced, re-examine later for evidence of missing party.

If a ***negotiated agreement,***

was a decision rule (unanimity, consensus, majority) discussed? If so, what was agreed upon?

when agreement was reached, was it: (check one)

- ☐ Unanimous.
- ☐ There were dissenters at table.
- ☐ There were parties not at table, who left or walked out on negotiations. Specify:
- ☐ There were significant parties not at table at all.
- ☐ Other. Explain: ....

....

If ***another outcome****, (legislation, administrative ruling, court ruling):*

Document voting results related to the outcome (i.e., legislative vote, vote of administrative body or decision of panel of judges. If the decision of a single individual (one judge, an agency head, etc.) note name, title, etc.).

B. VERIFIABLE TERMS

This criterion seeks to verify that consensus existed on the terms of the outcome and that reaching the outcome was publicly confirmed.

OUTCOME REACHED

SOURCES TO CHECK: Media, internet — Multiple sources advised.

BEST TIME TO ASSESS: After agreement is announced.

Was the outcome (check all that apply):

☐ Written and formally signed

☐ Oral (Specify who articulated the agreement, title or affiliation, date, setting.)

☐ Terms accessible to interested parties and the public? (or confidential). If yes, note sources.

☐ Terms published in media or posted in public forum (Federal Register or state equivalent, libraries, etc.)

C. PUBLIC ACKNOWLEDGMENT OF OUTCOME

This criterion verifies that reaching of the outcome was publicly confirmed. [Note: record *reactions* of stakeholders to outcome in category IV, Relationship of Parties to Outcome.]

SOURCES TO CHECK: Media, internet — local access to press may be helpful.

BEST TIME TO ASSESS: After outcome made public.

Describe:

- Media coverage (note geographic breadth of coverage — local paper, regional papers, "national" papers, such as the *New York Times*)
- Internet
- Newsletters (e.g., Western States Water Council)
- Professional Journals

D. RATIFICATION

This criterion assesses whether relevant governing bodies and/or courts formally approved outcomes. Court rulings may require follow up legislation to be implemented. Legislation may require court approval. Assume any signatory to a negotiated agreement would have to formally ratify it (e.g., vote of board, referendum or resolution). Such action can occur before or after agreement is reached.

1. Governing Bodies:

SOURCES TO CHECK: Media, local access helpful

BEST TIME TO ASSESS: Any time after, but easier at time outcome made public

Check off governing bodies (councils, boards) needed to ratify the outcome:

Constituency	Ratification-Date/Forum[9]	Type of Action[10]
☐ U.S. Congress		
☐ State legislature		
☐ Tribal councils		
☐ City council		
☐ Water district board (Irrigation districts, conservancy districts, etc.)		
☐ Board of Directors of utility or private firm		
☐ County Board of Supervisors/Commissioners		

2. Judicial Approval

Often this is required of legislation that was passed to address a conflict that has been the subject of litigation. Was judicial oversight required?

OUTCOME REACHED

Sources to Check: Media, local access helpful, diverse sources not needed. One source confirming each is adequate.

Best Time to Assess: After agreement is announced.

Which court, if any, needs to approve the outcome? (federal, state, tribal courts?)

Jurisdiction: ....

For ***court rulings****:*

Sometimes a higher level of court is asked to review a lower level ruling. If so, note here, with dates of hearings and dates and citation for final ruling:

....

For ***legislation, agency actions, and negotiated agreements****:*

A court may be asked to rule on an outcome's legal legitimacy. If so, note here. Also, a negotiated agreement may require court review and approval. For instance, a negotiated agreement about sharing water may require court approval and incorporation by the courts into a water decree.

....

[Note: Subsequent government action to comply with a specific outcome should be discussed under category V, Compliance. If a prior court ruling exists with which this outcome must comply, discuss compliance with that ruling under criteria III.G.1, Legal Feasibility.]

Also, a prior ruling may have motivated outcomes, with which they need to comply. If so, note prior ruling and legal citation here.

....

Criteria Category I

[Note: Any outcome (legislation, administrative action, negotiated agreement) may be subject to litigation and appeals to higher courts. Note legal citations to such subsequent litigation here. Also note legal challenges under criterion III.F, Feasibility/ Realism; IV.B.3, Compliance with Outcome Over Time; IV.D.2, Stability/Durability, if relevant.]

II CRITERIA CATEGORY II – PROCESS QUALITY

A. *PROCEDURALLY JUST*

This criterion seeks to ascertain parties' and others' *perceptions* of the justice of the process, i.e., was it fair, balanced, complete, thorough in the sense both of issues and parties, and not compromised by time constraints or power imbalances. This includes both parties' perceptions of fairness of the procedure and parties' satisfaction with the procedure, which research[11] suggests is combined together by raters.

[Note: Distinguish these measures of procedural fairness, etc. from "outcome fairness" which should be reflected under category III, Outcome Quality. Distinguish comments and perceptions about the outcome from those about the process.]

SOURCES TO CHECK: Media (local papers helpful), texts of agreements, stakeholder newsletters, web sites, and promotional materials. Public speeches by stakeholder representatives or convener may be helpful. Consider if certain parties may be less adept at using media to make their views known, and search for alternative venues. This criterion requires thorough information search on each and all aspects listed below. Must use multiple sources to get full range of views from diverse stakeholders. Do ***not*** base on "overall impression" or one source. Ideally, survey or interview parties during process and/or after agreement reached with procedural justice instrument.[12]

BEST TIME TO ASSESS Directly following the process, when it is prominent in the media. Follow up verification may require a lag for it to appear in newsletters, web sites, etc.

The first five dimensions of procedural justice are all essential to the construct.[13] Check for information on ***all*** dimensions below; note if no information available or not applicable to case being analyzed.

- procedure used was: (fair, useful, unfair, slow, rushed, etc.)
- all interests were taken into account [Interests should have been outlined in "cover sheet" section.]
- time for decision-making/negotiation/action was adequate/inadequate
- perception that citizen views or minority views were given equal consideration
- perception that specific views or certain people's views carried more weight
- perception that emotional issues were/were not given time
- perception that underlying concerns were unearthed

B. PROCEDURALLY ACCESSIBLE AND INCLUSIVE

This criterion seeks to ascertain the actual availability of three components that contribute to perceptions of procedural justice: first, did opportunities for public participation exist? Second, did the public have access to information on upcoming participation opportunities? Third, did the public have access to substantive and technical information on issues? The first asks for details (dates, timing, location, attendance, effectiveness) on any public hearings, town meetings, surveys, hotlines, citizen boards, or other forms of public outreach and polling. The second component asks for notes on attempts to notify the public, and the nature of the contact medium. The third component more specifically addresses public access to information on issues.

Sources to Check: Sometimes easily accessible (if an administrative rule), sometimes may require persistence to obtain (local presence may be helpful). If public participation is mandated, formal records on process are required. Check media, agency files and documents, promotional materials, agency web sites (may summarize public meetings), agency press releases on file, Federal Register; also Environmental Impact Reports (EIR) and Environmental Impact Statements (EIS). EIS and EIR are required to be publicly archived, e.g. put in public libraries (See appendix B). Sources such as advertisements, public postings, radio broadcasts, etc. will require local investigation.

Best Time to Assess: During process is best. (Certain forms of public notice may be difficult to locate after the fact). However, if public participation is mandated, formal records of process are kept.

1. Public Notice

For processes leading to the outcome itself: What were the public notice requirements that apply to the process being analyzed (e.g., hearings require public notice; EIS process requires public participation)? Certain processes, such as negotiation, may be private or even secret and therefore public notice requirements do not apply.

For subsequent ratification processes: Public hearings may come later if a legislative body ratifies agreement. What were public notice requirements for any subsequent ratification? Note such subsequent opportunities for public involvement also. Check for various attempts to notify public:

- media
- notifying community leaders
- public signage
- public notices in library, or media
- announced in Federal Register
- other efforts to notify public (e.g., Special Master's newsletter to claimants in general stream adjudication)

2. Public Participation

As in item #1 above, (a) analyze for specific process that is the focus, and (b) analyze for any ratification processes that occur subsequent to the outcome.

Were citizens given the opportunity to express their views before an outcome was achieved? Note dates of, timing of, location of, and comments regarding attendance, effectiveness of:

- public hearings
- town meetings
- surveys
- hotlines
- citizen boards
- submitted comments, written (e.g. letters) and oral.

For subsequent ratification processes:
Were citizens given the opportunity to express their views during the ratification process?

Note dates of, timing of, location of, and comments regarding attendance, effectiveness of:

- public hearings
- town meetings
- surveys
- hotlines
- citizen board

For court processes: Note —

- how broadly the court allowed testimony — from many affected parties or just a few direct litigants?
- which parties filed amicus ("friend of the court") briefs? What interests did they represent?

3. Access to Technical and Substantive Information on Issues

Indicate parties' access to technical and substantive information on issues.

Information accessibility among stakeholders: (Direct participants in process.) Did stakeholders share technical information or keep it confidential? Was it only in the hands of one, or a few, parties?

Information accessible to public: Was information available in public archives at the time of the process? (technical reports in public libraries, websites, etc.)
Public policy assistance in acquisition of information: Note where public policies may have assisted the public or stakeholders in gaining information, e.g. Federal Freedom of Information Act, open meeting laws, open record laws, "sunshine laws," Federal Advisory Committee Act of 1990.

4. Public Education

Were attempts actively made to educate the general public about the problem, including its scientific and technical issues (beyond merely making information accessible, as in item 3 above) as well as its general context? Examples: Public forums to explain; utility distributes flyers. [If these efforts represent a significant change in how agencies deal with the public, note this under criterion V.E, Transformation.]

C. REASONABLE PROCESS COSTS

This criterion examines costs associated with the *process being analyzed*, with cost information organized into three categories according to who bears the costs. Process costs are considered "reasonable" when they are (or are perceived as) in proportion to the magnitude of the conflict being addressed and the assets at stake. For a complex multi-party dispute over a large river basin involving millions of dollars in water rights, land and economic activities, it would not be unreasonable to spend many hours and hundreds of thousands of dollars on a process designed to resolve the conflict. However, these same expenditures would likely be seen as "unreasonable" for a two- or three-party dispute involving neighboring irrigators.

Only costs incurred as part of the process are to be noted here. Costs and cost-sharing agreements regarding the outcome (agreement/ruling) and its implementation should be noted in criterion III.A, Cost-Effective Implementation. See appendix C for specific instructions of collecting and recording cost information.

Costs are broadly defined to include monetary expenditures, staff time, and other resources dedicated to the process. The process to be analyzed is the one that led to the outcome that is the subject of the research — e.g., the process of negotiating an agreement, of drafting and enacting legislation, of litigating or of promulgating an administrative rule.

Researchers should note actual data on costs of process, as well as perceptions expressed regarding the costs of process and whether they were "reasonable." Actual or perceived agreements (if any) on splitting the costs of the process also should be noted.

SOURCES TO CHECK: Media, newsletters, communications among and within stakeholder groups, stakeholder, and court records on staff time expended, mediation or other records on meeting frequency and duration. (Interviews with parties may be useful.) Annual reports/reports to stakeholders, constituencies, congressional staff in Washington, D.C., and in local offices, congressional committee and staff reports, GAO reports (www.GAO.gov), lobbying costs.

BEST TIME TO ASSESS: Examine information generated while process is ongoing and for a short time after an outcome is reached.

1. *Costs to Stakeholders Who Participated in Process*

   List each party. Describe actual monetary, time (actual time spent in meetings, for instance), and other costs for each party (e.g., costs of mediation services, lodging and travel to participate in process, attorney's fees). Document cost-sharing agreements among the parties and actual contributions to process costs made per these agreements. Note perceptions also, such as "this process was so time consuming" or "travel expenses to participate really added up."

2. *Costs Borne by Taxpayers*

   Court staff time, public agency staff time, and expenses — for agencies not considered above as stakeholders (e.g., judge serving as facilitator), legislative appropriations to support process, comments on public/taxpayer costs, staff hours, number of people devoted to process.

3. *Costs Borne by Others*

   Often other parties (not direct participants, not supported by tax dollars) bear expenses to influence the process. For example, water or power rate increases to cover process costs, fee increases to resource users, costs to parties excluded from this process [being analyzed] in their attempts to influence process (i.e., lobbying, media campaigns…).

CRITERIA CATEGORY II

4. Splitting Process Costs

Describe formal arrangements (if any) on how process costs will be shared among the parties and information on actual contributions by the parties to process costs. These contributions can take varied forms: payments to mediators or facilitators, or technical experts; dedication of staff time to the effort; use of buildings and equipment; etc.

CRITERIA CATEGORY III – OUTCOME QUALITY

A. *COST-EFFECTIVE IMPLEMENTATION*

This criterion focuses on the costs of implementing the terms of the agreement, court ruling, legislation, administrative action, or other ECR outcome. It collects information that may be used comparatively to assess whether or not an agreement /ruling took a cost-effective approach to resolving the technical problems that are part of the conflict and to implementing the terms of the agreement. Implementation is considered cost effective if the actions taken were undertaken in a manner that considered and minimized the costs of accomplishing what was required. For instance, in a case that required an additional 10,000 acre feet of water to be made available for environmental restoration, cost-effective implementation would produce that water in the least-costly manner.

Note: Cost information is to be organized as: (1) costs to parties participating in the process, (2) costs to the public (agencies, courts, other costs paid through taxes), and (3) costs to others (not participating in the process, not paid by tax dollars), such as water utility ratepayers. Actual costs incurred, projected costs *and* perceptions regarding implementation, costs are to be reported under this criterion. Expenditures of staff, time, money, resources, water, and other assets all "count" as costs. Only costs attributable to implementing the outcome (i.e., the legislation, court ruling, agreement, etc.) are to be analyzed here. Costs of the *process* go under criterion II.C, Reasonable Process Costs. Researchers need to carefully distinguish between costs attributable to the outcome and more general changes in costs that would have occurred anyway. (Comments on costs that specifically imply fairness, or lack thereof, should also be noted under criterion IV.A, Satisfaction/Fairness.)

SOURCES TO CHECK: Consult Environmental Impact Statements (EIS) if available, agency reports analyzing costs, parties internal cost analyses, Federal Office of Management and Budget (OMB) or General Accounting Office (GAO) reports, if available. See appendix B for website information.

BEST TIME TO ASSESS: Once terms of outcome are available, although actual costs may not be known until implementation is complete. Parties' perceptions of future costs for implementation are best ascertained when outcome is achieved and there is media coverage.

1. *Summary of Cost-Sharing Arrangements*

   - How are costs of implementing the outcome allocated among parties; including staff time, water, and money? (Example: The court directed Party A to pay $10 million to Party B, in four $2.5 million annual payments with the first due on Jan. 1, 1999).
   - What guiding principles, if any, were used to determine cost sharing? (e.g. wealth of different parties, ability to levy taxes)
   - What specific cost sharing arrangements apply? (Example: Party A agreed to contribute 1,000 acre-feet per year of water, Party B to donate 500 acres of land and Party C to pay $5 million.)
   - Financial aid to be contributed by government, other parties.
   - Prepare a table of costs allocated under the terms of the outcome, to each party, to the degree to which such information is available.

2. *Costs to Parties Who Participated in the Process*

- Actual costs incurred since outcome — in past, ***not*** process costs.
  - Actual data on dollar figures (e.g., dam under construction — party A and B have each already contributed $20 million).
  - Comment on costs already incurred. [Note comments that imply fairness or lack thereof under category IV, Relationship of Parties to Outcome.]
- Direct participants' projected costs still to be incurred as a result of the outcome (agreement/ruling).
- Perceptions regarding costs to parties (note source and affiliation of source quoted)

3. *Costs Borne by Taxpayers*

- Costs incurred by public agencies that are not parties, by courts, by regulators
- Costs of court time
  - Actual costs incurred since agreement/ruling – in past; not process costs.
    - ❑ Actual data on dollar figures
    - ❑ Comments on costs already incurred. [Note comments that imply fairness or lack thereof under category IV, Relationship of Parties to Outcome.]
  - Projected costs still to be incurred as a result of the outcome (agreement/ruling).
  - Perceptions regarding costs to taxpayers (note source and affiliation of source quoted).

4. *Costs Borne by Others* (not direct participants, not taxpayer costs)

Direct parties in a conflict may negotiate an agreement that not only affects their own well-being but also affects others who were not party to the negotiations. Costs may be imposed on others, in order to ease the burden placed on the direct parties to the agreement. Shifting implementation costs onto other parties (such as taxpayers and ratepayers) should be noted here.

- Actual costs incurred since agreement — in past; not process costs.
  - Actual data on dollar figures.
  - Comments on costs already incurred.
- Projected costs still to be incurred as a result of the outcome — agreement/ruling (e.g., impacts on ratepayers in form of change in water or electric power rates).
- Perceptions regarding costs to others (note source and affiliation of source quoted).

5. Costs Considered in Implementation Decisions

Cite evidence that implementation decisions considered costs and selected lower cost alternatives.

- Cite opinions expressed by stakeholders and observers.
- Cite examples of actual implementation decisions.

B. PERCEIVED ECONOMIC EFFICIENCY

Economic efficiency, as used in cost-benefit analysis to evaluate projects and policies, means that the benefits of a specific activity (such as building a dam or regulating a pollutant) outweigh the costs. In order to rigorously assess economic efficiency for the outcome of an environmental conflict, it would be necessary to describe all of the relevant costs and benefits and to quantify them in dollars. However, less formally, economic efficiency asks "was it worthwhile?" "Are the costs justified by the benefits?" This criterion addresses the "was it worthwhile" issue from the perspective of stakeholders and observers.

1. Perceptions of Parties Participating in Process

- Cite parties' perceptions regarding benefits to themselves from outcome and benefits to other stakeholders.
- Record statements that compare benefits received to costs expended or that comment on "was it worthwhile?"

2. Perceptions of Observers

- Cite descriptions of benefits to parties, communities, larger society.
- Cite any attempts to quantify and compose benefits and costs and the results of these analysis.
- Cite opinions expressed by others on "was it worthwhile?"

C. FINANCIAL FEASIBILITY/SUSTAINABILITY

How will the money to pay for implementation of the outcome be obtained? This criterion assesses how the agreement addresses issues of securing funding for implementation and ensuring that economic incentives encourage compliance and support implementation.

While the previous criterion III.A, Costs of Implementing Outcome, addresses costliness of implementation, this criterion focuses on how the money will be obtained. Researchers should focus on actual or planned financial arrangements. Financial arrangements specify who pays for what, monthly or annual payment obligations, and the time period over which payments are made. Researchers must carefully distinguish financial considerations related to the outcome from those that would have occurred regardless of the outcome.

Comments on *perceived fairness* of cost sharing should be recorded elsewhere, under criterion IV.A.5, Satisfaction/Fairness (for stakeholders), or under category III, Outcome Quality (for public perceptions). Economic implications for viability of communities should be noted under criterion III.D, Cultural Sustainability/Community Self-Determination.

SOURCES TO CHECK: Terms of the agreement/ruling/legislation, commentary on these terms, media archives.

BEST TIME TO ASSESS: After agreement/ruling/legislation is made public.

1. *Plans to meet financial obligations:* How does each party with a financial obligation plan to obtain the money to meet its obligation? (Examples: County will raise property taxes ten percent; Utility will raise water rates five percent; State will issue $10 million in bonds.)

2. *Ability to meet financial obligations:* Is there evidence that these parties actually have the ability to pay their share of the costs? (Examples: City A already has raised taxes to pay its share; Irrigation district members voted down measure to raise water rates to cover their costs.)

3. *Long-term feasibility of financial obligations:* How are costs spread over time? A common tactic to achieve an agreement on cost sharing is to shift costs away from current water users, tax payers, and rate payers to the future. Deferring financial costs, cleanup costs, or endangered species mitigation costs (by not assigning them to current parties in the agreement) is an example of a strategy that may not be feasible when the time comes to cover those deferred costs.

4. *Incentive Compatibility:* Are the pricing incentives in the outcome consistent with achieving the goals of the outcome? Do water prices, electricity prices, and other prices support implementation of the agreement/ruling/legislation? (Example: Were water prices changed to promote conservation for cases where conservation was a key goal?)

5. *Loans:* What types of loans are involved, or are planned? (Examples: Federal loan to City to build water treatment plant; State loan to farmers to install water conservation devices.)

6. *Unfunded Mandates:* Are there "unfunded mandates"? — Does outcome require monitoring or other follow-up by agencies, but not provide funding for such follow-up? (Example: Court orders state agency to conduct water quality monitoring, but state does not appropriate money for new staff needed.)

7. *Assumptions regarding ability to follow through:* Note unrealistic assumptions about other parties' ability to follow through (personnel, resources, etc.)

D. *Cultural Sustainability/ Community Self-Determination*

This criterion asks for a record of communities affected by the agreement/ruling and an assessment of the types of potential effects. These include demographic and economic effects, such as changes in patterns of jobs, income, taxes, etc. These also include changes in patterns of ownership, changes in decision-making authority or jurisdiction, and changes in the social or cultural "lifeways" of the impacted communities or the relative balance of these lifeways (the "cultural mix"). Examples of "lifeways" include irrigated farming, ranching, community gardens, Native American cultural practices, Hispanic ditch associations (acequias), small town life. Expressed concerns about community and cultural impacts ***and*** actual indicators of impacts both should be reported.

Note: Researchers must attempt to separate out changes produced by the agreement/ruling analyzed from changes that would have occurred even without this specific outcome. Only the former should be discussed here. Carefully note sources for each comment/perception (i.e., potential effects will occur from whose perspective?). If the source of information for any of these effects is one of the parties, the need to check other sources (triangulation) is heightened.

Multipliers are an economic tool used to measure how economic impacts spread through a local economy. Suppose that an outcome being analyzed has caused the loss of 10,000 acres of irrigated farmland and a decline of $900,000 in farm income. This loss will have a ripple effect in the county which can be calculated using multipliers. The county income multiplier for farming is used to calculate changes in overall county income due to a drop in farm income. An employment multiplier is used to calculate how many job losses will occur in the county due to that loss in farm income. A business activity (or output) multiplier is used to calculate how much local business revenues will decrease due to the loss in farm income. Multipliers are developed by public agencies and are available on a county-wide basis for rural areas and on a metropolitan area scale for cities. If you use information that relies on multipliers in your case study write up, carefully note the source, the type of multiplier used, the size of the multiplier used, and the magnitude of economic impacts estimated using that multiplier.

SOURCES TO CHECK: Outcome itself, journals, stakeholder newsletters, publications of national and state water resource associations, promotional videos, and professional newsletters. Media, EIR, EIS (discuss local economic impacts), local government or tribal authority studies.

BEST TIME TO ASSESS: Potential for these effects and perceptions of these are accessible immediately upon agreement/ruling; actual changes will require time passage.

1. *Affected Communities*

   List all communities, towns, counties, Native American reservations, agricultural/irrigation districts affected by outcome (bearing consequences of outcome).

2. *Potential Effects*

   Be careful to distinguish comments/perceptions from actual data on impacts.

   Demographic

   - population loss/gain — if perceived as linked to agreement/ruling (e.g., immigration in and out)
   - crime rate (e.g., casinos or high growth comes to town and people perceive is due to outcome analyzed)

CRITERIA CATEGORY III

- change in residency patterns within the community (e.g., more outsiders moving in, break-up of neighborhoods, ethnic enclaves)

Economic

- jobs/unemployment rates
- tax revenues/tax base
- property values
- income patterns/poverty rates (change in average income and differential between wealthy and poor — income gaps)
- change in ownership of land or water (e.g., less local ownership of water rights, land)
- cost of housing
- cost of living (community gets "yuppified")

Community/local self-determination

- check all sources for expressions of concern over affected communities' ability to determine their own future
- tribal sovereignty — Check sources for concern over tribal government giving up power/control to outsiders
- legal recourse — Check for concern over limitations on community's future ability to litigate in order to protect themselves or to affect change
- decision-making authority/jurisdiction (loss or gain of local control). Examples: Decision making switched to city or federal agency; State agency took over water management from local district
- evidence of changes in control over economic future
- changes in local access to land, water, other resource availability/access
- evidence of changes in control over social future

Cultural impacts

- changes in cultural mix, loss/change in "way of life"
- change in access to cultural sites (ceremonial sites, burial grounds), control over artifacts and sites

E. ENVIRONMENTAL SUSTAINABILIITY

Sustainability suggests practices that allow for preservation of current resources in such a way that future generations will have comparable resources available to them. This criterion assesses the degree to which the outcome considers drought, environmental factors and other natural contingencies, either through direct language in the agreement /ruling or through participation of environmental advocates (agencies and organizations) in crafting the outcome and its implementation. The criterion also asks what natural resources are committed for implementation, over what time frame, and with what environmental impacts.

SOURCES TO CHECK: Outcome (ruling, agreement, etc.) itself, comments by parties reported in media. Media may also note participation by environmental groups.

BEST TIME TO ASSESS: Outcome and media coverage is best assessed after outcome is made public. However, assessing "adequacy" will require time passage.

1. Environmental Impacts

Note: Certain outcomes are required to explicitly address environmental impacts. See appendix B for notes on NEPA,

Clean Water Act, and Endangered Species Act, etc., and information on potential requirements.

Does outcome address existing problem? Was agreement/ruling intended to address existing environmental problem? Describe. If so, did the agreement/ruling address this environmental problem successfully?

Does outcome address environmental impact? Did (and how did) parties address environmental impact that may result from the agreement/ruling, such as endangered species concerns?

Does the outcome shift an environmental problem to a new location or new set of affected interests? An example is an outcome that cleans up water quality in one stretch of a river, but creates a water shortage or water quality problem for downstream water users.

Does outcome provide for resource acquisition?

2. *Natural Contingencies*

- Does Outcome consider how parties will jointly address natural contingencies (e.g., how will parties address water shortages during drought)?
- What natural contingencies could affect implementation of the Outcome? Have these contingencies been taken into account in the preparation of outcome (e.g., drought, flooding, earthquake interruptions of water supply, fire, non-native species invasion, pest infestations)?

Note: Processes and forums established in the agreement or ruling to deal with uncertainties should be noted under criterion IV. D.3, Stability/Durability.

OUTCOME QUALITY

3. Projected Resource Use

- What changes in natural resources availability or quality could affect the Outcome (agreement, ruling, etc.)? (Example: federal water project will contribute 10,000 AF(acre-feet) per year to fisheries water needs.)
- What are the projections of resource use over time under the Outcome?
  - Land, acquired from whom? Any environmental impacts?
  - Water, from what sources? Any environmental impacts?
  - Electric power, from what sources? Any environmental impacts?

4. Consideration of Environmental Concerns

Environmental issues may not always have an advocate during the process, particularly if there are no local environmental groups with resources to be present at all relevant disputes. Note crossover with criterion II.A, Procedurally Just.

- Did the parties crafting the Outcome (legislation, administrative action, negotiated agreement) contact and negotiate with environmental agencies and try to get assurance on Endangered Species Act compliance and regulatory compliance?
- Were environmental groups active in the process and/or were public agency environmental watchdogs active in the process?
- Were there local environmental advocates present?
  - If not, were there other groups active in the negotiations and advocating environmental sustainability?

CRITERIA CATEGORY III

F. CLARITY OF OUTCOME

Agreements, rulings, and other Outcomes that lack clarity may face problems in implementation. This criterion assesses whether the agreement/ruling was clearly worded and performance standards were specified. It assesses the presence of misunderstandings and differences in interpretation (if any), and examines Outcome language for ambiguity and checks Outcome and implementation for well-defined baselines and performance standards (e.g., water use, stream levels, conservation efforts).

Note: Consider tradeoffs with flexibility. Comments on lack of clarity may also have relevance for criterion IV.C, Flexibility, as parties make trade-offs between concretizing details and leaving room for flexibility.

SOURCES TO CHECK: Outcome itself, independent commentaries and reviews (e.g., in law reviews)

BEST TIME TO ASSESS: When Outcome is made public and as implementation proceeds. Ideally, best assessed after implementation has begun but can be assessed in a predictive way from agreement and commentaries. Baselines and performance standards can be assessed any time after outcome is made public.

1. Language of Outcome

Note: It is not necessary to review the terms of the Outcome as a whole. However, check for the presence of certain items. This information may be available through secondary sources (e.g., commentaries).

Performance standards/baseline conditions

Check Outcome for wording having to do with performance standards or baselines (minimum flows, lake levels, etc.)

Were baseline conditions used as reference points in the Outcome well defined? Examples: If a party agreed to cut back water use 20%, what baseline will they cut back from? If a party agreed to double the available habitat for a species, what base amount of habitat are they going to double?

Other examples to consider include hydrologic conditions (stream levels), water use limits and conservation efforts (e.g., 10% reduction in water use)

- Were performance standards and timelines specified (e.g., build the dam, canal or water conservation infrastructure by specific date)?
- Any evidence of varying interpretations of standards during and after signing or completion of the agreement/ruling?

Discrepant versions of outcome

Note any mention of "the X version" or "the Y version" (different recordings of terms), or differing translations due to more than one language.

2. *Perceptions of Ambiguity*

Note any comments on ambiguity of the Outcome, agreement, or ruling. — according to which parties? — note from outside observers?

CRITERIA CATEGORY III

3. Confusion/Controversy During Implementation

Were there misunderstandings regarding the Outcome required, or differences in interpretation? During implementation, did ambiguities or varying interpretations come to light?

G. *FEASIBILITY/REALISM*

This criterion addresses whether the Outcome is realistic in its assumptions and can be implemented in a practical sense, given legal, political, and technical considerations. Does it consider the legal and political context? Are the scientific and technical assumptions valid?

Note: Researchers are not asked to make legal, political, or scientific assessments themselves, but rather to note discussions of such types of feasibility in media and other sources.

SOURCES TO CHECK: Media, reviews of and commentary on Outcome itself.

BEST TIME TO ASSESS: After Outcome is announced, for things specified in the outcome. For things that must be formulated later (e.g. monitoring teams) assessment can only occur then.

1. Legal Feasibility

- consistency with existing legislation and administrative policies
- consistency with applicable court rulings
- Was the Outcome challenged on legal grounds?
- How were problems fixed (e.g. amendments)?

2. Political Feasibility

- Did (will) legislature/Congress pass needed legislation?
- For each party, could Outcome be justified to their constituency?
- Was a monitoring team assembled?
  - Is the proposed representation for monitoring bodies realistic and feasible?
  - Feasibility of representativeness for group that monitors implementation (e.g., The outcome may require environmental representative on the monitoring team, but environmental representatives spread too thin. The outcome may specify community representatives to oversee technical decisions, but no compensation is provided for time involved so they are unlikely to be able to afford to participate.)
- Any indication that actions required for implementation may be politically difficult?

3. Scientific and Technical Feasibility

- Was the scientific and technical basis for the agreement/ruling perceived as credible *in the scientific/technical community*? Was the "science right"? (Note: *Parties*' perceptions should be described under criterion IV.A.4, Satisfaction/Fairness.)
- Note discussion of unrealistic assumptions about the supply of the resource or the effectiveness of technology in the future

H. PUBLIC ACCEPTABILITY

Apart from the stakeholders themselves, the general public may judge the Outcome. Many argue that an Outcome should receive "public scrutiny." This criterion assesses whether the Outcome was perceived as fair by the public and by political leaders.

CRITERIA CATEGORY III

Note: Comments by stakeholders on fairness should go under criteria II.A, Procedurally Just, or IV.A, Satisfaction/Fairness. Discuss public scrutiny by official bodies under I.D. Ratification.

SOURCES TO CHECK: Media, Newsletters of parties.

BEST TIME TO ASSESS: After outcome is announced.

1. *Did public perceive outcome as fair?*

2. *Did the general public and political leaders perceive outcome as feasible from legal, political, technical and financial perspective?*

I. EFFICIENT PROBLEM-SOLVING

Negotiation theory suggests that efficiencies can be created when the CR process allows for collaborative problem-solving. If parties can work together, they can recognize opportunities for mutual gain, and collaborate to "expand the pie." Exchanges can be made that benefit everyone without anyone losing anything ("elegant trades"[14]). An efficient agreement is one where parties have not missed opportunities for "elegant trades,"[15] and where parties have "created value" by problem-solving together.

1. *Perceptions of Parties Participating in Process*

- Cite parties' perceptions regarding new options, increased resources, or multipliers ("value") created through collaborative problem-solving
- Cite parties' accounts of "elegant trades" or joint gains
- Cite parties' perceptions that there were "missed opportunities" for joint gains, win-win solutions, elegant trades, etc.

2. Perceptions of Observers

- Cite descriptions of new options, increased resources, or multipliers ("value") created through collaborative problem-solving
- Cite observers accounts of "elegant trades"
- Cite observers' perceptions that there were "missed opportunities" for joint gains, win-win solutions, elegant trades, etc.

CRITERIA CATEGORY III

CRITERIA CATEGORY IV – RELATIONSHIP OF PARTIES TO OUTCOME

A. *SATISFACTION/FAIRNESS – AS ASSESSED BY PARTIES*

Research[16] has found that outcome satisfaction and outcome fairness are highly related in peoples' minds, and so they have been combined here. This criterion seeks to assess parties' perceptions of satisfaction with and fairness of the Outcome immediately upon completion or announcement, either overtly expressed or expressed through behavior, such as a through refusal to sign or endorse.

Note: Need to check across various stakeholders. Multiple source use is important here. Distinguish between expressions of satisfaction with Outcome (to be noted here) and those about its subsequent implementation. Note the latter under criterion IV.D, Stability/Durability. Note if comments are made by spokespeople or by individuals.

SOURCES TO CHECK: Media, wire services, stakeholder newsletters, web sites, and other stakeholder promotional material. Some parties may have completed or commissioned reports. Scholarly sources may include interview results.

BEST TIME TO ASSESS: After enough time has passed for parties' summaries or reports to be prepared, though media can provide an initial barometer of reactions when

Outcome is announced and also while formal responses are being prepared (e.g., formal summaries, EIS responses, court appeals). Interviews may be helpful here.

1. *Satisfaction/Dissatisfaction with outcome*

Check all sources for: expressions of satisfaction/dissatisfaction with provisions of the outcome (economic, emotional, symbolic), did parties get what they "deserved," " people benefited (or suffered) equally," comments on fairness of the outcome. Comments on process should be noted under criterion II.A., Procedurally Just.

- Boycotting: Did any parties at table refuse to sign, refuse to show up at press conferences, a signing ceremony, or other symbolic event after the outcome was announced? [If evidence of dissatisfaction, quote/cite carefully, listing reasons.]
- Legal recourse: Were there new lawsuits filed, or threats to litigate?
- Partial Satisfaction: Were there indications that outcome met some (but not all) of the parties's needs (i.e., only partial satisfaction)?

2. *Fairness of cost sharing*

Comments about fairness of cost-sharing arrangements and the burden on one party versus the other. Watch for language such as "We gave up so much!" (implies the party is bearing what they consider large and unfair costs — in terms of money, water, or other claims they relinquished); or "They (other party) really made out well!" (implies perception of "imbalance" in how benefits and costs are distributed). Any statements regarding benefits also provide insights about costs because **costs and benefits are two sides of the same coin**. For instance, a party who did not get a benefit they believe they deserve feels that they have incurred a cost – the foregone benefit. Conversely, parties who make out better than they expected (incur lower costs) feel as though they have benefited from an outcome.

3. *Perception of scientific/technical credibility*

Did parties perceive scientific/technical basis for Outcome as credible (e.g., ruling based on uninformed science)?

B. COMPLIANCE WITH OUTCOME OVER TIME

Agreements/rulings/legislation compel parties to engage in certain behaviors and refrain from other behaviors. This criterion assesses whether parties did indeed engage in, or refrain from, actions as prescribed by the Outcome. Indicators include any subsequent litigation initiated or threatened in order to bring a party into compliance, the subsequent renewal of mediation or negotiations due to perceived noncompliance, records of compliance kept by any monitoring entity, and the inclusion of any provisions in the Outcome for verifying compliance (procedures, mechanisms, entities).

Note: The researcher is not being asked to independently assess compliance. Also consider that dissenters who did not sign an agreement (or who opposed legislation or a ruling) may still be bound by it.

SOURCES TO CHECK: Media, journals, party or third party reports. Information on provisions and mechanisms for verifying compliance is detailed in acts and rules. Once monitoring bodies are established, their records may be available.

Best Time to Assess: Provisions in Outcome can be assessed immediately upon completion. The presence of provisions and/or mechanisms for verifying compliance can be addressed immediately. Subsequent litigation can be assessed shortly after an agreement or settlement is achieved and can continue over time. Records of regulatory or monitoring organizations or other records kept to verify compliance can only be assessed after time has passed.

1. *Provisions/Mechanisms in Outcome for Measuring Compliance*

   - Are there follow-up documents/procedures required to verify compliance?
   - Are there ways to verify compliance? Are there adequate ongoing mechanisms set up (and funded) to verify compliance?

2. *Compliance Record-Keeping*

   - Any internal record keeping by parties of compliance? Do they indicate compliance? Note degree of compliance/noncompliance and which parties.
   - Records of regulatory/monitoring organizations? Do these records indicate compliance?

3. *Subsequent Actions Related to Compliance*

   - Subsequent litigation initiated or threatened by parties that are bound by agreement? Note: Here only include litigation initiated or threatened in order to bring another party into compliance with ruling or agreement.) Litigation initiated because dissatisfied with ruling or agreement should be noted under criterion IV.A, Satisfaction/Fairness.

- Subsequent mediation, negotiating session due to perceived non-compliance? List specific reasons for new sessions.

C. *FLEXIBILITY*

While no outcomes can be written to anticipate all future contingencies, outcomes can be designed to be responsive and flexible. This criterion assesses an Outcome's ability to be adapted to changing conditions. Indicators assess details of any subsequent modifications, the process specified in the original Outcome for modification (if any), and any unachieved but desired modifications, particularly if the barrier to modification was in the Outcome itself.

SOURCES TO CHECK: Information on modification procedures available in Outcome itself. Subsequent modifications may be noted in media, parties' papers, newsletters, and web sites; also board minutes or commissioned reports.

BEST TIME TO ASSESS: Provisions in Outcome can be noted upon completion. Ability for Outcome to be adjusted to changing conditions can only be judged after time has passed.

1. Provisions/Mechanisms in Outcome

- Was a process to achieve modification specified in original Outcome (i.e., modification must be unanimous, must be in writing)?

- Were there obstacles to modifications in the Outcome itself (e.g., unanimity required for change; modification requires re-ratification by all stakeholders)?

2. *Subsequent Actions Related to Modification Outcome*

- Describe the process.
- Were terms of outcome modified?
- Who initiated the change?
- What circumstances prompted modification?
- Did all parties agree?
- Was process cumbersome/costly to get modifications?

D. STABILITY/DURABILITY

This criterion addresses the ability for the Outcome to persist over time. It includes two types of indicators: those indicators that look at characteristics in the Outcome itself and in any accompanying framework for implementation that may affect stability over time; and indicators that actually note evidence of stability or instability over time. Indicators in the first category include stability-promoting incentives in the Outcome such as penalties, deadlines, or benchmarks, identification of a party (or parties) as responsible for implementation, and provision of an ongoing forum for future conflict resolution. Indicators in the second category, noting actual instability, include non-compliance, resumed litigation or introduction of counteracting legislation, expressions of hostility, communication breakdown, and coercive behavior.

SOURCES TO CHECK: Implementation guidelines and other predictive indicators may be found in the outcome itself, or in agency press releases, legislative committee testimony and addenda or third party sources. Decreases or increases in stability over time may be reported in the media.

BEST TIME TO ASSESS: Predictive indicators can be measured when outcome is reached. Long-term indicators require time passage after outcome, before assessment.

Note: Asking implementation questions of court rulings may have questionable validity, as rulings often are not designed to address future implementation of their implications, but only to rule on questions of law. Also, stability may be inferred from favorable assessments in category V, Relationship Between Parties.

1. Provisions/Mechanisms in Outcome

Responsibility for implementation

Which party (or parties) have been assigned responsibility for implementation? Is there an identifiable implementation team or other mechanism for taking responsibility for adequate progress over there?

Ongoing forum for problems

Is there an ongoing forum the parties agreed to turn to when problems and conflicts arise?

Incentives built into outcome to promote stability

Check agreement for:

- penalties for non-compliance
- deadlines for specific actions to be completed
- positive rewards for progress; benchmarks
- carefully staged timing of meeting parties' needs to keep them "invested" in successful implementation?

RELATIONSHIP OF PARTIES TO OUTCOME

Breakdown particulars of timing as required by agreement: When are each party's needs met? At the outset? At the end of some period? Staged over the course of implementation?

- mechanisms for assuring inclusivity in future decision-making affecting terms and obligations.

2. *Subsequent Actions Related to Stability/Durability*

- If Outcome did not assign responsibility for implementation, were responsibilities subsequently assigned? If so, how, where, when, etc. (See *Responsibility for Implementation* under item 1, above.)
- If no forum for problems was specified in original outcome, was a forum subsequently established? (See *Ongoing Forum for Problems* under item 1, above.)

3. *Indicators of Instability over Time*

- non-compliance
- litigation is resumed months or years after agreement reached
- hostility expressed in press, or other public forum
- shifting alliances, solicitation of new allies (evidence of polarization)
- legislation introduced/proposed to counteract or undermine agreement
- communication breakdown
- coercive behavior to extract concessions, payoffs, from others

Note: Many indicators of long-term stability also may be found in category V, Relationship Between Parties.

CRITERIA CATEGORY IV

CRITERIA CATEGORY V – RELATIONSHIP BETWEEN PARTIES

A. *REDUCTION IN CONFLICT AND HOSTILITY*

A common measure of improvement in conflictual relationships is a reduction in hostility. This criterion captures a sense of whether the conflict is de-escalating or escalating, either in actions, rhetoric or in tone of communication. Various factors from the literature on conflict escalation, such as the presence or absence of the possibility for non-alignment (indicates level of polarization) also are included as indicators.[17]

Note: This analysis is most useful if a baseline from the past is presented for comparison. It is best to use multiple sources to present range of impression of this criterion.

SOURCES TO CHECK: Media, publishing from parties, web sites or newsletters, public reactions, documents, congressional testimony, secondary sources writing about this conflict

BEST TIME TO ASSESS: Assess at points both before or during agreement process and then again after Outcome is reached. It is often difficult to assess whether or not the conflict is escalating or deescalating upon settlement (such an assessment will likely be mixed), and waiting for time to pass may be useful.

Check the following indicators for evidence of escalation or de-escalation:

- Check rhetoric and tone in parties' communications. What type of rhetoric is "acceptable"? "Out of bounds"?
- Is language used to describe the conflict from public platforms and in the media one of reasonableness or one of escalated, hostile rhetoric? Note both before and after Outcome.
- Are people being pressured to take sides or express opinion, or is it accepted for people/parties to be "undecided" or "on the fence?" (The former indicates polarization). Are people concerned about being sanctioned?
- How broadly are the disputants' lives affected? (where they shop, drink coffee, etc.)
- Hostile actions (blockades, shouting matches, threats)
- Public initiatives, legislative actions used to escalate or de-escalate conflict (e.g., joint projects as indication of de-escalation and collaborative tone).
- Character assassination, public put-downs.
- Any comments on the role of a convenor/facilitator/mediator in facilitating positive relational change? Indicate direct discussions of relationship change, citing source(s).
- Describe channels or modes through which parties interacted. For instance, did they only interact through the court or through other channels as well?
- Note references to other parties' credibility and acting in a trustworthy manner. Note discussions of "acting in good /bad faith". Do parties perceive that other parties are doing (or will do) what they say they are doing (or will do)?

B. *IMPROVED RELATIONS*

Theorists have sought to conceptualize "peace" or "good relations" as something beyond the lack of hostilities.[18] What represents "good relations" in terms of the presence, rather than the absence, of something? This criterion seeks to capture changes in the way parties see and relate to one another that may reflect the essence of successful resolution.[19] To note change, one also must first note the nature of the original relationship as a baseline for comparison. Indicators to explore for change include discussions of the relationship itself, as well as the tone of communication among the parties (hostile, conciliatory), the effort parties expended to protect themselves, and their sense of trust as indicated by the necessity of lack of enforcement clauses or other formalities.

Note: The subjective nature of relationship assessment makes it highly dependent on the type and number of sources consulted. Multiple sources should be sought and each should be cited clearly. Researchers should cite evidence and not rely on their own inferences.

SOURCES TO CHECK:	Language of Outcome itself. Media, correspondence between parties, speeches by party representatives, web sites, promotional materials, third party reports.
BEST TIME TO ASSESS:	Assess early on for baseline and then again upon completion of outcome.

1. Media Evidence of Relationship Quality

Check general media for evidence of relationship quality (good, bad, mixed):

- What baseline or starting point is used in making comparisons over time? What was the relationship like between parties as they entered this process?
- Note evidence of change in relationship quality during or as a result of the process/Outcome of interest.
- Characterize relational barriers to future hostile actions.
- Characterize relationship currently (at time of analysis).

2. *Improved Understanding*

Note evidence of a "more realistic and more sympathetic understanding of other stakeholders' positions and what their response might be to various scenarios or options."[20]

3. *Improved Communication*

- Is there any correspondence between parties? Please characterize.
- Note various sources' observation on the tone of communications among parties.

Note: Good interpersonal relationships among representatives are necessary, but not sufficient.

4. *Trust/"Good Faith"/Climate*

- Evidence of trust:
  - How much effort/expense do parties exert to protect themselves from one another ("Trust" is a handshake instead of a written contract. Is this bad practice or just evidence of trust?)
  - No enforcement clauses
  - No written agreement (note, these two could indicate a "bad" or "incomplete" agreement)
  - Change in enforcement mechanisms to less formal or less frequent
- Did the process create "a climate in which side-by-side problem solving was possible"[21]? Were parties willing to share true priorities?
- Note discussion of parties' ability to "follow through" on commitments. Include any sense of the "tone" of such discussions. Did assessments of others' competence in good faith improve?
- Was atmosphere during process one of "reciprocity" (e.g., where "good faith" efforts were reciprocated, where favors were offered, where give and take was

uncalculated, but in general balanced)? Note: Increases in general level of reciprocity in larger community should be noted in category VI, Social Capital.

C. *Cognitive and Affective Shift*

This criterion is designed to provide evidence of the phenomenon that many practitioners (and even parties) note a shift in parties' framing of the conflict and/or the relationship. Indicators include noting the ways parties refer to one another and the way they describe or explain the other parties' behavior (pre- and post- agreement). Building on literature from family systems theories, it also includes a bit of narrative analysis of the way "stories" are told about the conflict — do narratives change (pre- to post-) in their description of causality, interactions, values, etc.

Sources to Check: Internal party documents, newsletters, commissioned reports, fundraising brochures, media reports on a "new" relationship or process. Often will appear that parties' report on their own changes (self-reflection) rather than noting change in others.

Best Time to Assess: Upon completion of Outcome, and possibly even during the resolution process if it refers to perceptions of a new process and a new relationship that leads to reaching a constructive outcome. For most of these indicators, note the baseline state (i.e., before the resolution process began or early in it) and compare this to the state after the Outcome.

- Note evidence of shift in parties' "framing" of the conflict or the relationship.
- Note evidence of shift in parties' affective responses to the conflict or relationship (e.g., angry, resigned, hopeful, excited, agitated).
- Adjectives used to explain other parties' behavior. Note before and after agreement. Note in primary and secondary sources adjectives used to describe others' behavior.
- Have parties changed in how they think/consider other parties (e.g., now may involve tribal party representatives or environmental representatives from beginning of process on subsequent issues)?
- Do parties or observers describe changes in attitudes? Changes in public values?
- Cite evidence of views reflecting a shift in parties' sense of their interdependence.
- To what do parties (or observers) attribute any shifts observed?
- Some further dimensions to use in exploring the narratives parties tell about the conflict.[22] Note changes in the storytellers construction of:
  - time (static vs. fluctuating, nouns vs. verbs, ahistoric vs. historic)
  - space (non-contextual vs. contextual)
  - causality (cause vs. effect)
  - interactions (intrapersonal vs. interpersonal, intentions vs. effects, symptom conflicts, roles vs. rules)
  - values (good intent vs. bad intent, sane vs. insane, legitimate vs. illegitimate)
  - telling-style (passive vs. active, interpretations vs. descriptions, incompetence vs. competence)

CRITERIA CATEGORY V

D. Ability to Resolve Subsequent Disputes

This criterion addresses the degree to which the relationship between parties is handling subsequent related conflict, such as problems with implementation of the Outcome. Indicators include evidence that problems are handled constructively, that an ongoing relationship has emerged in which it is possible to address future concerns, and possibly the emergence of an ongoing forum for conflict management. This considers the parties' subsequent joint "track record" in terms of actions rather than simply perceptions.

Sources to Check: Media reports on process or on subsequent actions, parties' press releases, public speeches, and third party reports.

Best Time to Assess: After time has elapsed. Must be subsequent to Outcome itself. May involve disputes arising during implementation.

Note: It is best not to rely solely on inference, but to document responses in verifiable sources (which may be contradictory). This section is for an actual track record, i.e. actions. Note perceptions in criteria V.C, Cognitive Shift and V.E, Transformation.

Note absence or presence of new problems:

- Evidence that problems which arise are handled in a constructive manner
- Successive rounds of negotiations? (This can indicate intractable problems or the comfort level of parties in dealing with one another).
- Calling a mediator back in to the process, exchanging representatives

- Did the Outcome itself make it necessary to have an ongoing relationship (e.g., negotiation of future agreements required by initial agreement)? Characterize this subsequent relationship.
- Cite evidence of use of ongoing forum for conflict management. **Note:** Actual use of this forum or creation of a new forum should be noted here. If a forum was agreed to in the original Outcome, this should be noted under criterion IV.D, Stability/Durability.
- Cite observations of parties' informal commitment to working things through "in relationship" rather than resorting to more hostile forms (as in going to the courts, or an "end run" to Congress).

E. TRANSFORMATION

Some argue that conflict presents an opportunity for individual and collective moral growth.[23] More specifically, this moral growth is toward a social vision that integrates individual freedom and social conscience, and integrates concerns over justice and rights with concerns about care and relationships.[24] This moral growth can occur if conflict resolution processes help people to change their old ways of operating and to achieve new understanding and new relationships through conflict. Indicators include evidence of empowerment (i.e., the parties' renewed sense of their own capacity to handle challenges), evidence of recognition (i.e., empathy for and acknowledgement of others' circumstances)[25] and evidence of other major shifts in perception, (e.g., of relationship context, of paradigm, of social and political context, of tools and solutions.

Note: *Perceptions* of ability to achieve results and resolve future challenges should be noted here. Assessment of actual "track record" should be noted under criterion V.D, Ability to Resolve Subsequent Disputes. Changes in perceptions of the relationship between parties should be noted in criterion V.C, Cognitive Shift.

CRITERIA CATEGORY V

SOURCES TO CHECK: Media, observers' reports, internal party documents, newsletters, commissioned reports, fundraising brochures, media reports on a "new" relationship or process. Often will appear as parties' reports on their own changes (self-reflection) rather than as noting change in others.

BEST TIME TO ASSESS: Upon completion of outcome, and possibly even during the resolution process if it refers to perceptions of a new process and a new relationship that leads to reaching a constructive outcome.

Empowerment: Cite examples of parties' recognition of new or renewed capacity to resolve current or future disputes, and achieve future joint results.

Transformation of perception of the conflict and relationship: Note mention of major shifts in perception of relationship, or larger (environmental, social and political) context of the conflict and the relationship. Language to watch for is on the order of the following examples - "a paradigm shift," "a sea change," "a new era," "a new ethos," etc.

Transformation of policies/procedures: Note any jettisoning of past tools, models, algorithms, frameworks, operating policies, procedures, and any calls for new ones.

Recognition of the other: Note evidence of empathy for and acknowledgment of other parties' circumstances and contextual constraints.

- Cite evidence of parties' understanding of the others' circumstances. Is there any evidence of empathy with other parties' experiences?
- Note examples where a party may have acknowledged the legitimacy of another party's concerns either to them directly or at least via a public format (media, public forum).

CRITERIA CATEGORY V

CRITERIA CATEGORY VI – SOCIAL CAPITAL

This category includes criteria that address positive changes that occur in the larger system in which this conflict is embedded: Changes that go beyond the relationships between these particular stakeholders and/or beyond the particular issues in this conflict. These changes are grouped loosely together as Social Capital. [Note: Changed relationships between same stakeholders on same issues should be noted in category V., Relationship Between Parties.]

Social capital, like other forms of capital, is a potential resource that must be drawn upon to realize its value. It is the capacity for individuals to command resources that comes from having social connections. However, social capital is not inherent to individuals; rather, it is a characteristic or possession of relationships and communities.[26] It has been defined as potential assistance relationships between people,[27] "generalized reciprocity,"[28] the capacity for individuals to command scarce resources by virtue of their membership in networks or broader social structures,[29] the aggregate of these actual or potential resources linked to a durable network,[30] or even the capability for trusting strangers.[31] Related literature on consensus-building[32] leads one to the conclusion that social capital can be seen as a system's (i.e., a community's) increased capacity for responding that comes from cooperation and coordination.

Several of these criteria are outlined below. Given the relatively new nature of the increased theoretical focus on the construct of social capital, these criteria reflect the varying ways the category of social capital has been defined. Though it may be difficult to tie the development of social capital to a particular conflict resolution process, monitoring for change on these indicators may at least suggest changes in the large system that are correlational.

SOURCES TO CHECK: Media, internal party documents, newsletters, announcements of trainings or other activities, annual reports, reports to stockholders or memberships.

BEST TIME TO ASSESS: Social capital clearly can be assessed after the catalyzing conflict resolution processes have been completed, but can often be discerned during the process itself. Certain facilitative conditions for social capital (such as the presence of networks and trust) may even exist prior to the conflict resolution process's initiation.

A. *ENHANCED CITIZEN CAPACITY TO DRAW ON COLLECTIVE POTENTIAL RESOURCES*

1. Aggregate of resources

Aggregate of actual and potential resources that can be drawn upon from one's network. [hard to measure amount, but ability to draw can be noted - see next few criteria]

2. Potential assistance relationships

New partnerships and projects between same parties.

3. Generalized Reciprocity

- Evidence that assistance is provided across groups (e.g., environmental group or local government helps mining company get loans to make needed improvements.
- Evidence that groups and people across groups are "scratching each others backs."
- Likeliness that one might call upon contacts in other groups for a favor.

CRITERIA CATEGORY VI

B. INCREASED COMMUNITY CAPACITY FOR ENVIRONMENTAL/POLICY DECISION-MAKING

1. Aggregate of resources

- Coordinate meeting times so that people can attend both groups' meetings
- Evidence that parties have "divided up" tasks (e.g., information gathering, publicity).
- Seek out common avenues, e.g. sharing a web site, sharing a public forum, sharing office space, sharing publicity outlets.
- Public education across conflict lines (e.g., writing in each other's newsletters, guest speaking, visits to schools — separate and/or together).
- New decision-making structure (e.g., commission)

2. Increased System Efficiency

- Information shared
- Meeting process in place; follows prior conflict resolution process
- Processes utilize time efficiently; quick responses to problems
- Use of prevention strategies and long-range planning (e.g., proactive identification of problems)
- Cost sharing and joint financial arrangements

3. Increased Capacity for Cooperation

- Evidence that can utilize:
  - New Communication (How?) Know how to communicate with those in the other groups.
  - New Networks (Who?) Know whom to speak to in other groups - have "connections"
  - New Knowledge (What?) Knowing "who does what" may be useful

SOCIAL CAPITAL

- Increased capacity for handling conflict (either within the community/system itself or with judicious use of conflict resolution professionals)
- Greater conflict resolution knowledge; skills
- Use of training in conflict resolution, team-building, etc.
- New and different people /groups involved in the decision-making system (e.g., previously under-represented groups)

4. *Increased System Capacity for Responding to External Challenges*

- Evidence that connections and relationships established during initial conflict resolution are reactivated.
- Subsequent dispute or debate over new/different issues shows previously involved groups working together at outset.

5. *Increased Information Flow*

- Is it seen as normative to share information rather than to withhold it?
- Prestige given to those who share information (Widner).
- Cite references made or evidence that more people "know" things (e.g., about water, land use, etc.).
- Increased information infrastructure.
  - Evidence that information is more broadly available.
  - Evidence that people share information, have more ways/capacity to share information.
  - New newsletters, websites, etc.
- Traditional power holding groups (e.g., government agencies, corporations) more communicative, send out community newsletters , hold open houses, hold tours, etc.

CRITERIA CATEGORY VI

C. *SOCIAL SYSTEM TRANSFORMATION*

1. Assistance and Support Provided to General Community

- Donations to community funds, events
- In-kind donations (e.g., irrigators provide seeds to public schools in nearby municipalities; environmental organization representatives do guest speaking at water department event).

2. More Resilient Social/Political/Economic System

- Economy rebounded more quickly to crisis
- Unified, coordinated response of diverse groups to social crisis (e.g., hate crime, natural disaster)

3. Increased Civic Discourse

- Town Meetings, etc.
- Face to face discussions
- Drama, humor on debated issues (rather than antagonism)
- More openness to dialogue among traditional power-holding entities (e.g., government, corporations, utilities)

4. Creating a learning system/ "double-loop learning"[33]

- Evidence that parties (and community) know how to organize themselves (collectively, not just within parties) — Town meetings called, community agenda set, etc.
- Ongoing links between community leaders and between community organizations (e.g., Urban League, Valley Planning Commission) where members jointly search out what they "need to know".

- Focus on prevention and proactivity (e.g., proactive data collection and monitoring; tax incentives for using new approaches).
- Organizations' agendas and missions are renegotiated

Items 5-8 do not so much define social capital, but rather both result from it and also facilitate its development.

5. *Enhanced Networks*
   (Putnam[34] says these should be "horizontal associations")

   - Stability/ Transience of community (numbers moving in and out)
   - Note change in awareness of how to contact others in other groups.
   - Increased linkages across conventional divides of race, class, profession, etc.

6. *Perceived Mutual Reciprocity/Assistance Relationships*

   Cite evidence for beliefs in existence of help and/or cooperation from other parties.

7. *Perceived Interdependence*

   Cite parties' observations of interdependence , common fate or linking, (e.g., "we're all in this together" "we'd all lose/win").

8. *General Trust*

   - Note any mention in media, editorials, etc., or any actual survey data showing trust in neighbors, schools, government, other institutions.
   - Note any mention of degree of trust in strangers and ability to spontaneously work together for common purposes.

CRITERIA CATEGORY VI

NOTES

[1] Sources leading to criteria development are not cited in this Guidebook. For full review and citations, see Chapter 2, this volume.
[2] Readers interested in more detail on these topics should refer to Chapter 2 and Chapter 9.
[3] See Chapter 8.
[4] See Chapter 2.
[5] See Chapter 3.
[6] Judith E. Innes, "Evaluating Consensus Building," in *Consensus Building Handbook*, eds. Lawrence Susskind, Sarah McKearnon, and Jennifer Thomas-Larmer, 631-675 (Thousand Oaks, CA: Sage Publications, 1999).
[7] A CDROM version of the Guidebook that allows for simply "filling in the blanks" is available from the first author.
[8] Comments and case analyses can be submitted to Tamra Pearson d'Estrée, Conflict Resolution, University of Denver, 2199 S. University, Denver, CO 80208, or Bonnie Colby, Dept. of Agricultural & Resource Economics, University of Arizona, PO Box 210023, Tucson, AZ 85721.
[9] Forum: City Council meeting, session of Congress, Board of Directors meeting, etc.
[10] Vote, referendum, resolution
[11] E. Allen Lind, and Tom R. Tyler, *The Social Psychology of Procedural Justice* (New York: Plenum Press, 1988).
[12] For example, Janice A. Roehl, "Measuring Perceptions of Procedural Justice," (doctoral dissertation, George Washington University 1988).
[13] Based on Janice A. Roehl, "Measuring Perceptions of Procedural Justice," 1988.
[14] Howard Raiffa, *The Art and Science of Negotiation* (Cambridge, MA: Harvard University Press, 1982).
[15] Ibid; see also Lawrence Susskind and Jeremy Cruikshank, *Breaking the Impasse: Consensual Approaches to Resolving Public Disputes* (New York: Basic Books, 1987).
[16] E. Allen Lind and Tom R. Tyler, *The Social Psychology of Procedural Justice*, 1988.
[17] Jeffery Z. Rubin, Dean G. Pruitt, and Sung Hee Kim, *Social Conflict: Escalation, Stalemate, and Settlement* (2nd ed.) (New York, Colin McGraw Hill, 1994).
[18] Adam Curle, *Making Peace* (London: Tavistock, 1971); Johan Galtung, *Peace by Peaceful Means: Peace and Conflict Development and Civilization* (Thousand Oaks, CA: Sage Publications, 1996).
[19] Further discussion of this criterion category may be found in Tamra Pearson d'Estrée, "Achievement of Relationship Change," in *The Promise and Performance of Environmental Conflict Resolution*, eds. Rosemary O'Leary and Lisa B. Bingham, 111-128 (Washington, DC: Resources for the Future, 2003).

[20] Judith E. Innes, "Evaluating Consensus Building," 1999.
[21] Lawrence Susskind and Jeremy Cruikshank, *Breaking the Impasse: Consensual Approaches to Resolving Public Disputes*, 1987.
[22] Carlos E. Sluzki, "Transformations: A Blueprint for Narrative Changes in Therapy," *Family Process* 31 (1992): 217-230.
[23] Robert A. Baruch Bush, and Joseph P. Folger, *The Promise of Mediation: Responding to Conflict Through Empowerment and Recognition* (San Francisco: Jossey-Bass, 1994).
[24] cf. V. Held, *Justice and Care: Essential Readings in Feminist Ethics* (Boulder, CO: Westview, 1995).
[25] Robert A. Baruch Bush, and Joseph P. Folger, *The Promise of Mediation: Responding to Conflict Through Empowerment and Recognition*, 1994.
[26] Shawn MacDonald, "Social Capital and Its Measurement" (Unpublished manuscript, George Mason University, 1999).
[27] James S. Coleman, "Social Capital in the Creation of Human Capital," *American Journal of Sociology* 94 (supplement), (1988): S95-S120; James S. Coleman, *Foundations of Social Theory* (Cambridge, MA: Harvard University Press, 1990).
[28] Robert D. Putnam, "The Prosperous Community: Social Capital and Public Life," *American Prospect* 13, (1993): 35-42; Robert D. Putnam, Bowling Alone: America's Declining Social Capital," *Journal of Democracy* 6 (1), (1995): 65-78.
[29] A. Portes, "Social Capital: Its Origins and Applications in Modern Sociology," *Annual Review of Sociology* 24 (1998).
[30] Pierre Bourdieu, "The Forms of Capital," in *Handbook of Theory and Research for the Sociology of Education*, ed. J. Richardson (Westport, CT: Greenwood Press, 1986).
[31] Francis Fukuyama, *Trust: Social Virtues and the Creation of Prosperity* (New York: Simon & Schuster, 1995).
[32] Judith E. Innes, "Evaluating Consensus Building," 1999.
[33] Chris R. Argyris, Robert D. Putnam, and D.M. Smith, *Action Science* (San Francisco: Jossey-Bass, 1985).
[34] Robert D. Putnam, "The Prosperous Community: Social Capital and Public Life," 1993.

NOTES

APPENDIX B

FEDERAL POLICIES AFFECTING ENVIRONMENTAL CONFLICT RESOLUTION

The Resource, Regulatory and Planning Frameworks

Kathryn Mazaika

A general overview of the federal laws and major policies provides a useful starting point to generate ideas and strategies for research. Every federal agency must respond to a host of environmental laws as it plans to implement projects. The National Environmental Policy Act (NEPA), specific sections of the Endangered Species Act (ESA), the Clean Water Act (CWA), the Clean Air Act (CAA), the Resource Conservation Recovery Act (RCRA), the Comprehensive Environmental Response, Compensation, and Liability Act (CERCLA) and their amendments are among them. These laws provide ready-resources for information to evaluate an environmental conflict resolution effort using the criteria described in this book. Among these laws, NEPA requires the broadest analysis (addressing issues covered by the other Acts) and the production of environmental documents for public review. In addition to the broader-based laws, organic acts for land and resource management agencies such as the National Forest Management Act (NFMA), the Federal Land Policy and Management Act (FLPMA), and the National Park Service Organic Act are other potential sources of information on public involvement, and the level of detail that documents will cover.

This appendix highlights specific issues to consider as you seek information for the ECR case analysis. While the laws and implementing regulations provide one framework through which a community or project must navigate, they too are subject to debate and reflect the differences in changing administrations and ideologies. Even though the basic legal frameworks remain unchanged, it is worthwhile to stay abreast of policy shifts in how they are implemented. Recent efforts to alter public input to forest projects, endangered species impacts, and mining are examples.[1] It is also important to bear in mind that there are many other potential sources of information within state and local laws and other regulations for the geographic area in which the project is located.

National Environmental Policy Act (NEPA) and Clean Air Act § 309

NEPA requires all federal agencies to prepare a detailed statement on proposed legislation, regulations, and other major federal actions significantly affecting the quality of the human environment. The statement should discuss: (1) the environmental impact of the proposed action; (2) unavoidable adverse environmental impacts; (3) alternatives to the proposed action; (4) the relationship between local short-term uses and the maintenance and enhancement of long-term productivity; and (5) any irreversible commitments of resources should the proposed action be implemented.[2] NEPA also requires all federal agencies to develop implementation procedures that guide how they consider the environmental impacts of their decision-making processes.[3] Section 309 of the Clean Air Act directs the Environmental Protection Agency to review and comment on these statements of environmental impact, and refer to the Council on Environmental Quality those proposals it deems unsatisfactory because of the impacts the project poses to public health and welfare or environmental quality.[4] There is greater potential to identify documents and activities related to the criteria when the project involves a federal sponsor. NEPA, the implementation regulations, and the Forty Most Asked Questions Concerning the NEPA regulations provide the framework for public participation and the baseline scope of the environmental impact analysis.[5] Where land management laws apply, specific elements of the analysis will also be addressed.

Because environmental impact statements and assessments (documents) and the review processes conducted to comply with NEPA require describing the affected environment and the potential impacts of a project, a variety of information can emerge that is responsive to the ECR review criteria. The scope of the public input and participation can provide clues into both how the outcome was reached and its quality. Ideas about the process quality may be evident in the way that various parties were able to express their concerns in public comments and how the project sponsor(s) addressed them. Sections within the "affected environment" of a document can provide a sense of the outcome quality because it will include discussions on impacts to resources (water, wildlife, air, land), culture, and social environment. The comments and response to comments sections will similarly provide insights into the public acceptability of a project. Discussions on cumulative and reasonably foreseeable direct and indirect impacts may provide insights into the criteria focusing on sustainability by extending into the future the impacts considered. Taken together the documents produced to comply with NEPA are a readily available public source of information when a project or process constitutes a major federal action (MFA) and may significantly affect the environment. MFAs could include timber sales, highway widenings, Park Service management plans, renegotiating water contracts, or land transfers.

Regulations and policies pertaining to NEPA change from time to time and a brief review of the latest policies is worth consideration. Two central websites (the Environmental Protection Agency's National Environmental Policy Act and Council on Environmental Quality (CEQ) in Washington, DC) provide general information as well as updates on changing policies.[6] A more thorough review of changes in case law, legislation, and regulations would be available through NEPAnet's Case Law Review, and legal research. Among the policies and guidance the EPA includes in reviews of environmental impact statements are "Environmental Justice in 309 Reviews" and "Pollution Prevention/Environmental Impact Checklists for NEPA/309 Reviewers". The CEQ also makes available to federal agencies a "Memorandum on Incorporating Pollution Prevention into NEPA Documentation". These documents may provide the case reviewer with a sense of the degree to which project proponents have considered pollution prevention and environmental justice, and the degree to which the review criteria have been addressed.

In addition to environmental assessments and impact statements, it is possible to locate similar NEPA-type documents produced in response to NEPA-type state legislation and policy. A comprehensive review of state laws, policies, and regulations addressing biodiversity can be found in Defenders of Wildlife's Saving Biodiversity.[7] This publication provides a list of the types of impact assessments states required as of 1995.[8] This document may be especially helpful in locating information for projects that do not include a federal component. Another resource on state law and policy is available online through the State Environmental Resource Center (SERC).[9] Through SERC's site it is possible to stay abreast of state policies and legislation, including trends.

Endangered Species Act

Three sections of the federal Endangered Species Act (ESA) may provide ideas when searching for potential sources of information pertaining to environmental sustainability.[10] Section 4 of the ESA outlines the criteria the Secretaries of Interior and Commerce should consider when determining whether to list a species as either threatened or endangered.[11] Factors such as threatened or actual modification or destruction of habitat or range; disease or predation; overuse for commercial, recreational, scientific or educational purposes; inadequacy of existing regulations; or other natural or manmade factors which threaten its continued existence may justify listing a species as threatened or endangered. The public can also petition the Secretary to either add or remove species from the list. Information pertaining to a species' listing would be available in the Federal Registrar.

Section 7 requires all federal agencies to consult with the Secretary and ensure that any actions they authorize, fund or carry out are not likely to jeopardize the continued existence of any threatened or endangered species, or destroy or adversely modify critical habitat of such species.[12] If a federal agency is uncertain about the presence of listed species in their project area, it will request the Secretary's assistance to determine the likelihood, and prepare a biological assessment should data suggest its presence. The Secretary will review the biological assessment and follow the consultation with a biological opinion of its findings. These procedures are detailed in a consultation handbook finalized by the Fish & Wildlife Service and the National Marine Fisheries Service.[13] If the impact is in conjunction with a major federal action, NEPA will also apply and, an environmental assessment or impact statement will also be available for public comment and review. These documents should be available for review from the U.S. Fish & Wildlife Service (FWS) or the National Marine Fisheries Service, who implement the ESA in the region where the project is located.

Section 10 grants the Secretary the authority to permit certain actions otherwise prohibited under the ESA.[14] Under this section, the Secretary through the FWS can permit the incidental taking of listed species as long as a habitat conservation plan accompanies the application for a permit. Parties other than federal agencies typically use this approach. The habitat conservation plan must discuss the likely impact, the steps taken to minimize and mitigate the impact as well as the funding available to support these plans, any alternative actions considered with the reasons that they were not chosen, and other measures the Secretary deems appropriate. Habitat conservation plans and environmental impact documents (written in response to issuing the Section 10 permit) would be available for public review and comment.

Both the Department of Interior and the U.S. Fish & Wildlife Service provide information through their websites in addition to information located in the Federal Register pertaining to species listings, habitat conservation plans and incidental take permits. The U.S. Fish & Wildlife Service maintains an Endangered Species Homepage that would be a good starting point for obtaining copies of the Endangered Species Act and relevant policies and procedures.[15] FWS also maintains a website on the endangered species Habitat Conservation Planning (HCP) program.[16] This comprehensive website includes an HCP overview, Frequently Asked Questions, Questions and Answers on the "No Surprises" Policy, "No Surprises" Myths and the Endangered Species Act, as well as contact information and other policies.

Two additional sources of information on state endangered species law and its scope are available in Defenders and in Goble et al.[17] These documents may be useful in developing an understanding of the requirements a state imposes, in addition to federal endangered species law.

Clean Water Act

The federal Clean Water Act provides similar opportunities for public participation as it regulates decisions affecting water resources. Three sections in particular may provide insights into the environmental sustainability of a decision, as well as opportunities for public input, and thus a measure of how these criteria might be addressed in an environmental conflict resolution effort.

Section 319 Nonpoint Source Pollution Control

Although the Clean Water Act is a delegated program, Congress recognized the need for greater federal leadership when it amended the Act in 1987 to include a section that addresses nonpoint source (NPS) water pollution.[18] Nonpoint source pollution has diffuse sources of water pollution that originate from many undefinable sources, and which normally include agricultural and urban runoff, in addition to runoff from construction activities and forestry practices.[19] Section 319 provides financial assistance in the form of grants to states, territories, and tribes to support nonpoint source programs. These funds might be used to provide technical assistance, develop and implement watershed demonstration projects, or assess the success of these projects. Through the nonpoint source water pollution grant program communities have come together to identify their watersheds, assess NPS impacts, and cooperatively develop plans to address these water quality problems.

Conference proceedings may be useful sources of information on watershed plans developed in the area where a conflict resolution analysis takes place. The EPA also provides access to information on this program through a Nonpoint Source Pollution website.[20] Through the Publications and Information Resources website one can obtain a copy of Section 319, which describes the program, and numerous other publications, including two volumes documenting successful 319 programs that produced a variety of solutions to nonpoint pollution. Through its libraries, the EPA also provides access to a wealth of documents relating to water quality programs.[21] One such document lists opportunities for public involvement in nonpoint source control.[22] Among the noted activities are volunteer monitoring, water conservation, household management, and forming community-based forums.

Section 303(d) Total Maximum Daily Loads (TMDLs) Program

Through Section 303(d) of the Clean Water Act the Environmental Protection Agency expects states to identify and rank water-quality limited waters. In these situations effluent limitations established elsewhere in the Act have not been sufficient for these water bodies to achieve applicable water quality standards. The integrated process of developing total maximum daily loads is intended to create a link between water quality standards assessment and water quality-based control actions. Through public notice and hearings the public has the opportunity to help identify and prioritize targeted water bodies and help identify the most feasible and implementable treatment strategies. Once again, through a website maintained by the EPA, it is possible to access information and updates about the TMDL Program.[23] New developments in the program as well as case studies and program guidance are available through this website.

Section 404 Wetlands and Waters of the United States

Section 404 of the Clean Water Act established programs to regulate the discharge of dredged and fill material into waters of the United States, including wetlands.[24] Activities that are regulated under this program include fills for development, water resource projects (such as dams and levees), infrastructure development (such as highways and airports), and conversion of wetlands to uplands for farming and forestry.[25] Under this program, no discharge of dredged or fill material is permitted if a practicable alternative exists that is less damaging to the aquatic environment or if the nation's waters would be significantly degraded. Permit applicants must show that they have: (1) taken steps to avoid impacts to wetlands, (2) minimized the potential impacts, where they were unavoidable, and (3) provided compensation for any remaining, unavoidable impacts through activities to restore or create wetlands. Individual permits are needed for activities posing potentially significant impacts; general permits issued at a nationwide or regional level allow activities that will have minimal adverse effects.

The Environmental Protection Agency and U.S. Army Corps of Engineers jointly administer Section 404 of the Clean Water Act. The EPA develops and interprets the environmental criteria for evaluating permit applications, determines the scope of geographic jurisdiction, approves and oversees State assumption, identifies activities that are exempt, reviews and comments on individual permit applications, has the authority to veto the Corps' permit decisions (Section 404[c]), can elevate specific cases (Section 404[q]) for review, and enforces Section 404 provisions. The Corps administers the day-to-day program (including individual permit decisions and wetlands jurisdictional determinations), develops policy and guidance, and enforces Section 404 provisions.

Within fifteen (15) days of receiving all information supporting an individual permit application, the Corps issues a public notice. The public notice should describe in the permit application the proposed activity, its location and potential environmental impacts. Public comments are invited for a 15 - 30 day period, depending on the proposed activity. The Corps and other interested federal and state agencies, organizations, and individuals review the application and comments, and the Corps determines whether an environmental impact statement is needed to permit the action. Citizens may request that the Corps hold a public hearing, though this is not normally done according to the EPA. More often, the Corps evaluates the permit application and issues an environmental assessment and statement of finding. These documents are then available for public review.

The public can also participate in volunteer wetlands monitoring programs.[26] The EPA expects to provide technical assistance to states, tribes, and NGOs by making skilled personnel available to carry out necessary wetlands monitoring tasks. One example of such an effort is Frogwatch USA, established in February 1999 to help researchers track populations of frogs and toads. The EPA's Volunteer Monitoring Home Page discusses methods and tools to monitor, assess, and report on the health of water resources, and provides access to software and automated information systems used to manage monitoring data.[27]

The Wetlands Walk Manual, available as an Adobe Acrobat file,[28] provides citizens with opportunities to learn about the value of wetlands and to collect information and data that helps identify trends in wetlands health and location.

Through the 404 permit process, public hearings, and review of environmental documents the public has the opportunity to participate in decisions affecting wetlands resources. Volunteer opportunities provide the chance to learn through hands-on experience. The opportunity for public input and to protect resources through these processes should provide some clues into both the availability of documentation and the potential depth of analysis one might expect to find as you review a conflict resolution effort.

Organic Acts & Land Management

Three acts and their implementing regulations are worth considering when reviewing environmental conflict resolution efforts. These acts, like NEPA, specify a process and framework through which environmental impacts are assessed, disclosed, and communicated to the public. Land management plans in conjunction with the environmental assessment or impact statement can provide a sense of how issues of environmental sustainability were considered, and the degree to which the public participated in the decision making processes around proposed land uses.

Forest Service Planning & the National Forest Management Act (NFMA)

The National Forest Management Act, its implementing regulations and the Forest Service's NEPA implementing regulations found in the Forest Service Manual and Handbook provide insights into the scope of analysis one can expect to find in land management plans and the accompanying environmental documents, as well as the depth of public involvement in crafting the proposed management directions.[29]

Forest planning and its accompanying environmental impact analyses have been the subject of much recent review and revision. In November 2000, the Department of Agriculture (Forest Service) promulgated national forest land resource management planning regulations that included expanded opportunities for public input and analyses for ecological, social and economic sustainability.[30] Later in December 2002, following a change in administrations, the Department of Agriculture proposed another modified set of rules for forest planning and management.[31] According to the Forest Service's website on NFMA's proposed planning rules the public comment period closed April 7, 2003. The website also notes that the Forest Service received nearly 200,000 comments, and is currently reviewing them to make adjustments to the planning rule.[32] Seeking updates on the status of these implementing regulations would be helpful here to follow how earlier expansions on public participation, and ecological, social, and economic sustainability fare.

Land and resource management planning regulations specify the content of forest plans, including the resources it should address, and ways to include the public in the planning process. The Forest Service's NEPA implementing procedures further specify a process for public input that determines the extent of public interest and plans for public involvement, including methods to inform the public as the process unfolds. To the maximum extent possible, the Forest Service seeks to integrate the forest planning requirements with that of NEPA. Among the common elements of NFMA and NEPA, respectively, are: (1) proposed action/purpose and need; (2) possible management practices/alternatives; (3) environmental effects/environmental impacts; (4) public participation/scoping and review of environmental documents. Therefore, when reviewing a conflict resolution effort, one can turn to both the land and resource management plans for details on the content and depth of resources considered, and the NEPA document for details and disclosure of environmental impacts and public comments. The EPA's review and comments on the environmental documents pursuant to Section 309 of the Clean Air Act, moreover, will provide a sense of the degree to which it finds the proposed plan and document adequately disclose and assess environmental impacts. Taken together, these documents and supporting public records can simplify one's search for the major issues and the degree to which they were addressed within these frameworks.

Bureau of Land Management & Federal Land Policy Management Act (FLPMA)

The Federal Land Policy and Management Act of 1976 created the Bureau of Land Management (BLM), and declared it the policy of the United States that: "...the public lands be retained in Federal ownership, unless as a result of the land use planning procedure provided in this Act, it is determined that disposal of a particular parcel will serve the national interest...."[33] FLPMA also gave the BLM planning authority that included a periodic inventory to assess public lands and their resources for current and emerging conditions.[34] Congress, moreover, made it clear that these public lands should be managed for multiple use and sustained yield in a manner consistent with other public land use planning.[35] Through its land use planning, acquisition and disposition authority BLM manages the lands that had previously remained in the public domain.

Land use planning regulations for resource management planning include: identifying issues, developing planning criteria, analyzing the management situation, formulating alternatives, estimating the impacts of the alternatives, selecting a preferred alternative, selecting a resource management plan, and monitoring and evaluating it.[36] These regulations clearly mirror NEPA requirements. Moreover, a policy statement earlier in the planning regulations specifies that, "approval of a resource management plan is considered a major Federal action significantly affecting the quality of the human environment. The environmental analysis of alternatives and the proposed plan shall be accomplished as part of the resource management planning process and, wherever possible, the proposed plan and related environmental impact statement shall be published in a single document."[37] This explicit policy attempts to integrate resource management planning and NEPA requirements to the greatest extent possible.

In addition to these planning regulations the BLM provides a handbook to guide its staff in preparing NEPA documents. The handbook is intended for use by BLM officials responsible for oversight of and compliance with the National Environmental Policy Act (NEPA) within their program area and the BLM personnel responsible for preparing NEPA documents.[38] The guidance provided in this handbook is intended to address both CEQ's Regulations for Implementing NEPA and the Department of the Interior's manual guidance on the National Environmental Policy Act of 1969 (516 DM 1-7).[39]

As one reviews conflict resolution efforts that have included issues under the BLM's jurisdiction, these laws and regulations may provide some insights into the types of documents that already exist, as well as chronicle the scope of public involvement and the resources considered within this planning framework.

National Park Service & Organic Act of 1916

The National Park Service Organic Act of 1916 created the National Park Service and the national park system.[40] The Park Service prepares general management plans intended to preserve natural and cultural resources and guide recreational uses in units within the park system. Management plans should: include measures to preserve resources, and assess the needs and impacts of development associated with public visitation, visitor carrying capacities and potential modifications to boundaries of the park service unit.[41] The Park Service maintains websites that provide helpful and up-to-date background on their planning process, current and completed planning efforts, planning policy, a sourcebook, and its NEPA implementing procedures.[42] Through these documents, one can access information about the planning framework, and identify any general management plans that may be relevant to a conflict resolution analysis.

NOTES

[1] Defenders of Wildlife, "Weakening the National Environmental Policy Act: How the Bush Administration Uses the Judicial System to Weaken Environmental Protection," (Washington,DC:Defenders ofWildlife, 2003), http://www.defenders.org/publications/nepareport.pdf; Zachary Coile, "House OKs Bush Plan for 'Healthy Forests,'" *San Francisco Chronicle*, May 21, 2003; Matthew Daly, "Bush and Environment: Getting His Way By Settling Lawsuits," *AP Wire*, April 18, 2003; John Heilprin, "Interior Department Tries to Accelerate Research into Oil and Gas Drilling in Rockies," *AP Wire*, April 18, 2002; Charles Levendosky, "Bush Turns BLM Into Energy Machine," *High Country News*, March 18, 2002.

[2] 42 U.S.C. § 4332 (2)(C)

[3] 42 U.S.C. § 4332 (2)(B)

[4] 42 U.S.C. § 7609

[5] 40 C.F.R. Parts 1500-1508; 46 Fed. Reg. 18026-18038 (3/23/81).

[6] U.S. Environmental Protection Agency, National Environmental Policy Act (http://www.epa.gov/Compliance/resources/policies/nepa/index.html). Last Updated April 8, 2003; CEQ NEPANet (http://ceq.eh.doe.gov/nepa/nepanet.htm).

[7] Defenders of Wildlife, *Saving Biodiversity,* (Albuquerque, NM: Defenders of Wildlife, 1996) (also available online at http://www.defenders.org/pb-bst00.html).

[8] Ibid., pp. 36, 37.

[9] State Environmental Resource Center. http://www.serconline.org/

[10] 16 U.S.C. § 1531- § 1544, as amended.

[11] 16 U.S.C. § 1533

[12] 16 U.S.C. § 1536

[13] U.S. Fish & Wildlife Service and National Marine Fisheries Service, "Procedures for Conducting Consultation and Conference Activities under Section 7 Handbook of the Endangered Species Act," March 1998.

[14] 16 U.S.C. § 1539

[15] U.S. Fish &WildlifeService, "Endangered Species Homepage," (http://endangered.fws.gov/).

[16] U.S. Fish & Wildlife Service, "Endangered Species Habitat Conservation Planning," (http://endangered.fws.gov/hcp/index.html).

[17] Defenders of Wildlife, *Saving Biodiversity*, 1996; Dale D.Goble, Susan M. George, Kathryn Mazaika, J. Michael Scott, and Jason Karl, "Local and National Protection of Endangered Species: An Assessment," *Environmental Science & Policy* 2 (1999): 43-59.

[18] 33 U.S.C. § 1329

[19] U.S. Environmental Protection Agency, Region 10, "A Glossary of Watershed Related Terms," September 1994.

[20] U.S. Environmental Protection Agency, Office of Water, "Nonpoint Source Pollution," (http://www.epa.gov/owow/nps/), March 14, 2003.

[21] U.S. Environmental Protection Agency, "Online Library System," (http://www.epa.gov/natlibra/ols.htm), updated April 15, 2002.

[22] U.S. Environmental Protection Agency, *Nonpoint Pointers Pointer No. 2* (EPA-841-F-96-004B), March 1996 (http://www.epa.gov/owow/ nps/facts/point2.htm).

[23] U.S. Environmental Protection Agency, Office of Water, "Total Maximum Daily Load Program," (http://www.epa.gov/OWOW/tmdl/docs.html), February 24, 1998.

[24] 33 U.S.C. § 1344

[25] U.S. Environmental Protection Agency, Office of Wetlands, Oceans, Watersheds, "Section 404 of the Clean Water Act: An Overview," (http://www.epa.gov/owow/wetlands/facts /fact10.html), last updated January 16, 2003.

[26] U.S. Environmental Protection Agency, Office of Wetlands, Oceans, Watersheds, "Wetlands – Monitoring & Assessment," (http://www.epa.gov/owow/wetlands/monitor/#vol), last updated March 31, 2003.

[27] U.S. Environmental Protection Agency, Office of Water, "Monitoring and Assessing Water Quality," (http://www.epa.gov/owow/monitoring/index.html), last updated March 27, 2003.

[28] http://www.epa.gov/owow/wetlands/wetwalk.pdf

[29] National Forest Management Act, 16 U.S.C. § 1600-14; National Forest System Land and Resource Management Planning, 36 C.F.R. Part 219; Forest Service Manual, Chapter 1950, Environmental Policies and Procedures; Forest Service Handbook, Chapter 1909.15, Environmental Policies and Procedures Handbook.

[30] 65 Fed. Reg. 67514

[31] 67 Fed. Reg. 72770

[32] "The 2002 Proposed NFMA Rule," (http://www.fs.fed.us/emc/nfma/index3.html), last Modified July 2003.

[33] 43 U.S.C. § 1701 (a)(1)

[34] 43 U.S.C. § 1711 (a)

[35] 43 U.S.C. § 1701 (a)(7)

[36] 43 C.F.R. Parts 1610.4-1 to 1610.4-9

---

[37] 43 C.F.R Part 1601.0-6

[38] H-1790-1 – National Environmental Policy Act Handbook (http://www.blm.gov/nhp/efoia/wo/handbook/h1790-1.html).

[39] Id. At 4

[40] National Park Service Organic Act of 1916, 16 U.S.C. § 1-18f

[41] 16 U.S.C. § 1a-7

[42] National Park Service – Park Planning (http://planning.den.nps.gov/default.cfm) Last Modified May 5, 2003

APPENDIX C

APPLYING THE ECONOMIC AND FINANCIAL CRITERIA

This appendix provides some concrete guidance on applying the economic and financial criteria to the case studies. These criteria are II.C, Process Costs; III.A, Cost-Effective Implementation; III.B, Perceived Economic Efficiency; III.C, Financial Feasibility/Sustainability; III.D, Cultural and Community Sustainability; and III.E, Environmental Sustainability.

Why Consider Economic and Financial Aspects?

Policymakers and the public require accountability for the manner in which environmental conflicts are resolved. Public agencies often are stakeholders in conflicts, public resources are expended in grappling with conflicts and issues of public interest – such as air and water quality, endangered species, and management of public lands – frequently are the subject of the disputes. Public officials want to know how much money, time, and other resources were expended, whether the costs incurred were justified by the positive outcomes of the dispute resolution process and whether the best possible process was used.

In order to fully address these questions, it would be necessary to document the costs incurred by all parties to the conflict, the costs imposed on taxpayers and other indirect parties, and the costs yet to be borne in the implementation of the outcome. Moreover, in order to determine whether the process and the outcome were worthwhile, it would be necessary to identify all the current and future benefits, express them in dollars, and then weigh the benefits against the costs. If the benefits were found to outweigh the costs, then the process and outcome could be characterized as worthwhile in the sense that they returned more in benefits than was expended in costs. If one wishes to rigorously document that the best possible process was used for the case, it is necessary to estimate the costs and benefits of an alternative process and the likely outcome (i.e., litigation compared to a negotiated agreement). This sort of hypothetical comparison may be possible in some cases, but is fraught with difficulties. Another approach is to examine actual costs and benefits in parallel sets of similar cases that were resolved through different processes.

Such rigorous inquiry is not possible with limited data, the absence of comparable data across similar cases resolved by different mechanisms, and the limited resources to study the matter. Consequently, this Guidebook takes the strategy of collecting available data on costs and characterizing benefits largely in descriptive terms, except where monetary measures of benefits can be obtained from other studies. Over time, with a sufficiently large number of carefully documented cases, it will be possible to say more about the costs and benefits of litigation, mediation, administrative actions, legislative remedies, and other resolution methods.

Suggestions for Collecting and Recording Information

Record anecdotal evidence on costs and benefits (including quotes by affected parties or observers on their perceptions) with careful notes on the source, the individual being quoted, and their title and institutional affiliation. Perceptions may not be consistent with factual evidence, but it is still interesting for our purposes to learn how parties perceive the economic and financial aspects of resolving disputes and implementing outcomes.

There are numerous mechanisms through which parties incur costs and raise funds to financially support a process or to implement an outcome. These include, but are not limited to:

- legislative appropriations
- changes in tax, water and electricity rates
- direct grants and donations
- contributions of land, water, equipment
- contributions of staff time and expertise directed to specific purpose
- bonds issued by public entities, such as cities, counties, irrigation districts
- loans provided at favorable interest rates or with favorable repayment provisions
- forgiveness of existing debts or restructuring of loan repayment.

If you are planning to contact parties to obtain economic and financial information, be aware that much of this may be confidential data and not readily shared. Moreover, many organizations do not compile data on how much staff time and money has gone into a specific case. More likely, a manager may simply be able to comment that "three of my staff have each devoted 50 percent of their time to this case for the past two years; a hydrologist, a biologist, and an attorney." This is useful information. Try to get estimates of the percentage of staff members' workload devoted to the case, over how many months/years, and the profession and title of the staff members involved. Salary levels are useful in converting time into monetary figures. Salary data may be obtainable from public records for public agency staff but may be confidential information for businesses and non-governmental organizations.

Refrain from asking organizations to compile detailed cost and financial information that have not already been collected for other purposes prior to your inquiry. This task can be very time consuming and such requests should only be made under special circumstances or when the organization indicates a willingness to compile the information.

The costs of a dispute continuing unresolved become the benefits of achieving resolution IF they are costs that the parties no longer have to incur once an outcome has been achieved. Costs of the dispute continuing are also important because they affect the stakeholders' incentives to resolve the dispute. In particular, the distribution of the costs affects incentives to settle. If the costs of the dispute are high and affect all parties, then they all have incentive to settle. However, costs are usually unevenly distributed. Some parties benefit from the status quo and face relatively low costs. They prefer to delay reaching a settlement and to wear other parties down by imposing high costs on them. Examples of this include irrigation districts that benefit from current water allocations which can impose high costs on environmental interests by impeding settlement while a species declines. Likewise, non-Indian water users may delay settlement to put off the day when water they have been using is allocated to a tribe.

Examples of the types of benefits that may be generated by resolving a dispute include:

improved economic output, productivity
- additional crops planted
- increased fish harvests
- new subdivisions built

enhanced certainty and ability to plan
- planning for urban growth

- endangered species recovery planning

environmental improvements
- habitat restoration
- better recreation quality
- increased population of endangered species

political/organizational benefits
- better cooperation between states, tribes, cities, irrigation districts
- good press for stakeholders and the process/outcome
- improved organizational morale and credibility
- increased confidence in the system to solve problems
- enhanced quality of community life
- decreased anxiety and tension.

Operationalizing the Economic Criteria □ Economic Terminology

Below are economic terms, that will be useful as you work through the economic and financial criteria. Using this terminology in case analysis will help create some consistency across case studies.

Baseline

The baseline is a crucial concept in examining the costs and benefits of a process or an outcome. We seek to identify effects specifically caused by the process and outcome being analyzed. However, there are likely to be other changes that can not be attributed to the specific process and outcome we are studying, but which affect the same region and resources. For instance, a water district may vote to increase water rates in the communities we are studying for reasons unrelated to the outcome we are examining. The increased water rate should not be described as an effect of the outcome, even though it is occurring during the time frame of the process/outcome we analyze. As another example, a specific case may examine a cutback in logging, which is causing job losses and is associated with the outcome being analyzed. At the same time and in the same area, a military base may be closed (for reasons unrelated to the process/outcome), also causing loss of jobs. We need to separate the economic effects of the base closure from those impacts which are properly attributable to the process and outcome we are analyzing.

In order to be clear about the effects of the cases we analyze vis a vis changes occurring due to other factors, we need to define a "baseline" for our cases. The baseline is the conditions that would exist without the process and outcome we analyze. Effects due to the process/outcome are those that would not have occurred without that process and outcome. This is called the "with and without" principle. It attributes to the case only those effects that would not have occurred anyway. For instance, if a fish species is declining and the outcome we are analyzing provides 100,000 acre feet per year of additional water for fish recovery (water that would not otherwise have been available for this purpose), then improvements in fishery conditions linked to the new 100,000 acre feet can properly be described as benefits of the outcome. But suppose that unusually favorable rainfall in one year brings another 200,000 acre feet for the fish. The outcome we are studying can't take credit for that.

It is not easy to isolate the impacts of a specific process and outcome from all of the other events that are affecting the resources and the region involved in a case study. In reporting the cases, be careful to note where multiple factors contributed to changes in environmental, economic, and social conditions. Search for studies that attempt to carefully trace the effects of the process and outcome being analyzed and distinguish them from other forces at work in the area. If your sources of information are not clear in this regard, please note this in your case write up.

Accounting Stance

The choice of accounting stance is a decision made by the research team regarding how widely (across time, layers of parties, and geographically) to count cost, benefits, and other impacts in writing up cases. These choices may differ for each case, but they should always be explicitly stated.

Here is an example of an explicit statement regarding accounting stance: "The conflict over the Middle River has been ongoing since the early 1900s, but in this analysis we examine the period 1960 to the present. The primary parties are farmers, cities, and anglers. Boaters also are affected by the conflict, but were not key players in the process. Consequently, there is little information on impacts on boating, and we do not include boater impacts in our detailed analysis. While the conflict does affect river management in several downstream states, the primary impacts are in Nebraska. We do not assess effects in downstream states because there is little information and these effects are peripheral to the process and outcome we are analyzing, which occurred within Nebraska. We count costs and benefits to federal taxpayers, but examination of all other economic impacts is limited to Nebraska."

As the example illustrates, each case must clearly state the time period, geographical area, and the range of parties considered in the analysis. Provide a brief explanation of the reasons for excluding some time periods, regions, and parties; e.g., not central to the case, limited information.

Opportunity Costs

An opportunity cost is not a direct expense paid out of some party's pocket. Rather, it is an opportunity (flow of expected benefits) that is given up in order to obtain something else. While subtle as a cost concept (compared, for instance, to the cost of constructing a dam), perceptions about opportunity costs are powerful influences on parties. For instance, an irrigation district that gives up some of the water it has been using in order to achieve a negotiated settlement (with a tribe that claims the water for their own purposes) will view the future stream of farm profits that were foregone as a genuine cost, even though it is not money spent out of pocket, but rather disappointed expectations about the future. Every decision has opportunity costs such as what a party could have done with the land, water, money, etc. if they were not committed as specified in the outcome.

References

AEA Task Force on Guiding Principles for Evaluators. 1995. Guiding principles for evaluators. *New Directions for Program Evaluation* (Summer): 19-34.

Allen., J. C. 2001. Community conflict resolution: The development of social capital within an interactional field. *Journal of Socio-Economics* 30: 119-120.

Argyris, Chris, Robert Putnam, and D. M. Smith. 1985. *Action science.* San Francisco: Jossey-Bass.

Arnold, Craig Anthony, and Leigh A Jewell. 2001. Litigations bounded effectiveness and the real public trust doctrine: The aftermath of the Mono Lake case. *8 Hasting West-Northwest Journal of Environmental Law and Policy* I. (Fall).

Babbitt, E., and Tamra Pearson d'Estrée. 1996. An Israeli-Palestinian women's workshop: Application of the interactive problem-solving approach. In *Managing global chaos: Sources of and responses to international conflict,* eds. C. Crocker, F.O. Hampson, and P. Aall, 521-529 Washington, DC: U.S. Institute of Peace.

Barrett-Howard, E., and Tom R Tyler. 1986. Procedural justice as a criterion in allocation decisions. *Journal of Personality and Social Psychology* 50:296–304.

Baucom, Donald H., and Norman Epstein, 1991. Will the real cognitive-behavioral marital therapy please stand up? *Journal of Family Psychology* 4:394-401.

Beierle, Thomas C., and Jerrell C. Cayford. 2002. Democracy in practice: Public participation in environmental decisions. Washington, DC: Resources for the Future Press.

Bennett, Lynne Lewis, Charles W. Howe, and James Shope. 2000. The interstate river compact as a water allocation mechanism: Efficiency aspects. *American Journal of Agricultural Economics*, 82(4) (November): 1006.

Bingham, Gail. 1994. Findings and recommendations on convening second round Truckee/Carson settlement negotiations. Washington DC: RESOLVE Inc.

Bingham, Gail. 1986. *Resolving environmental disputes: A decade of experience.* Washington, DC: The Conservation Foundation.

Bourdieu, Pierre. 1986. The forms of capital. In *Handbook of theory and research for the sociology of education*, ed. J. Richardson. Westport, CT: Greenwood Press.

Bourdeaux, C., Rosemary O'Leary, and R. Thornburgh. 2001. Control, communication, and power: A study of the use of alternative dispute resolution of enforcement actions at the U.S. Environmental Protection Agency. *Negotiation Journal* 17(2): 175-191.

Bratt, Duane. 1997. Assessing the success of UN peacekeeping operations. *International Peacekeeping 3*:64-81.

Bruce, Fredrick R. 1993. Salvaged water: the failed critical assumption underlying the Pecos River Compact. *Natural Resources Journal* 33, no. 1 (Winter): 217-228.

Brunet, Edward. 1987. Questioning the quality of alternative dispute resolution. *Tulane Law Review* 1, no. 62:1-56.

Buckle, Leonard G., and Suzann R. Thomas-Buckle. 1986. Placing environmental mediation in context: Lessons from failed mediations. *Environmental Impact Assessment Review* 6 (March): 55–70.

Bush, Robert A Baruch, and Joseph P. Folger. 1994. *The promise of mediation: Responding to conflict through empowerment and recognition.* San Francisco: Jossey-Bass.

Bush, Robert A. Baruch. 1989. Defining quality in dispute resolution: Taxonomies and anti-taxonomies of quality agreements. *Denver University Law Review* 66:335–380.

———. 1995. Report on the assessment of the Hewlett Foundation's Centers for "Theory Building" on conflict resolution. Hewlett Foundation.

California State Water Resources Control Board-Division of Water Rights. 1993-1994. *Draft and final environmental impact report for the review of the Mono Basin water rights of the City of Los Angeles.*

Camplair, Christopher W., and Arnold L. Stolberg. 1990. Benefit of court-sponsored divorce mediation: A study of outcomes and influences on success. Publisher unknown.

Capra, Fritjof. 1996. *The web of life: A new scientific understanding of living systems.* New York: Anchor Books.

Carpenter, Susan, and W. J. D.Kennedy. 1988. *Managing public disputes: A practical guide to handling conflict and reaching agreements.* San Francisco: Jossey-Bass.

Carstarphen, Berenike. Undated. O.H.M. shift happens: Transformations during small group interventions in protracted social conflicts. Unpublished doctoral dissertation, George Mason University.

Clagett, M.P. 2002. Environmental ADR and negotiated rule and policy making: criticisms of The Institute For Environmental Conflict Resolution and The Environmental Protection Agency. *Tulane Environmental Law Review* 15:409-417.

Cobourn, J. 1999. Integrated watershed management on the Truckee River in Nevada. *Journal of the American Water Resources Association* 35, no. 3:623-32.

Colby, Bonnie G., and Gail Bingham. 1997. Economic components of success in resolving environment policy disputes. *Resolve Newsletter* 28. Washington, DC.

Coleman, James S. 1988. Social capital in the creation of human capital. *American Journal of Sociology* 94 (supplement): S95-S120.

Coleman, James S. 1990. *Foundations of social theory.* Cambridge, MA: Harvard University Press.

Committee on Western Water Management, Water Science and Technology Board, Commission on Engineering and Technical Systems, Board on Agriculture, National Research Council. 1992. Water transfers in the west. 119-136. Washington, D.C.: National Academy Press.

Conner, Mike. 2001. *Congressional Briefing: Background and Status of Indian Land and Water Claims.* Department of Interior Ad Hoc Group on Indian Water Rights.

Consortium on Negotiation and Conflict Resolution. 1994. *The facility issues negotiation process of the Georgia Comprehensive Solid Waste Management Act: An evaluation with recommendations.* Atlanta, GA: Author.

Cook, Stuart W. 1984. The 1954 social science statement and school desegregation: A reply to Gerard. *American Psychologist* 39:819-832.
Corburn, J. 1996. *Pursuing justice in environmental decision-making: Deliberative democracy and consensus building.* Unpublished Master's thesis. Massachusetts Institute of Technology.
Curle, Adam. 1971. *Making peace.* London: Tavistock Press.
Davis, G., and Roberts, M. 1988. *Access to agreement.* Milton Keyes, UK: Open University Press.
d'Estrée, Tamra Pearson. 2003. Achievement of relationship change. In *The promise and performance of environmental conflict resolution,* eds. Rosemary O'Leary and Lisa B. Bingham, 111-128. Washington, DC: Resources for the Future.
Dillon, P. A., and R.E. Emery. 1996. Divorce mediation and resolution of child custody disputes: Long-term effects. *American Journal of Orthopsychiatry* 66:131–140.
Dotson, A. Bruce. No date. *Defining success in environmental and public policy negotiations.* Unpublished manuscript.
Dukes, Franklin. 1993. Public conflict resolution: A transformative approach *Negotiation Journal* 9(1): 45-57
Easter, K. W. 1993. Economic failure plagues public irrigation: An assurance problem. *Water Resource and Research* 29:1913-1222.
Edwards, Harry T. 1986. Commentary: Alternative dispute resolution: Panacea or anathema? *Harvard Law Review* 99:668-684.
Emerson, Kirk. April, 1996. *A critique of environmental dispute resolution research.* Presentation to the Conflict Analysis and Resolution Working Group Seminar, University of Arizona.
Fisher, Ronald J. 1997. *Interactive conflict resolution.* New York: Syracuse University Press.
Fisher, Roger, William Ury, and Bruce Patton. 1991. *Getting to Yes, 2nd Ed.* Penguin Books.
Fukuyama, Francis. 1995. *Trust: Social virtues and the creation of prosperity.* New York: Free Press.
Galtung, Johan. 1996. *Peace by peaceful means: Peace and conflict, development and civilization.* Thousand Oaks, CA: Sage.
Grant, Douglas L. 2003. Interstate water allocation compacts: When the virtue of permanence becomes the vice of inflexibility. *74 University of Colorado Law Review* 105 (Winter).
Gray, Brian E. 2002. The property right in water. *9 Hasting West-Northwest Journal of Environmental Law and Policy* I. Fall.
Gregg, Frank, Doug Kenney, Kathryn Mutz, and Teresa Rice. 1998. *The state role in western watershed initiatives.* Boulder: Natural Resources Law Center, University of Colorado.
Gulliver, P.H. 1979. *Disputes and negotiations.* New York: Academic Press.
Gwartney, P.A., F. Fessenden, and Gail Landt. 2002. Measuring the long-term impact of a community conflict resolution process: A case study using content analysis of public documents. *Negotiation Journal* 18(1): 51-74.
Hall, Emlen G. 2002. *High and dry: the Texas-New Mexico struggle for the Pecos River.* Albuquerque, NM: University of New Mexico Press.
Hammond, Kenneth R., Lewis O. Harvey, & Reid Hastie. 1992. Making better use of knowledge, separating truth from justice. *Psychological Science* 3(2): 80-87.
Hart, John, and Nancy Fouquet. 1996. *Storm over Mono: The Mono Lake battle and the California water future.* Berkeley, CA: University of California Press.
Held, Virginia. 1995. *Justice and care: Essential readings in feminist ethics.* Boulder, CO: Westview.
Hollis, Martin. 1994. *The philosophy of social science.* Cambridge: Cambridge University Press.

Innes, Judith E. 1999. Evaluating consensus building. In *Consensus building handbook,* eds. Lawrence Susskind, Sarah McKearnon, and Jennifer Thomas-Larmer, 631-675. Thousand Oaks, CA: Sage Publications.

Johnston, Joseph R., Linda E. G. Campbell, and Mary C. Tall. 1985. Impasses to the resolution of custody and visitation disputes. *American Journal of Orthopsychiatry* 55:112-119.

Kakalik, James S., Terence Dunworth, Laural A. Hill, Daniel McCaffrey, Marian, Oshiro, Nicholas M. Pace, and Mary E. Vaiana. 1996. *An evaluation of mediation and early neutral evaluation under the Civil Justice Reform Act.* Report produced by the Institute for Civil Justice, Rand, Santa Monica, CA.

Keilitz, Susan L., Harry W. K. Daley, and Roger A. Hanson. 1992. *Multi-state assessment of divorce mediation and traditional court processing.* Project report for the State Justice Institute, Williamsburg, VA.

Kelly, J. B. 1990. *Mediated and adversarial divorce resolution processes: An analysis of post-divorce outcomes.* Final report prepared for the Fund for Research in Dispute Resolution.

Kelman, Herbert C. 1992. Informal mediation by the scholar-practitioner. In *International Mediation: A Multi-level Approach to Conflict Management*, eds. Jacob Berkowitz and Jeffrey Z. Rubin, 64-96. London: Macmillan.

Kerwin, Cornelius, and Laura Langbein. 1995. *An evaluation of negotiated rulemaking at the environmental protection agency Phase I.* Washington, DC: Administrative Conference of the United States.

Kloppenberg, Lisa A. 2002. Implementation of court-annexed environmental mediation: The district of Oregon pilot project. *Ohio State Journal on Dispute Resolution* 17, no.3:559-596.

Lamb, B.L., N. Burkardt, and J.G. Taylor. 2001. The importance of defining technical issues in interagency environmental negotiations. *Public Works Management & Policy* 5(3): 218-223.

Lang, M.D., and A. Taylor. 2000. *The making of a mediator: Developing artistry in practice.* San Francisco: Jossey Bass.

Lederach, John Paul. 1995. *Preparing for peace: Conflict transformation across cultures.* Syracuse, NY: Syracuse University Press.

———. 1997. *Building peace: Sustainable reconciliation in divided societies.* Washington D.C.: U.S. Institute of Peace Press.

Lewicki, Roy, Barbara Gray, and Michael L.P .Elliott, eds. 2003. *Making sense of intractable environmental conflicts: Concepts and cases.* Washington, DC: Island Press.

Leventhal, G. S. 1980. What should be done with equity theory? In *Social exchange: Advances in theory and research,* eds. K. J. Gergen, M. S. Greenberg, and R. H. Weiss. New York: Plenum Press.

Lind, E. Allen and Tom R. Tyler. 1988. *The social psychology of procedural justice.* New York: Plenum Press.

MacDonald, Shawn. 1999. *Social capital and its measurement.* George Mason University. Unpublished manuscript.

McCrory, John P. 1981. Environmental mediation—another piece for the puzzle. *Vermont Law Review* 49 (Spring): 77-79.

Meierding, N. 1993. Does mediation work? A survey of long-term satisfaction and durability rates for privately mediated agreements. *Mediation quarterly* 11:157–170.

Menkel-Meadow, Carrie. 1991. Pursuing settlement in an adversary culture: A tale of innovation co-opted or 'the Law of ADR'. *Florida State University Law Review* 19(1): 1-46

———. 1995. The many ways of mediation: the transformation of traditions, ideologies, paradigms, and practices. *Negotiation Journal* 11:217-242.

Moore, Christopher W. 1987. *The mediation process: Practical strategies for resolving conflict.* San Francisco, CA: Jossey-Bass.

Morris, Douglas D. 1996. Alabama-Tombigbee Rivers Coalition v. Department of Interior: Giving sabers to a "toothless tiger" the Federal Advisory Committee Act. *Environmental Law* 26(1), (Spring): 393-417.

Nabatchi, Tina, and Lisa B. Bingham. 2002. Expanding Our Models of Justice in Dispute Resolution: A Field Test of the Contribution of Interactional Justice. Paper presented at the annual meeting of the International Association of Conflict Management.

National Institute for Dispute Resolution. 1996. *Final report: Fund for research on dispute resolution.* Washington DC: Author.

Neuman, J. C. 1996. Run river run: Mediation of a water rights dispute keeps fish and farmers happy for a time. *University of Colorado Law Review* 67 (Spring): 259–339.

Newman, Cathy . 1993. The Pecos: River of hard-won dreams. *National Geographic* (September):

Oregon Department of Justice. 2001. Collaborative dispute resolution pilot project. A report submitted January 30, 2001 to the Honorable Gene Derfler, Senate President, The Honorable Mark Simmons, House Speaker, and The Honorable Members of the Legislature.

Ozawa, Connie. 1996. Science in environmental conflicts. *Sociological Perspectives* 39(2): 219-230.

Orr, Patricia . 2003. ECR cost effectiveness: Evidence from the field. Briefing Tucson, AZ: U.S. Institute for Environmental Conflict Resolution, April 16, 2003.

Patton, M.Q. 1997. *Utilization-focused evaluation.* Thousand Oaks, CA: Sage Publications, Inc.

Pearson, Jessica. 1994. Family mediation. In *A report on current research findings – implications for courts and future research needs*, ed. S. Keilitz. National Symposium on Court-connected Dispute Resolution Research, State Justice Institute, 1993.

Pearson, Jessica and Nancy Thoennes. 1989. Divorce mediation: Reflections on a decade of research. In *Mediation research*, eds. K. Kressel and Dean Pruitt. 9-30. San Francisco, CA: Jossey-Bass.

Pharris, Jim, Mary Sue Wilson, and Alan Reichman 2002. *Federal and Indian Reserved Water Rights: A Report to the Washington State Legislature by the Office of the Attorney General.* Olympia, WA: Attorney General of Washington.

Policy Consensus Initiative. 1998. States mediating change: Using consensus tools in new ways. Portland, OR: Policy Consensus Initiative.

Portes, A. 1998. Social capital: Its origins and applications in modern sociology. *Annual Review of Sociology* 24.

Pratt, Jeremy. Truckee-Carson River Basin study: Final report. Seattle: Clearwater Consulting Corporation.

Pruitt, Dean G., Robert S. Pierce, Neil B. McGillicuddy, Gary L. Welton, and Lynn M. Castrianno. 1993. Long-term success in mediation. *Law and Human Behavior* 17: 313–330.

Putnam, Robert D. 1995. Bowling alone: America's declining social capital. *Journal of Democracy 6* (1): 65-78.

Putnam, Robert D. 1993a. The prosperous community: Social capital and public life. *American Prospect* 13:35-42.

Raab. Jonathan. 1988. The politics of environmental dispute resolution. *Policy Studies Journal* 16:585-601.

———. 1994. *Using consensus building to improve utility regulation.* American Council for an Energy Efficient Economy: Washington, D.C.

Raiffa, Howard. 1982. *The Art and Science of Negotiation.* Cambridge, MA: Harvard University Press.

Regan, Patrick. 1996. Conditions of successful third-party intervention in interstate conflicts. *Journal of Conflict Resolution* 40:336-359.

Resolve. 1997. *What is success in public policy dispute resolution? Building bridges between theory and practice.* A roundtable sponsored by Resolve and the National Institute for Dispute Resolution, Washington, DC.

Roehl, Janice A. 1988. *Measuring perceptions of procedural justice.* Doctoral dissertation, George Washington University.

Ross, M. H. and J. Rothman. 1999. *Theory and practice in ethnic conflict management: Theorizing success and failure.* New York: St. Martin's Press, Inc.

Rubin, Jeffrey Z., Dean G. Pruitt, and Sung Hee Kim. 1994. *Social conflict: Escalation, stalemate, and settlement, 2nd ed.* New York: Colin McGraw Hill.

Runge, C. F. 1984. Institutions and the free rider: The assurance problem in collective action. *Journal of Politics* 46:154-181.

Saposnek, D. T. 1983. *Mediating child custody disputes: A systematic guide for family therapists, court counselors, attorneys, and judges.* San Francisco, CA: Jossey-Bass.

Schön, Donald A. 1983. *The Reflective Practitioner.* New York: Basic Books.

Sluzki, Carlos E. 1992. Transformations: A blueprint for narrative changes in therapy. *Family Process* 31:217-230.

Southeast Negotiation Network. 1993. *Carpet policy dialogue assessment.* Atlanta, GA: Author.

Stephan, Walter G. and J.C. Brigham. 1985. Intergroup contact: Introduction. *Journal of Social Issues* 41: 1-8.

Straus, Murray A. 1979. Measuring intrafamily conflict and violence: The conflict tactics (CT) scales. *Journal of Marriage and the Family* 41:75–86.

Susskind, Lawrence, and Jeremy Cruikshank. 1987. *Breaking the Impasse: Consensual Approaches to Resolving Public Disputes.* New York: Basic Books.

Susskind, Lawrence E., P. Levy, and Jennifer Thomas-Lerner, *Negotiating Environmental Agreements: How to Avoid Escalating Confrontation, Needless Costs, and Unnecessary Litigation.* Washington, DC: Island Press, 2000.

Susskind, Lawrence E., Ravi K. Jain, and Andrew O. Martyniuk. 2001. *Better environmental policy studies: How to design and conduct more effective analyses.* Washington, DC: Island Press.

Susskind, Lawrence, Sarah McKearnon, and Jennifer Thomas-Larmer. eds. 1999. *Consensus building handbook.* Thousand Oaks, CA: Sage Publications.

Susskind, Lawrence and Connie Ozawa, 1983. Mediated negotiations in the public sector. *American Behavioral Scientist* 27(no. 2): 255–79.

Susskind, Lawrence, Mieke van der Wansem, and Armand Ciccarelli. 2000. An analysis of recent experience with land use mediation – Overview of the consensus building institute's study. In *Mediation land use disputes pros and cons.* Lincoln Institute of Land Policy.

Tyler, Tom R. 1989. The psychology of procedural justice: A test of the group-value model. *Journal of Personality and Social Psychology* 57:830–838.

———. 1991. The influence of decision makers' goals on their concerns about procedural justice. *Journal of Applied Social Psychology* 21 (no. 20): 1629-1658.

Tyler, Tom R. and E. Griffin. 1991. The influence of decision-maker goals on resource allocation decisions. *21 Journal of Applied Social Psychology*:1629-1658.

US Dept of Interior and California Dept of Water Resources. 1998. Draft environmental impact statement/environmental impact report for the Truckee River Operating Agreement. US Dept of Interior and California Dept of Water Resources:5-7.

Ury, William, J. M. Brett, and Stephen B. Goldberg, 1988. *Getting disputes resolved: Designing systems to cut the costs of conflict.* San Francisco: Jossey-Bass.

Wissler, Roselle L. 2002. Court-connected mediation in general civil cases: What we know from empirical research. *Ohio State Journal on Dispute Resolution* 17 (3): 641-704.

Walker, J.A. 1989. Family conciliation in Great Britain: From research to practice to research," *Mediation Quarterly* 24: 34.

Walker, J. A. 1986. Assessment in divorce conciliation: Issues and practice. *Mediation Quarterly* 11:43-56.

Western Governors Association. 1991. *Park City principles*. Document produced following series of three workshops developed by Western Governors Association and Western States Water Council. http://www.westgov.org/wga/initiatives/enlibra.

Index